Tropical Conservation Biology

Tropical Conservation Biology

Navjot S. Sodhi
National University of Singapore

Barry W. Brook
University of Adelaide

Corey J. A. Bradshaw
Charles Darwin University

Blackwell
Publishing

BLACKWELL PUBLISHING

350 Main Street, Malden, MA 02148-5020, USA
9600 Garsington Road, Oxford OX4 2DQ, UK
550 Swanston Street, Carlton, Victoria 3053, Australia

The right of Navjot S. Sodhi, Barry W. Brook, and Corey J. A. Bradshaw to be identified as the Authors of this Work has been asserted in accordance with the UK Copyright, Designs, and Patents Act 1988.

First published 2007 by Blackwell Publishing Ltd

1   2007

Library of Congress Cataloging-in-Publication Data

Sodhi, Navjot S.
Tropical conservation biology / Navjot S. Sodhi, Barry W. Brook, Corey J. A. Bradshaw.
    p. cm.
Includes bibliographical references and index.
ISBN 978-1-4051-5073-6 (pbk. : alk. paper) 1. Conservation biology–Tropics–Textbooks. I. Brook, Barry W. II. Bradshaw, Corey J. A. III. Title.
QH77.T78.S63 2007
333.95'160913–dc22
        2007009687

A catalogue record for this title is available from the British Library.

Set in 10.5 pt on 12 pt Sabon
by Prepress Projects Ltd

The publisher's policy is to use permanent paper from mills that operate a sustainable forestry policy, and which has been manufactured from pulp processed using acid-free and elementary chlorine-free practices. Furthermore, the publisher ensures that the text paper and cover board used have met acceptable environmental accreditation standards.

For further information on
Blackwell Publishing, visit our website:

www.blackwellpublishing.com

Contents

Preface

Habitat loss and fragmentation, overexploitation of species and environmental degradation are operating on a massive scale worldwide, and these are the processes fuelling the current extinction crisis. These human impacts on the biosphere predict a grim future for global biodiversity, especially considering that in many regions the rates of destruction and modification are accelerating. Although the temperate regions of the earth suffered severely from human encroachments in the ancient and recent past, the present-day and future biodiversity crisis loom largest in the tropics. This is especially important in terms of global biodiversity conservation because it is in the tropics that the largest tracts of still-pristine habitat, the greatest species diversity and the richest centres of endemism are found. The mounting threats to tropical biodiversity require conservation practitioners to make urgent conservation decisions. As such, it is critical to document the current state of tropical biodiversity, determine possible ways to protect it and synthesize the vast body of scientific research relevant to tropical regions in a concise, yet comprehensive, format. It was with these aims that we wrote *Tropical Conservation Biology* – the first book of its kind to focus exclusively on conservation issues in the tropics. A brief summary of the contents by chapter follows:

Chapter 1: Diminishing habitats in regions of high biodiversity. We report on the loss of tropical habitats across the tropics (e.g. deforestation rates). We also highlight the drivers of habitat loss such as human population expansion. Finally, we identify the areas in immediate need of conservation action by elucidating the concept of biodiversity hotspots.

Chapter 2: Invaluable losses. We discuss the utilitarian role of tropical forests in maintaining soil stability, regulation of hydrological processes and climate and carbon sequestration, especially in relation to the importance of ecosystem servicing offered by nature to humanity (e.g. pollination and catchment protection). We also highlight the evolutionary uniqueness, irreplaceability and intrinsic value of nature that is being eroded by habitat and species loss.

Chapter 3: Broken homes: tropical biotas in fragmented landscapes. Owing to massive deforestation and forest degradation, fragmentation is one of the major drivers of species loss in the tropics. We discuss the theoretical premises upon which the importance of habitat fragmentation are based (e.g. negative edge effects and meta-population dynamics). We then exemplify various concepts using empirical information and a wide range of examples.

Chapter 4: Burning down the house. Prolonged drought and poor land use decisions have made many tropical landscapes vulnerable to fire, thus creating a negative feedback cycle that is damaging at both local (e.g. habitat loss) and global scales (e.g. atmospheric pollution). We describe the effects of fire on tropical biotas and how it interacts with other drivers to exacerbate threats.

Chapter 5: Alien invaders. We report on the documented impacts of invasive species on tropical biotas. We describe the process of invasion and the factors that affect 'invasiveness' (of both the invading organism and the invaded habitat). We then described a range of documented impacts of invasive species in tropical regions.

Chapter 6: Human uses and abuses of tropical biodiversity. Tropical biodiversity is under heavy threat from anthropogenic overexploitation (e.g. harvest for food or live specimens for the pet trade). For example, wild ('bush') meat hunting is imperilling many tropical species as expanding human populations in these regions seek new or long-favoured sources of protein and potentially profitable new avenues for trade. Here we highlight the effects of human exploitation on tropical biodiversity and the unsustainability of current practices.

Chapter 7: Threats in three dimensions: tropical aquatic conservation. This chapter expands on the major marine and freshwater conservation issues plaguing tropical regions. We focus on the impacts of overfishing, water pollution and climate change on the astonishing biodiversity supported by marine and freshwater ecosystems.

Chapter 8: Climate change: feeling the tropical heat. We summarize the main lines of evidence for the biotic response to climate change in the tropical realm – past, present and future. Examples are drawn from studies on local populations through to investigations of pantropical ecosystems. Finally, we discuss options for mitigating the worst of climate change's predicted effects on tropical biodiversity.

Chapter 9: Lost without a trace: the tropical extinction crisis. We first discuss the empirical evidence for and controversies surrounding current and predictive extinction estimates. We then provide case studies of both local population and species extinctions from the tropics and extract generalities of extinction trends in this region. We also discuss the species traits correlated to extinction proneness or vulnerability of species to decline due to human encroachment in mega-diverse tropical ecosystems.

Chapter 10: Lights at the end of the tunnel: conservation options and challenges. We make pragmatic recommendations for the protection of existing biodiversity. We report on the state, adequacy and complementarity of current protected areas, and discuss the minimum preserved areas required to protect adequately biodiversity at national, regional and global scales. We stress the need to consider the social issues (e.g. human hunger) in order to achieve effective conservation objectives by highlighting examples of conservation successes in different regions of the tropics.

We have also provided a short biography of some of the world's most eminent conservation scientists within a special Spotlight section in each chapter: Norman Myers (Chapter 1), Gretchen Daily (Chapter 2), William Laurance (Chapter 3), Mark Cochrane (Chapter 4), Daniel Simberloff (Chapter 5), Bruce Campbell (Chapter 6), Daniel Pauly (Chapter 7), Stephen Schneider (Chapter 8), Stuart Pimm (Chapter 9) and Peter Raven (Chapter 10). These biographies are followed by a brief set of questions and answers that focus on some of the most pertinent and pressing issues in tropical conservation biology today. It is our intention that readers of *Tropical Conservation Biology* will benefit from the knowledge and be inspired by the passion of these renowned conservation experts.

In summary, we have chosen to emphasize the biological aspects of biodiversity conservation, rather than the social or political processes that drive human behaviour. Regardless, the book should not be viewed as a resource solely for biologists – it also contains important information on how natural resource managers, politicians and policy makers can mitigate many of the negative effects of an increasingly human-dominated planet. We have strived overall to present the book's content so that it will appeal to advanced undergraduate and postgraduate students, scientists and managers with an interest in tropical conservation biology. It is with this general objective that we hope you will find *Tropical Conservation Biology* interesting, useful and instructive.

Acknowledgements

Funding for research for this book was provided by the National University of Singapore (R-154-000-264-112), Charles Darwin University and the Australian Research Council. David Bowman, Bruce Campbell, Robin Chazdon, Richard Corlett, Tony Cunningham, Robert Dunn, Tim Flannery, Lian Pin Koh, Kelvin Peh, Mark Meekan, Hugh Tan and Cagan Sekercioglu reviewed individual chapters and provided helpful and insightful recommendations for improvement. David Bickford read the entire manuscript and made very valuable suggestions. Matthew Lim provided invaluable assistance in collating material and references, and preparing figures and diagrams. Karah Wertz helped in designing the book's cover. The scientists featured in each chapter's Spotlight took the time to provide their expert opinions and biographies: Bruce Campbell, Mark Cochrane, Gretchen Daily, William Laurance, Norman Myers, Daniel Pauly, Stuart Pimm, Peter Raven, Stephen Schneider and Daniel Simberloff. Our sincerest thanks to all listed above, and our families for their support.

1

Diminishing Habitats in Regions of High Biodiversity

In this chapter, we review the loss of native habitats across the tropics: the region that lies between the Tropics of Cancer and Capricorn, i.e. 23.5° north and south of the equator (Figure 1.1). The word 'tropics' is derived from the Greek word *tropos* meaning 'turn'. The average annual temperature of the tropics is higher and the seasonal change in temperature is less pronounced than in other parts of the world because the tropical zone receives the rays of the sun more directly than areas at higher latitudes. The seasons in the tropics are marked not by large temperature fluctuations, but by the combination of winds taking water from the oceans and creating seasonal rains, called monsoons, over the eastern coasts. Several different climatic types can be distinguished within the tropical belt. Distance from the ocean, prevailing wind conditions and elevation are all contributing elements. Tropical highland climates, which have some characteristics of temperate climates, also occur where high mountain ranges are located. The tropics are the world's largest reservoirs of humid forests (Amazon, Congo Basin and New Guinea); the immense vegetative growth of these lush 'rain forests' is attributable to the monsoon rains. Owing to decreasing rainfall towards the northern and southern limits of the tropics, climatic conditions favour low-latitude savanna, steppe and desert biomes. High temperatures and abundant rainfall make rubber (*Hevea brasiliensis*), tea (*Camellia* sp.), coffee (*Coffea robusta* and *C. arabica*), cocoa (*Theobroma cacao*), spices, bananas (*Musa* spp.), pineapples (*Ananas* sp.), oils and timber the leading agricultural exports of tropical countries.

Ironically, the tropical region where two-thirds of the world's biodiversity is found is also the backdrop of massive contemporary loss of native habitats, mimicking the historical land conversion witnessed over the past few centuries in Europe and the temperate regions of North America and Australia. As a result of this mega-rich biodiversity and unprecedented loss of habitats, the tropical region has obviously attracted a high level of interest from conservationists. We first present an overview of habitat loss, namely in rain forests, mangroves and tropical savannas and limestone karsts, and then follow with a report on the

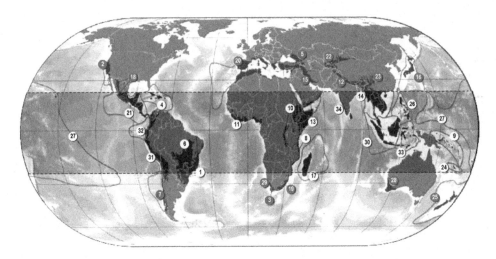

Figure 1.1 A map of the world showing the tropics and the distribution of 'biodiversity hotspots' outside (grey circles) and within (white circles) the tropics (shaded region): 1, Atlantic forest; 2, California floristic province; 3, Cape Floristic region; 4, Caribbean islands; 5, Caucasus; 6, Cerrado; 7, Chilean winter rainfall – Valdivian forests; 8, coastal forests of eastern Africa; 9, East Melanesian islands; 10, eastern Afromontane; 11, Guinean forests of west Africa; 12, Himalaya; 13, Horn of Africa; 14, Indo-Burma; 15, Irano-Anatolian; 16, Japan; 17, Madagascar and Indian Ocean islands; 18, Madrean pine–oak woodlands; 19, Maputaland–Pondoland–Albany; 20, Mediterranean basin; 21, Mesoamerica; 22, mountains of Central Asia; 23, mountains of southwest China; 24, New Caledonia; 25, New Zealand; 26, Philippines; 27, Polynesia–Micronesia; 28, southwest Australia; 29, Succulent Karoo; 30, Sundaland; 31, tropical Andes; 32, Tumbes–Chocó-Magdalena; 33, Wallacea; 34, western Ghats and Sri Lanka. (After conservation.org. Copyright, Conservation International.)

known and postulated drivers of native habitat loss in the tropics. Finally, we identify the areas in immediate need of conservation action by discussing the concept of biodiversity hotspots.

1.1   Loss of native habitats

If human impact on the natural environment continues unabated at its present rate or increases in severity, then by the turn of the century the resulting changes in land use will have exerted a profound and irreversible effect on tropical biodiversity (Sala *et al.* 2000). Habitat loss will probably have far greater effects on terrestrial ecosystems in the tropics than other drivers such as climate change, elevated carbon dioxide (CO_2) levels and invasive species (Sala *et al.* 2000). However, among these factors there are likely to be large and complex interactions that exacerbate the foreseen problems. Rain forest loss, degradation and fragmentation are the most widely publicized examples of habitat loss in the tropics; indeed, human activities, such as logging, are degrading and destroying tropical rain forests at a rate that has no historical precedence (Jang *et al.* 1996; Whitmore 1997; W.F. Laurance 1999). Given that the vast majority of the earth's terrestrial biodiversity is harboured in these threatened and little-studied biomes (E.O. Wilson 1988; Myers *et al.* 2000; Sodhi and Liow 2000), they are critical for conservation.

1.1.1   Rain forest depletion

Tropical forests cover 7% of the earth's land surface, yet they support over 50% of described species, plus a large number of undescribed taxa (W.F. Laurance 1999; Dirzo and Raven 2003). They are also critical for global carbon and energy cycles [Intergovernmental Panel for Climate Change (IPCC) 2002]; therefore, tropical forests are not only crucial for biodiversity conservation, they also play pivotal roles in moderating global climate change. Despite this importance, more than 40% of original tropical forests have been cleared in Asia alone (Wright 2005).

The United Nations Food and Agriculture Organization (FAO) has reported that countries with the largest annual net forest losses between 2000 and 2005 are all situated in the tropics (FAO 2005). These countries include Brazil, Indonesia, Sudan and Myanmar, and they collectively lost 8.2 million hectares (ha) of forest every year between 2000 and 2005 (FAO 2005). W.F. Laurance (1999) used data provided by the FAO (1993) on forest cover change from 1980 to 1990 and estimated that 15.4 million ha of tropical forest is destroyed every year, with an additional 5.6 million ha being degraded through activities such as selective logging. Overall, an average of 1.2% of existing tropical forests is degraded or destroyed every year (Whitmore 1997; W.F. Laurance 1999). In terms of absolute loss of area, forest conversion is the highest in the neotropics (South and Central America: 10 million ha/year), followed by Asia (6 million ha/ year) and Africa (5 million ha/year). However, if we consider forest conversion relative to the existing forest cover in the region, Asia clearly tops the list (W.F. Laurance 1999) (Figure 1.2), with 1.5 million ha of forest removed each year from the four main Indonesian islands of Sumatra, Kalimantan (Indonesian Borneo), Sulawesi and West Papua (Indonesian New Guinea) alone (DeFries *et al.* 2002). Even the so-called 'protected forests' in the tropics are not safe from

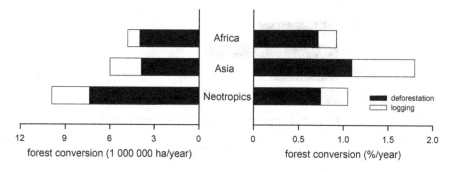

Figure 1.2   Relative and absolute rates of forest conversion in the major tropical regions throughout the decade of the 1980s. (After W.F. Laurance 1999. Copyright, Elsevier.)

plunder – of 198 protected areas surveyed, 25% have been losing forests within their administrative boundaries since the 1980s (DeFries *et al.* 2005).

Worryingly high as they are, whether the FAO values are accurate is controversial because they may fail to include catastrophic events (such as the vast 1997–98 forest fires that occurred in Indonesia) and may erroneously include forestry plantations as native forest cover (Matthews 2001; Achard *et al.* 2002). Deploying remotely sensed satellite imagery, Achard *et al.* (2002) reported that tropical forest loss may be much lower (5.8 million ha/year) than FAO estimates. Yet even Achard *et al.*'s estimates have being questioned. It has been argued that their lower estimates of forest loss may be unrepresentative, owing to the fact that they sampled only 6.5% of the humid tropics (Fearnside and Laurance 2003). Nevertheless, despite the different methods used, Achard *et al.* (2002) also found, as reported earlier by W.F. Laurance (1999), that rates of deforestation and forest degradation are highest in Asia (Figure 1.3).

The expansion of agriculture in the humid tropics is the main culprit in this devastating forest loss, with more than 3 million ha of forest converted annually by this activity (Achard *et al.* 2002). Although native forest loss in tropical Latin America seems to be decelerating, in a particularly disconcerting trend it continues to accelerate in tropical Asia (Matthews 2001) (Figure 1.4). This trend is further corroborated by a satellite imagery study of Latin America, tropical Africa and Asia by M.C. Hansen and DeFries (2004), who reported that deforestation appears to be accelerating in the last two regions (Figure 1.5). Depending on factors such as soil fertility and proximity to remnant forests (i.e. seed source), forest regeneration can proceed in abandoned areas following disturbance (Chazdon 2003). These secondary forests could be crucial for global carbon cycles and conservation of some forest biota (Wright 2005). However, about a quarter of regenerating forests are also being lost in tropical Asia and Africa (M.C. Hansen and DeFries 2005), with an increase of this forest type found only in Latin America (Figure 1.5).

There has also been controversy as to whether the global deforestation rate is subsiding over time. FAO data show that globally there has been a 31% decrease in the deforestation rate over the past two decades. But a study by M.C. Hansen and DeFries (2004) showed that deforestation has, in reality, accelerated by

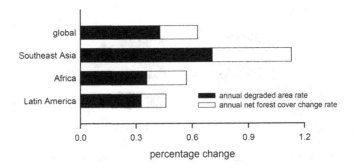

Figure 1.3 Mean annual estimates of deforestation in the humid tropics from 1990 to 1997. (Data derived from Achard *et al.* 2002.)

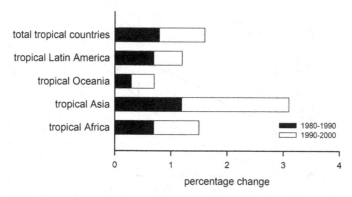

Figure 1.4 Worsening deforestation rates in all tropical regions except Latin America. (Data derived from Matthews 2001.)

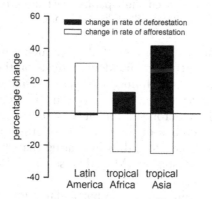

Figure 1.5 Positive increase in rate of deforestation (filled bars) and decrease in rate of afforestation (blank bars) in tropical Africa and Asia, with only Latin America showing opposite trends. (Data derived from M.C. Hansen and DeFries 2004.)

the same amount during the same period. Wright and Muller-Landau (2006) argue that the deforestation rate will slow down in the future due to a decrease in human population growth and increasing migration to urban centres. They argue that such changes in human demographics will be conducive to forest regeneration. However, Brook *et al.* (2006a) dispute this scenario because of decoupling between rural and urban human populations (Figure 1.6); even if deceleration does occur, it may be a little too late to stop the mass extinction of biodiversity in the tropics caused by the momentum of past habitat loss (see Chapter 9).

Globally, 0.8% of native tropical forests (primary and secondary forests, excluding plantations) are likely to be lost each year (Matthews 2001), and in countries plagued by civil war, such as Burundi and Rwanda, the rates of loss can be much higher (Table 1.1). Perhaps most dramatically, it has been estimated that, by 2010, human actions will have caused almost complete destruction of native lowland (< 1000 m elevation) forests from the hyper-biodiverse regions

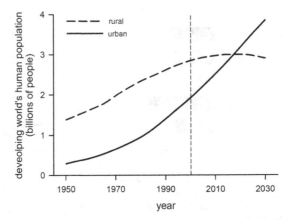

Figure 1.6 Trends in rural and urban human population growth in the world's developing countries. Vertical dashed line separates past and projected figures. (After Brook *et al.* 2006a. Copyright, Blackwell Publishing Limited.)

of Sumatra and Kalimantan (Jepson *et al.* 2001). Such a massive loss of habitat will almost certainly have profound incidental effects on the region's spectacular megafauna, such as the Sumatran rhinoceros (*Dicerorhinus sumatrensis*), Sumatran tiger (*Panthera tigris sumatrae*) and Asian elephant (*Elephas maximus*). Tropical countries with the highest deforestation rates (> 0.4% deforestation annually) usually have a large percentage of the remaining dense forests (> 80% canopy cover) near deforestation activities, thus indicating that these areas also remain highly vulnerable to deforestation (M.C. Hansen and DeFries 2004) (Figure 1.7).

The lowland tropical rain forests are particularly imperilled owing to their ready accessibility to an expanding human population and their increasing conversion to logging concessions, agricultural land and urban areas (Kummer and Turner 1994). In addition to this widespread forest type, other forest types also existing in the tropics are being destroyed (Whitmore 1997; Figure 1.8). For example, montane/submontane (usually > 1000 m elevation) rain (cloud) forests provide timber, fuel wood, soil and catchment protection. This forest type makes up 12% of the existing tropical forests worldwide, and it is currently being cleared at a rate twice that of the global average [Long 1994; Whitmore 1997; IUCN (The World Conservation Union) 2000]. In fact, montane forests are lost at a relatively higher annual rate than lowland tropical forests (Figure 1.9; Whitmore 1997). Montane forests, because of their unique environmental conditions (e.g. cooler temperatures), support a high degree of endemism. For example, the proportion of endemic moths is at least twice as high in montane forests than in their lowland counterparts in Malaysian Borneo (Chey 2000). Montane forests have a low recovery potential following disturbance (Ohsawa 1995; Soh *et al.* 2006). However, despite their fragility and high endemism, human activities continue to threaten these vulnerable forests (Ohsawa 1995; IUCN 2000).

Eighty-five per cent of global forest loss occurs in tropical rain forests (Whitmore 1997). However, the tropics also contain seasonal, dry or monsoonal

Table 1.1 Summary information on forests in tropical countries that have lost most of their original forests, showing land area, original and current natural forest area and change in forest area. Data derived from the United Nations Food and Agriculture Organization (FAO), United Nations Environment Programme, World Conservation Monitoring Centre (WCMC) and World Resources Institute (WRI). Countries are arranged (alphabetically) according to positive or negative annual percentage change in forest area ('total forest').

Country	Land area (000 ha)	Original forest area (000 ha) (% of land area)	Current natural forest area (000 ha) (% of original forest area) WRI (2000)	FAO (2005) primary forest	Mean annual % change in forest area 'Natural forest' – WRI (1990–2000)	'Total forest' – FAO (2000–2005)
Benin	11 262	1802 (16.0)	2538 (140.8)	na	-2.2	-2.5
Burundi	2783	1280 (46.0)	21 (1.6)	0 (0.0)	-8.9	-5.2
Comoros	186	112 (60.2)	6 (5.4)	0 (0.0)	-4.1	-7.4
Ghana	23 854	15 744 (66.0)	6259 (39.8)	353 (2.2)	-1.6	-2
Honduras	11 209	11 209 (100.0)	5335 (47.6)	1512 (13.5)	-1.1	-3.1
Mauritania	102 552	0 (0.0)	293 (na)	na	-3	-3.4
Nigeria	92 377	41 570 (45.0)	12 824 (30.8)	326 (0.8)	-2.5	-3.3
Philippines	30 000	28 500 (95.0)	5036 (17.7)	829 (2.9)	-1.9	-2.1
Togo	5679	1874 (33.0)	472 (25.2)	0 (0.0)	-3.1	-4.5
Uganda	24 124	16 873 (69.9)	4147 (24.6)	na	-1.8	-2.2
Cape Verde	403	na	0	na	na	0.4
Costa Rica	5110	5008 (98.0)	1790 (35.7)	180 (3.6)	-1.3	0.1
Côte d'Ivoire	32 246	na	na	625 (na)	na	0.1
Cuba	11 086	9977 (90.0)	1867 (18.7)	na	0.1	2.2
Gambia	1130	441 (39.0)	479 (108.6)	na	na	0.4
Palau	46	na	na	na	na	0.4
Rwanda	2634	948 (36.0)	46 (4.9)	0 (0.0)	-7.8	6.9
St. Vincent and Grenadines	39	17 (43.6)	6 (35.3)	na	-1.6	0.8
Cape Verde	403	na	0	na	na	0.4

na, not available.

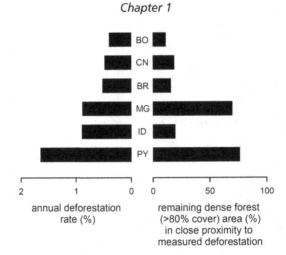

Figure 1.7 Deforestation rates and remaining forest area for countries with a clearing rate of greater than 0.4% per year. Country codes: BO, Bolivia; BR, Brazil; CN, People's Republic of China; ID, Indonesia; MG, Madagascar; PY, Paraguay. (Data derived from M.C. Hansen and DeFries 2004.)

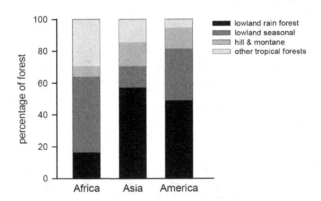

Figure 1.8 Percentage occurrence of different forest types across different regions. (Data derived from Whitmore 1997.)

deciduous forests. These generally lie below 1000 m elevation in regions such as Central America, Madagascar and Asia (Thailand) (Ruangpanit 1995; W.F. Laurance 1999), and constitute 33% of the existing tropical forests in the world (Whitmore 1997). Owing to their proximity to human habitation, seasonal forests also suffer a similar fate as lowland rain forests. In fact, seasonal forests are often grouped, for convenience, with rain forests, and are thus included in some of the regional deforestation calculations (e.g. Achard *et al.* 2002). Indeed, it is estimated that seasonal forests are being lost at the highest rate of any forest type (Figure 1.9) (Whitmore 1997). In some regions, such as Central America and Madagascar, more than 96% of these forests have already been destroyed (Krammer 1997; A. P. Smith 1997; W.F. Laurance 1999).

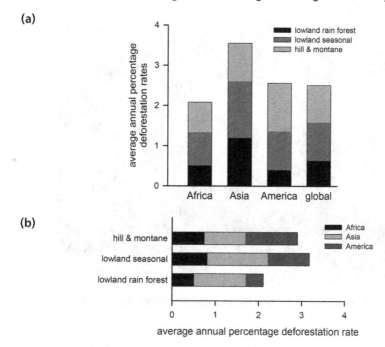

Figure 1.9 Average annual deforestation rates of different forest types at regional (a) and global scales (b), with lowland seasonal forests being lost at the highest rate globally. (Data derived from Whitmore 1997.)

1.1.2 Mangrove loss

Mangrove forests (growing in saline coastal environments) represent another unique tropical ecosystem. Mangroves are juxtaposed between land and sea and found within 25° north and south of the equator, covering approximately 8% of the world's coastline in 112 countries (Figures 1.10 and 1.11) (Adeel and Pomeroy 2002). In addition to direct overharvesting of trees, mangroves face threats from pollution, siltation, coastal development, aquaculture development and boating and shipping (Adeel and Pomeroy 2002). Traditionally, mangroves have been undervalued and largely considered to be useless swamps or wasteland (Liow 2000; Adeel and Pomeroy 2002). However, as with other forest types, mangroves support extensive biodiversity and contribute to varied ecosystem functions. For example, the presence of mangroves may enhance fish, shrimp and prawn catches (Baran and Hambrey 1998), producing an estimated US$66 to almost US$3000 of fisheries-related annual income from 1 ha of mangrove (Baran and Hambrey 1998). Although this estimate may be inflated because it does not include fisheries yield exclusively reliant on mangroves, it does show that human livelihoods can depend on this habitat type. Because of these indirect benefits to human well-being, conversion of mangroves for aquaculture may generate around 70% less revenue from the overall system than the pristine

Figure 1.10   A tropical mangrove swamp in Singapore. (Photo by Hugh Tan.)

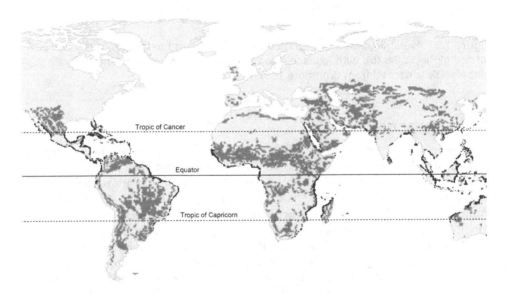

Figure 1.11   Global distribution of mangroves (black) and tropical savannas and grassland (dark grey). [After United Nations Environment Programme – World Conservation Monitoring Centre (UNEP-WCMC) online 2007. With permission.]

state (Balmford *et al.* 2002). Despite their environmental and economic benefits, mangroves are currently being lost at a rate of 2–8% per year. Between 4% and over 60% of the original mangrove cover has been lost in different tropical regions (Figure 1.12) (Valiela *et al.* 2001; Adeel and Pomeroy 2002).

In Southeast Asia, Singapore epitomizes mangrove destruction and conversion: mangrove forest cover amounted to 6334 ha (63% of original) in 1953, but by 1993 this had declined to only 6.5% (Hilton and Manning 1995), with a further reduction to 4% projected by 2030 (Figure 1.13). The primary driver of this massive destruction of mangrove forests in Singapore has been coastal development associated with urban expansion and industrialization (Hilton and Manning 1995). The mangrove loss in Singapore has also resulted in biotic losses. For example, at least four mangrove plant species (e.g. *Barringtonia conoidea*) have been extirpated from the island (Liow 2000). On a positive note, mangrove cover in the Central American nation of Costa Rica increased by 6% between 1983 and 1990 as a result of regeneration and plantations (Adeel and Pomeroy 2002).

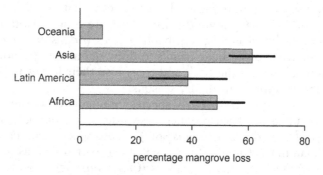

Figure 1.12 Percentage of original mangrove forest loss in different regions. Error bars represent standard error and are missing from Oceania because it is represented only by Papua New Guinea. (Data derived from Adeel and Pomeroy 2002.)

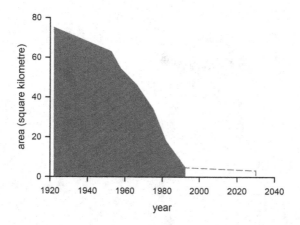

Figure 1.13 Decrease in area of mangrove forest in Singapore, with projected estimates up to 2030. (After Hilton and Manning 1995. Copyright, Cambridge University Press.)

1.1.3 Loss of tropical savannas

Tropical savannas or grasslands are associated with a highly seasonal climate of a prolonged dry and a shorter wet season (Figures 1.11 and 1.14). This type of extreme climate is expected to produce some form of forest or woodland, but soil conditions (e.g. low soil fertility) or disturbance prevent the establishment of dominant tree cover in these areas. Indeed, savannas are shaped (and now managed) by fire and grazing pressures of mega-herbivores. The African savannas are the best known, covering much of central and southern Africa. However, savannas also cover large areas of Central and South America (pampas), western India and northern Australia (Figure 1.11). Savannas are characterized by continuous cover of perennial (a plant persisting for several years) grasses, often reaching 3 m at maturity. Savannas may have an open canopy of drought-, fire- or browse-resistant trees. The dominant layer of these trees distinguishes the type of savanna, for example acacia savanna and pine savanna. Both tree and grass species in savannas have underground root systems that allow them to survive prolonged periods of drought and/or fire. There are other adaptations to resist the stress imposed by droughts, e.g. baobab trees (*Adansonia digitata*; Figure 1.15) in Africa have evolved huge trunks to draw out and store moisture during drought, perennial grasses die back and trees lose their leaves to reduce water loss through transpiration during dry seasons.

The world's greatest diversity (> 40 species) of ungulates (hoofed mammals) is found in the savannas of Africa and includes wildebeest (*Connochaetes taurinus*), oryx (*Oryx gazella*) and zebra (*Equus* spp.). These species-rich communities of large-bodied mammals attract a diverse set of carnivores such as lions (*Panthera leo*), cheetahs (*Acinonyx jubatus*), jackals (*Canis adustus*) and hyenas (*Crocuta crocuta*). Termites are also abundant in tropical savannas; they feed on decomposing animal and plant remains and thus are important for maintaining soil fertility. In

Figure 1.14 A tropical savanna in Kenya. (Photo by Chuck Bargeron, University of Georgia, www.forestryimages.org.)

Figure 1.15 Baobab trees (*Adansonia digitata*) in Kenya. [Photo by Robert L. Anderson, United States Department of Agriculture (USDA) Forest Service, forestryimages.org.]

fact, trees growing nearer to termite mounds in western Zimbabwe have higher nutrients and are preferred as browse by elephants (*Loxodonta africana*) (Holdo and McDowell 2004). Additionally, termite nesting mounds provide shelter for other animals and food for anteaters (*Myrmecophaga* spp.) and pangolins (*Manis* spp.).

The Cerrado region of Brazil exemplifies the importance of, and threats to, the savannas. These savannas cover 21% of Brazil and are exceptionally rich in endemic species (Figure 1.16) (Klink and Machado 2005). Despite this richness, 55% of the 2 million km² of forest cover in the Cerrado has been transformed or cleared for human use, with the main threats being soil erosion, uncontrolled fire and the spread of exotic grasses (Klink and Machado 2005). An issue of particular conservation concern is that 20% of threatened endemic species do not occur in any of the region's protected areas (Klink and Machado 2005). A similar predicament is unfolding in the savannas of the Serengeti National Park in Tanzania (Sinclair *et al.* 2002). Some 5% of the grasslands in this region are each year converted for human use, such as agriculture. Compared with native savannas, only 28% of all bird species, and 50% of insectivorous and granivorous species, are found in agricultural areas (Sinclair *et al.* 2002). That study suggested that continuing human encroachment in this ecosystem would negatively affect bird species and perhaps some larger vertebrate species, such as lions and cheetahs. Worldwide, 50% of the tropical and subtropical savannas have already been sequestered for use by humans, with further projected losses of about 20% by the year 2050 (Figure 1.17) (Millennium Ecosystem Assessment 2005).

1.1.4   Loss of limestone karsts

Limestone karsts are sedimentary rock outcrops composed primarily of calcium carbonate (Figure 1.18). They were formed millions of years ago by calcium-secreting marine organisms (e.g. corals) and were subsequently lifted above the sea level by tectonic movements. The complex terrain of karsts translates into high species diversity because many of the species can be site-endemic owing to

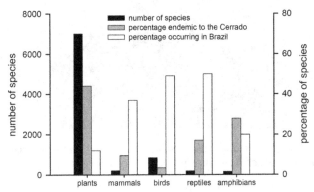

Figure 1.16   Number of species with per cent endemism and proportion of estimated species richness in the savannas of Brazil. (Data derived from Klink and Machado 2005.)

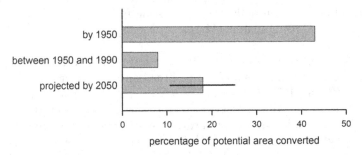

Figure 1.17   Loss of savannas over the past century and projected conversion by 2050. Bars represent medium certainty, and the error bar represents a range of values from four different Millennium Ecosystem Assessment scenarios. (After Millennium Ecosystem Assessment 2005. With permission.)

a high degree of isolation among the karsts. For example, 21% of 1216 karst-associated plant species are endemic to Peninsular Malaysia, with 11% of these being site endemics (Chin 1977). Similarly, 80% of Malaysian landsnails live on karsts, with many of them occurring only on individual karsts (Clements *et al.* 2006). Karsts should have high conservation priority because they are home to 143 globally threatened species, and 18 karst-dominated species have already been lost from Peninsular Malaysia due to habitat loss (Kiew 1991; Clements *et al.* 2006). In addition, karsts are valuable to humanity because of their rainwater storage abilities and thus play a role in maintaining the hydrological integrity of catchments. They also are magnets for tourists – karsts in Sarawak (Malaysia) generate at least US$80 000 annually in tourism-related revenues. Despite this, tropical karsts are currently being quarried heavily for limestone to be used in over 100 commercial products such as cement. More than 400 million tonnes of limestone were quarried from karsts in the tropics over a period of 5 years (Figure 1.18). Clearly, there is a need for more conservation attention to this imperilled ecosystem.

In addition to loss, fragmentation and degradation of once pristine areas, tropical habitats are affected by drivers such as climate change, invasive species, overexploitation (e.g. hunting) and pollution (Figure 1.19; Millennium

Figure 1.18 (a) Pristine tower karsts in Sarawak, East Malaysia, and (b) the mean (± standard error) annual limestone quarrying rates of four major tropical areas (1999–2003). (After Clements *et al.* 2006, Copyright, American Institute of Biological Sciences. Photo by Reuben Clements.)

Ecosystem Assessment 2005). These elements will be covered in later sections and chapters.

1.2   Drivers of habitat loss

Direct causes of deforestation (and loss of other habitats) are multifarious, and include slash-and-burn clearing associated with swidden agriculture, selective logging, cattle ranching, plantations, permanent agriculture, fuel wood collection and transmigration. These drivers can act individually or in concert. In the tropics, the main proximate drivers of deforestation are agriculture, followed by timber extraction and infrastructure (urban) expansion (Geist and Lambin 2002). However, the precise causes of deforestation are underpinned by complex and geographically variable factors. For instance, some governments have little choice but to sell forests as logging concessions to alleviate foreign debt (Bawa and Dayanandan 1997). Below we discuss in some detail the various drivers of habitat loss in the tropics.

1.2.1   Human population pressure

The increasing demand for, and consumption of, natural resources by humans show no sign of abating. Rapid economic development, population expansion and poverty are key drivers of land conversion (Giri *et al.* 2003; Jha and Bawa 2006). Demographic growth is particularly steep in tropical countries, where

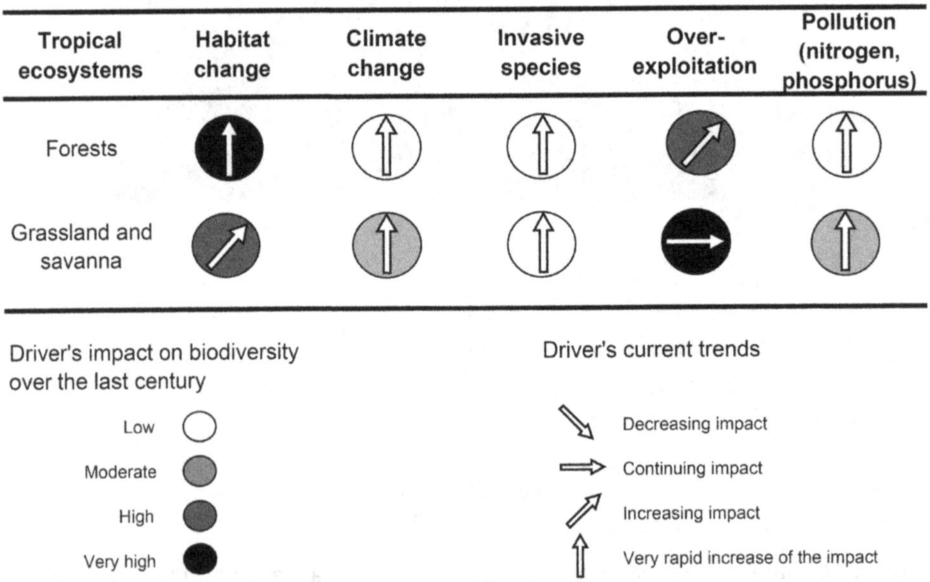

Figure 1.19 Factors that affect tropical habitats and their current trends. (After Millennium Ecosystem Assessment 2005. With permission.)

the size of the human population has increased by 3.1 billion between 1950 and 2000, and is projected to grow by another 2 billion before 2030 (United Nations 2004). Within the next 100 years, as many as 11 billion people may inhabit the planet, a number that will be difficult to sustain (Palmer *et al.* 2004). Urbanization will greatly expand in the future, with expectations that more than half of the world's total human population will be living in cities by 2030 (see Figure 1.6) (Palmer *et al.* 2004). Expanding human populations, and their specific actions (e.g. land conversion), exert substantial pressure on natural resources and native biodiversity (Cardillo *et al.* 2004). Poor policy choices can also exacerbate environmental destruction (Jha and Bawa 2006).

It would be immeasurably informative, from both a scientific and management perspective, if we could hypothetically excise a representative tropical country, allow it to fulfil its economic potential, and document the consequent loss of natural habitats and biodiversity, all within a greatly accelerated time frame. It is both depressing and fortunate that the Southeast Asian island nation of Singapore provides exactly such an ecological worst-case scenario for the tropics. Singapore has experienced an exponential population growth from around 150 subsistence-economy villagers around 1819 to 4 million people in 2001 (Corlett 1992; World Bank 2003). In particular, within the past few decades, Singapore has transformed itself from a Third World country of squatters and slums to a developed metropolis of economic prosperity and thus has been widely regarded, by the regional developing countries, as the ideal economic model. However, the success of Singapore came at a hefty price, one that was unfortunately paid for most particularly with its biodiversity (Brook *et al.* 2003a). The island has suffered massive deforestation, initially from the cultivation of short-term cash crops (e.g. gambier: *Uncaria gambir*, rubber), and subsequently from urbanization and industrialization (Corlett 1992). With this destruction of habitats (rain forest, swamp forest and mangroves) has come the extirpation of at least a third of the island's known biodiversity. Similar environmental scenarios are now unfolding in other tropical countries, often at much greater geographic scales (Jepson *et al.* 2001).

As the human population of tropical countries continues to grow, enormous pressures will be placed on their natural resources (World Bank 2003). The current trend in the tropics suggests that forest loss will likely increase in tandem with human population density, and in some cases due to economic expansion as well (Figure 1.20 for all tropical countries). The point to note, however, is that the pressure of dense human populations represents only one of the factors in habitat loss because, even in areas with relatively few human residents in the tropics, there can be widespread loss of natural forests (Whitmore 1997).

A burgeoning human population, even with an increasing concentration of people in urban areas, means more mouths to feed, and agriculture is the main factor in land conversion in the tropics (Figure 1.21), with its estimated contribution to annual tropical forest losses being as high as 90% (Hardter *et al.* 1997; Achard *et al.* 2002). In Asia, 100 million ha of land was converted to cropland between 1880 and 1980. Over these 100 years, the area of land converted to agriculture expanded by 86% in this region (Flint 1994; Richards and Flint 1994), largely at the expense of its forests (Flint 1994; Bawa and Dayanandan 1997) (Figure

(a)

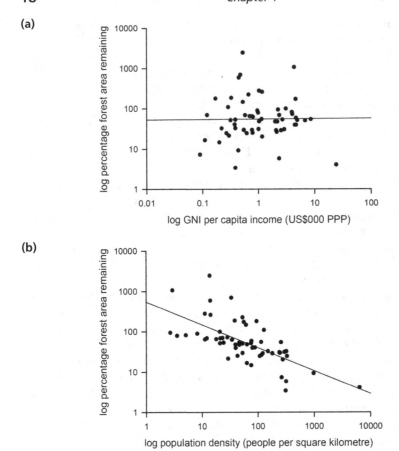

(b)

Figure 1.20 Socioeconomic correlates of percentage forest loss. The proportion of forest area remaining in tropical countries did not correlate with (a) gross national income (GNI) but correlated negatively with (b) population density. Percentage of forest area remaining is defined as the proportion of current total forest area over estimated forest cover about 8000 years ago assuming current climatic conditions. PPP, purchasing power parity. (Data derived from www.earthtrends.wri.org.)

1.22). Production of soya bean (*Glycine max*) has increased 100-fold since 1961 in Argentina and Brazil, largely for export to China (Donald 2004). This has resulted in a severe shrinking of Cerrado grasslands (Donald 2004).

Globally, over the last three decades, agricultural areas have doubled from 50 million ha to 100 million ha (Niesten *et al.* 2004), and now cover a quarter of earth's land surface (Millennium Ecosystem Assessment 2005). By 2030, it is predicted that an additional 120 million ha of agricultural land will be needed by developing countries to support their increased populations (M. Jenkins 2003); therefore, land clearing for agriculture is almost certain to continue at a rapid pace. In addition, in many areas of the tropics, factors such as low soil fertility and high levels of erosion mean that land conditions are not particularly conducive to sustainable agriculture, thus promoting a cycle of forest destruction. Farmers

Figure 1.21 Agriculture in the tropics. (Photo by Cagan Sekercioglu, naturalphotos.com.)

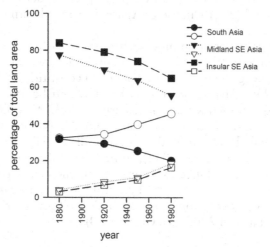

Figure 1.22 Change in cultivated area and forest cover of South and Southeast Asian countries (filled symbols: forest cover; unfilled symbols: cultivated areas). (After Flint 1994. Copyright, Elsevier.)

have to burn forest vegetation to release nutrients that enhance soil fertility and, as a result of the region's characteristically high rainfall, these nutrients are usually washed away rapidly, making the soil less fertile; in as little as 3 years, the ground is no longer capable of supporting crops (Härdter *et al.* 1997). Farmers are then forced, because of the reduced soil fertility, to look for new forested areas to clear and burn.

Agricultural plantations (large-scale, export-oriented monocropping) are also on the rise in the tropics. Areas under plantation doubled from 4% to 8% between the 1970s and 1984 (Hartemink 2005) – a trend that has most likely continued since. For instance, the area under oil palm (*Elaeis guineenis*) cultivation increased from 150 000 ha in the 1970s to over 3 million ha in 1998 (Hartemink 2005). Similar trends have been found for cocoa plantations in the Ivory Coast and for sugar cane (*Saccharum officinarum*) plantations in India and Brazil. In Brazil, soya bean (*Glycine* spp.) cultivation has increased by 13 million ha in the past 30 years, mostly at the expense of its wooded savannas (Niesten *et al.* 2004); these agricultural plantations contribute substantially to the gross national product (GNP). Forest conversion to cropland in the Brazilian Amazon is directly correlated with soya bean prices, indicating that high crop prices in the international market can promote deforestation (D.C. Morton *et al.* 2006). In the 1980s, exports from plantations contributed 22% to the GNP of the Ivory Coast, and rubber accounted for 10% of GNP exports in Malaysia (Hartemink 2005). However, these plantations can release pollutants, cause soil erosion and declining soil fertility and can be poor in sequestering carbon and promoting biodiversity values (Hartemink 2005).

Cultivation of coca and poppies for the illicit drug trade may be the cause of almost half of the deforestation in neotropical countries (Aldhous 2006). Biologists and governmental officials trying to prevent deforestation as a result of illicit drug demand (mostly in the USA) are threatened with violence (Aldhous 2006).

Massive resettlement programmes, such as those in Indonesia, Thailand, the Philippines, Burundi and Rwanda, also facilitate deforestation (R.L. Bryant *et al.* 1993; www.fao.org). Fuel wood is thought to constitute 85% of total energy consumption in West Africa, and this has resulted in heavy deforestation and occasional shortages of such wood, particularly in Niger, Nigeria and Togo (www.fao.org). Civil war in tropical countries has compromised conservation because illegal logging money has been used to fund conflict (Talbott and Brown 1998; Draulans and van Krunkelsven 2002), although war can sometimes facilitate forest regeneration, for instance by deterring commercial operations (Hecht *et al.* 2006). Last, but not least, the liberal granting of forestry concessions, largely though cronyism and corruption, does not bode well for the remaining tropical forests (Geist and Lambin 2002; see Chapter 10).

1.2.2 Perverse subsidies

Some government actions inadvertently promote poor land use practices. Subsidies designed to promote agricultural production also facilitate land clearing (Barbier 1993; A.N. James *et al.* 1999). These subsidies are called 'perverse' because, over the long term, they adversely affect the economy and the environment (Myers 1998). For example, worldwide, citizens pay US$950–1450 billion annually to subsidise fisheries and the timber and oil palm industries (Myers 1998; van Beers and de Moor 1999). The largest amount of money (US$345 billion/year) goes to agriculture and fisheries (van Beers and van den Bergh 2001). Global ocean fisheries cost US$100 billion a year, of which only US$80 billion is recovered through sales – the shortfall is made up by government subsidies (Myers 1998). This has resulted in the depletion of fisheries stocks to unsustainable levels, bankruptcy of fishing businesses and a high unemployment rate in fisheries communities. Despite this problem, the European Union (EU) has been increasing subsidies to its fishing fleets in West Africa, thus artificially inflating profitability against the backdrop of declining fishing stocks (Brashares *et al.* 2004). In Brazil, the depletion of soils, forests and fisheries as a result of perverse subsidies to agriculture has depressed its economic growth potential by up to 30% (Myers 1998).

Coffee cultivation provides a useful illustration of the nature of perverse subsidies in agriculture. During the 1990s, Asian governments with the support of the International Development Bank promoted intensive coffee cultivation in countries such as Indonesia and Vietnam, hence elevating Indonesia to the world's fourth largest coffee exporter and the second largest producer of coffee after Vietnam. Unfortunately, this short-term economic gain resulted in massive forest conversion, and eventually proved to be economically unsustainable due to overproduction and the subsequent collapse of world prices (O'Brien and Kinnaird 2003). Despite it being clear that coffee production is not economically attractive in many situations and is detrimental to biodiversity, the Indonesian government plans to expand coffee production further. O'Brien and Kinnaird (2003) recommended that the adoption of coffee cultivation be strongly discouraged in protected areas and that strident attempts should be made to curtail deforestation for this type of cultivation. Certainly, organizations such as the International Coffee Organization need to play a bigger role in promoting a balance between coffee production and biodiversity needs.

In addition to agriculture, governmental financing of road construction can act as an indirect subsidy by facilitating logging (Flint 1994; van Beers and van den Bergh 2001) and can lead to an increase in the hunting of wild animals by providing easier access to the forests (see Chapter 6). Further, underpricing of timber and subsidies of private harvesting (e.g. low logging fees and taxes) promote deforestation (Barbier 1993). For example, in the Philippines, timber revenues collected by the government during the late 1980s were six times lower than the prevailing market value (US$39 versus US$250 million; Barbier 1993). More than 46% of the money generated by timber concessions in Malaysia and Indonesia since the early 1990s remains uncollected (www.fao.org), suggesting that logging does not benefit the local people economically.

1.2.3 Commercial logging

Commercial logging is also another common direct driver of deforestation, especially in the tropics (Geist and Lambin 2002). Trees are felled for sale as timber, timber products (e.g. woodchips), or pulp. Between 1981 and 1990, 5.6 million ha of tropical forest was logged for timber (Whitmore 1997). Such logging activities predominantly occur in the primary forests (Figure 1.23; Whitmore 1997). Forestry industries account for 3% of the world's gross economic output, or approximately US$330 billion annually (Sizer and Plouvier 2000). The annual total worldwide consumption of wood is around 3 billion cubic metres, with about half of that consumed as firewood. The Asia-Pacific region now drives log exports in the tropics, encompassing 67% of the total volume (Sizer and Plouvier 2000; Figure 1.24).

Over the past few decades, Indonesia, Brazil and Malaysia have logged more than half of the world's commercially viable tropical timber, with Thailand and India also in the top five producers [ITTO (International Tropical Timber Organization) 2003; Figure 1.25]. The high level of commercial logging is also of concern in countries with currently vast intact tropical forests, such as Gabon, Cameroon and Papua New Guinea (Figure 1.26), as there is little possibility of further protection of forests in these countries. In many of these countries,

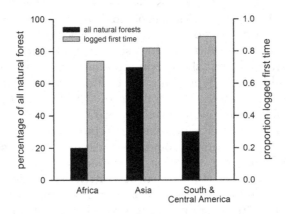

Figure 1.23 Proportion of remaining natural forest areas and percentage of primary forests area logged for the first time in those forests. (Data derived from Whitmore 1997.)

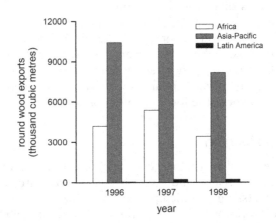

Figure 1.24 Log exports of round wood from different regions from 1996 to 1998. (Data derived from Sizer and Plouvier 2000.)

timber harvesting occurs in the absence of any management plan that might seek sustainability [Centre for International Forestry Research (CIFOR) 2004].

Japan, South Korea and the People's Republic of China are the main importers of tropical timber products (Sizer and Plouvier 2000). Forest products exported by developing countries are usually subject to low export duties, particularly on unprocessed logs (Barbier 1993). Trade liberation, through the removal of export

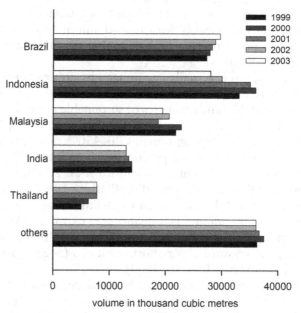

Figure 1.25 Tropical timber producers from 1999 to 2003. (Data derived from International Tropical Timber Organisation 2003.)

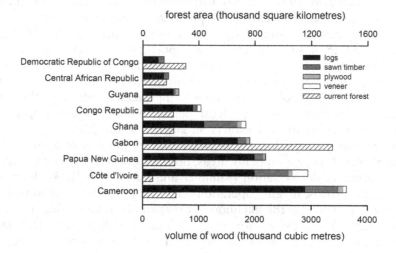

Figure 1.26 Production of various commercial logging materials from the top nine countries and their remaining natural forest area. (Data derived from Sizer and Plouvier 2000.)

restrictions, may increase log exports (up to fourfold in the Philippines) and thus cause a further acceleration in deforestation (Barbier 1993). In addition, current bans on commercial logging within India and the People's Republic of China promote higher demand for the supply of tropical wood from other countries (www.birdlife.net). Between 1997 and 2002, China's forest product imports soared by 75%, from US$6 to 11 billion (Sun *et al.* 2004). Over 70% of China's forest products are supplied by countries in the Asia-Pacific region (Katsigris *et al.* 2004). Many supplying countries are rife with unsustainable harvesting, illegal logging and concomitant negative impacts on human livelihoods. Greater attention by governments, market leaders and international organizations is needed to address the problems that tropical timber export creates in the source countries (Katsigris *et al.* 2004).

Moreover, deforestation is actually encouraged by governments in the tropics because of high international demand for tropical wood and wood products (Kummer and Turner 1994), and examples of sustainable natural forest use in the tropics are difficult to find (Bowles *et al.* 1998; Putz *et al.* 2000; W.F. Laurance *et al.* 2001a). Unsustainable logging remains 20–450% more profitable, at least over the short term, than sustainable practices such as fruit collection and meat production (Bowles *et al.* 1998; but see Balmford *et al.* 2002). Logging in many tropical countries is still done by clear-felling (Barbier 1993), yet even 'selective logging' can be very wasteful, resulting in the felling of an average of 25 non-commercial trees for every commercial-quality tree extracted (Myers 1991) and 40–50% of the canopy cover being destroyed (Cochrane 2003). Further, logging concessions are typically awarded for short durations (5–25 years) that tend to promote logging in new areas (Barbier 1993).

Old-growth forests are also being targeted and depleted across the tropics (Whitmore 1997). The highest percentage of logging in neotropical old-growth forests occurred from 1981 to 1990 (see Figure 1.23; Whitmore 1997), and in many areas of Southeast Asia old-growth forests are being logged to oblivion (Barbier 1993; Whitmore 1997). Clearly, stiffer guidelines are needed to regulate, and perhaps curtail, mass tropical timber production and export. These might include mandatory timber certifications, better economic incentives to encourage sustainable harvesting, substantial reviews of export duties and laws and a more widespread promotion and education of environmental issues (see Chapter 10). Furthermore, because they are often the only institutions in remote areas, logging companies can themselves assist in environmental protection, although there are instances of officials wanting to stop illegal logging being intimidated by arson, attempts at bribery and murder (Jepson *et al.* 2001; Aldhous 2006).

Agriculture and commercial logging may have differing impacts on deforestation rates in the tropics; however, in many areas, both logging and agriculture act in concert to exacerbate deforestation (Kummer and Turner 1994). As described earlier, by creating roads, logging operations enhance physical access to forests. This greater access increases the likelihood of invasion by humans (e.g. hunters, farmers and miners) and exotic organisms associated with humans (e.g. rats, dogs, cats), and is a cause of considerable concern for the long-term prospects of tropical biodiversity (W.F. Laurance *et al.* 2001a).

1.2.4   Weak governance

Anaemic national institutions and poor enforcement of legislation remain a major hindrance to curtailing tropical deforestation (W.F. Laurance 1999). Liberal granting of forest concessions, non-existent or poor forestry practices, weak governance structures and political corruption all work to maintain high deforestation rates in developing countries (R.L. Bryant *et al.* 1993; Geist and Lambin 2002; R. J. Smith *et al.* 2003). Illegal logging (timber harvesting, transportation and trade in violation of national laws) and encroachment into nature reserves remain a problem across the tropics (www.fao.org/forestry; T. Whitten *et al.* 2001; DeFries *et al.* 2005). For example, it is reported that illegal and possibly unsustainable logging remains rampant in Indonesia with the implicit backing of politicians, businesses and the military (Kinnaird and O'Brien 2001; T. Whitten *et al.* 2001; Stibig and Malingreau 2003; see Chapter 10). Certain people implicated in illegal logging continue to retain prominent political positions (T. Whitten *et al.* 2001). Forest policies are not sufficiently well developed to protect the remaining forests adequately and to stem the growth of industries based on forest exploitation. For instance, Indonesia's paper and pulp industry has grown sevenfold since the 1980s, with a similar expansion in the production of plywood. Further, the oil palm plantation areas and resettlement plans for some native people have placed even greater pressure on these forests (Stibig and Malingreau 2003).

Sadly, the events occurring in Indonesia are not anomalous – they in fact epitomize disturbing actions across most of the tropics. Illegal logging costs the timber industry up to US$15 billion annually (S. Johnson 2003). Seneca Creek Associates and Wood Resources International (2004) estimate that up to 10% of forestry trade worldwide is based on illicit timber products. This rampant illegal activity in the tropics makes up between 20% and 90% of forest production and trade in some tropical countries (Figure 1.27; Seneca Creek Associates and Wood Resources International 2004), with corruption seeming to facilitate logging (Figure 1.28). The construction of roads in many tropical areas further enables illegal logging (Barbier 1993).

In many tropical nations, forests are managed by government departments, whose main interest seems to be to increase commercial logging and the export of timber products without any consideration for the conservation of biodiversity (Byron and Waugh 1988). Furthermore, these departments remain understaffed and heavily politically influenced (Barbier 1993), so that logging companies identified as having poor environmental practices (e.g. promoting forest fires) can continue to conduct unscrutinized 'business as usual' (T. Whitten *et al.* 2001). Rampant corruption and illegal activities hinder proper management of forests in many parts of the tropics (Kummer and Turner 1994; see Chapter 10). For example, corruption has been a major contributing factor to the heavy deforestation of the Philippines (Kummer and Turner 1994), and this is very likely to be also a major factor in furthering deforestation in other tropical countries. The few but well-publicized struggles by some native groups (e.g. Penan of Sarawak) to halt logging have generally been in vain (R.L. Bryant *et al.* 1993).

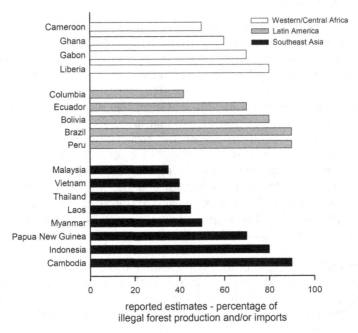

Figure 1.27 Proportion of illegal production of wood products and/or imports of different countries across various tropical regions. (Data derived from Seneca Creek Associates and Wood Resources International 2004.)

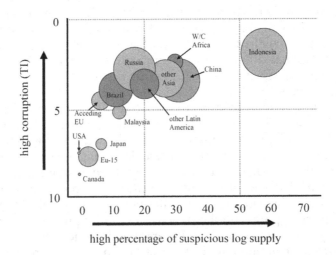

Figure 1.28 Relationship of estimates of suspicious log supply and the corruption index of selected countries. Size of bubbles represents volume of suspect roundwood, including imports. (After Seneca Creek Associates and Wood Resources International 2004. With permission.)

1.3   Biodiversity hotspots

Because there are finite economic and logistical resources available for conservation, wisdom dictates that their value in retaining biodiversity must be maximized. Therefore, it is critical to identify priority areas where conservation needs are the greatest, in order to achieve the most significant payoffs in terms of the global preservation of species and communities. On this basis, Myers (1988) identified 10 terrestrial 'hotspots' where levels of vascular plant endemism were exceptional but which were marred by massive habitat loss and thus should be accorded the highest conservation priority. To these, Myers later added another eight hotspots – these 18 hotspots support 20% of the earth's plant species within just 0.5% of its land surface (Myers 1990).

Critics of the hotspots approach to identifying conservation priorities have argued that there are poor complementarities among taxonomic groups, e.g. hotspots for plants may not be hotspots for butterflies (Reid 1998). In 2000, Myers *et al.* (see Spotlight 1: Norman Myers) expanded this paradigm of conservation planning by including, in addition to vascular plants, key vertebrate groups (amphibians, reptiles, birds and mammals) in their analyses. Vascular plants and vertebrate groups were the best candidates for such analyses because the available data on the distribution and conservation status were sufficient to make robust assessments. Myers *et al.*'s (2000) now famous analyses identified 25 terrestrial biodiversity hotspots globally – defined as regions that harbour a high diversity of endemic species from plant and animal taxa and, simultaneously, have been heavily impacted and altered by human activities. The criteria for designating a hotspot were that, out of the 300 000 known vascular plant species in the world, at least 0.5% or 1500 vascular plant species should be endemic to it, and that it should have lost at least 70% of its primary vegetation (habitats rich in endemic species). Collectively, these 25 biodiversity hotspots have exceptionally high endemism and harbour some 44% of world's plant species and 35% of its vertebrate species. These hotspot areas collectively have lost 88% of their primary vegetation, thus restricting these endemic species to only 1.4% of the earth's land surface. Sixteen (64%) of these hotspots are in the tropics. The five 'hottest' hotspots (highest levels of endemicity coupled with the greatest threats) are all in the tropics – the tropical Andes, Sundaland, Madagascar, Brazil's Atlantic Forest and the Caribbean. These five together contain 20% of all vascular plant and 16% of vertebrate species, yet cover a mere 0.4% of the earth's land surface.

Myers *et al.* (2000) argued that conservation in these 25 biodiversity hotspots is imperative if we are to counteract the mass extinction crisis that is now unfolding (see Chapter 9). The threat to biodiversity hotspots from humans is imminent – 20% of 1.1 billion people inhabit these areas (Cincotta *et al.* 2000). More worryingly, between 1995 and 2000, the human population expanded by 1.8% annually in these hotspots, which is a much higher rate than the average annual expansion for the whole world of 1.3% (Cincotta *et al.* 2000). This trend suggests that, in these hotspots, human-induced environmental changes are likely to proceed more rapidly than in other parts of the world (Cincotta *et al.* 2000; Shi *et al.* 2005). Conservation intervention is, therefore, absolutely necessary in these areas. Myers *et al.* (2000) calculated that, on average, US$20 million per

Spotlight 1: Norman Myers

Biography

My graduate education was based on systems ecology and resource economics, with lots of demography, sociology, ethics, forestry, and lengthy etc. thrown in. By the time I had completed my PhD at Berkeley, I was solidly disposed to specialize in being a generalist. I also decided that I was not a team player, and that I would be better off as a lone-wolf consultant in environment and development. I strongly recommend to anybody embarking on a career to find one that keeps their options open. In a former age it was OK to say at age 20 that you wanted to spend the next 50 years being a lawyer or a doctor or something of that sort. But today things are different. Within just another 10 years, the world will have changed out of sight, and you will have changed too, so you might well encounter a need to change horses in mid-career. I actually entered postgraduate school at the advanced age of 35, having been a colonial officer for my first career, a high-school teacher for my second, a professional photographer for my third and a journalist/book writer for my fourth. After Berkeley I became a consultant. At age 72, I am pondering what I could try for a sixth career.

Major publications

Myers, N. (1976) An expanded approach to the problem of disappearing species. *Science* **193**, 198–202.

Myers, N. (1996) The biodiversity crisis and the future of evolution. *The Environmentalist* **16**, 37–47.

Myers, N. (1998) Lifting the veil on perverse subsidies. *Nature* **392**, 327–328.

Myers, N. and Kent, J. (2003) New consumers: The influence of affluence on the environment. *Proceedings of the National Academy of Sciences of the USA* **100**, 4963–4968.

Myers, N. and Kent, J. (2005) *The New Atlas of Plant Management*. University of California Press, Berkeley, CA.

Myers, N., Mittermeier, R. A., Mittermeier, C. G., da Fonseca, G. A. B. and Kent, J. (2000) Biodiversity hotspots for conservation priorities. *Nature* **403**, 853–858.

Questions and answers

Have 'biodiversity hotspots' proven to be a useful concept for applied conservation?

Yes, the biodiversity hotspots thesis has (if I might indulge my immodesty) proved to be an especially useful concept for applied conservation. At any rate, it has attracted funding to the tune of US$850 million from the World Bank, Conservation International, the MacArthur and Moore Foundations and numerous NGOs. For fully two decades before I first formulated the hotspots thesis in the mid-1980s, I had been struck that conservation bodies had been spreading their all-too-inadequate funds in terms of a bit for this species, a bit for that species, and so on, and not really making a big enough impression with any species (for the most part at least). There were no logically derived priorities in play. Note that my hotspots thesis is but one way of postulating a priority ranking, and there are a lot of others, despite protests from some quarters that I was seeking a monopoly over conservation options.

Which tropical hotspots are in most urgent need of protection and management?

The hotspots in most urgent need of protection and management are Madagascar, Sundaland, the Atlantic coastal forest of Brazil, the Caribbean, Indo-Burma, Western Ghats/Sri Lanka, Eastern Arc, the coastal forests of Tanzania/Kenya and the Mediterranean Basin.

Do you think that the current media focus on climate change is shifting emphasis away from the more immediate, direct threats to biodiversity such as deforestation?

No. The current media focus on climate change should surely be complementary to long-standing and more immediate threats such as tropical deforestation. But note that all major environmental issues of today are intricately interlinked; for instance, deforestation can often increase the warming effect of climate change (bare earth absorbs more heat than thick vegetation).

What are the most urgent research problems now facing tropical conservation biology?

I consider that the urgent research problems facing tropical conservation biology are: (1) What are the socioeconomic and politicocultural factors that serve as root causes of deforestation (e.g. perverse subsidies)? (2) What are some interdependencies at work, e.g. how far do forest conservation and reforestation in temperate and boreal zones merely shift logging pressure onto tropical forests? (3) Which sectors of tropical biotas could serve as evolutionary hotspots, i.e. communities that can foster 'bounce-back' processes, notably speciation, when the current biotic crisis has played itself out?

In your opinion, in what condition will tropical ecosystems be at the end of the twenty-first century?

I fear that by the end of the twenty-first century tropical ecosystems will be badly battered, at best, owing to population pressures, socioeconomic forces and political incompetence and/or ignorance and/or corruption. But year 2100 is far too distant – as is 2050 – for one to make any informed or rational prognosis in detail.

year is needed to safeguard each hotspot – which amounts to a total of about US$500 million annually – a value many orders of magnitude lower than the US$1.4 trillion that global citizens spend in subsidies to degrade environments and economies alike (Myers 1999).

Conservation International has expanded Myer *et al.*'s (2000) analyses and now identifies 34 biodiversity hotspots (www.biodiversityhotspots.org). These hotspots once encompassed 16% of the earth's land surface, but the collective loss of 86% of vegetation means that their land coverage has now been reduced to 2.3%. Harbouring 50% and 42% of world's vascular plant and vertebrate species, respectively, these hotspots are clearly exceptionally rich in biodiversity. Out of the 34 hotspots (see Figure 1.1), 20 are located within the tropics, and biotas in these hotpots are severely imperilled. Seventy-seven per cent (1367 of 1770), 73% (898 of 1213) and 51% (568 of 1101) threatened amphibian, bird and mammal species are endemic to one or more of these declining hotspots (Figure 1.29).

Care should be taken when delineating biodiversity hotspots, as illustrated by Orme *et al.* (2005) using data collected on breeding birds. They showed that, depending on which of three different aspects of avian diversity was used as a basis of determination (i.e. overall species richness, threatened species richness and endemic species richness), alternative hotspot configurations could be generated that were very different spatially. Only 2.5% of hotspots were common to all three aspects (indices) of avian diversity. Orme *et al.* (2005) postulated that these disparities could have originated from differences in rates of speciation, past extinctions and anthropogenic influences that affect measures of diversity. Their study shows that conservation efforts will be aided if targets for conservation are clearly defined before delineating hotspots. For example, threatened or endemic (or, preferably, both) status should take precedence over other diversity indices if species conservation is the goal. This also is the most logical method because of the urgent need to stem the global biodiversity crisis (see Chapter 9).

With similar goals to Myers *et al.* (2000) and Conservation International, BirdLife International has identified 218 Endemic Bird Areas (EBAs) worldwide (Stattersfield *et al.* 1998). Out of 10 000 bird species, 2500 have restricted ranges (< 50 000 km²). In addition to being endemic to an EBA, half of all range-restricted bird species are threatened, making them a high priority for conservation. The EBAs identified have distributions of two or more range-restricted species that overlap, making them richer in endemic bird species by comparison with the rest of the planet. Collectively, these EBAs occupy 4.3% of the earth's land surface, with 77% of them located in the tropics, affirming the paramount conservation value of this region. Indeed, the countries with the highest number of EBAs are

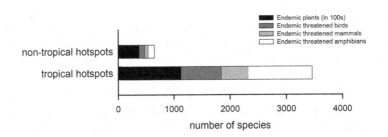

Figure 1.29 The total number of endemic plants and threatened endemic animals between tropical and non-tropical biodiversity hotspots. (Data derived from www. biodiversityhotspots.org.)

all tropical: Indonesia (24), Mexico (18), Peru (16), Brazil (15) and Colombia (14). Eighty-three per cent of EBAs are located in forested areas, particularly tropical forests. BirdLife International considers habitat conservation in EBAs critical for maintaining global avian biodiversity.

Critics claim that, instead of allocating conservation priorities based on taxonomic groups, conservation should target a range of habitats within terrestrial, freshwater and marine ecosystems on the basis that these protect vital ecological processes (e.g. seed dispersal) and ecosystem services (e.g. carbon sequestration; Olson and Dinerstein 1998). The World Wide Fund for Nature (WWF) has developed an approach known as the Global 200, in which 238 eco-regions are prioritized for conservation because they represent all ecosystems and habitat types present on earth (Olson and Dinerstein 1998; www.panda.org). An eco-region is defined as a unit of land or water harbouring geographically distinct species assemblages and environmental conditions. Eco-regions are prioritized based on a formulaic combination of indices of relevance to biodiversity conservation: their species richness, endemism, taxonomic uniqueness, unusual ecological or evolutionary phenomena and global rarity. Of these 238 eco-regions, 143 are terrestrial, 53 freshwater and 43 marine. Terrestrial realms dominate the critical eco-regions because they have higher endemism than aquatic realms (Olson and Dinerstein 1998). However, this could also be a reflection of the gaps in information regarding aquatic realms. The aim of this mode of conservation planning is that biomes containing unique ecosystems and species assemblages such as tundra are not ignored in favour of species-rich tropical forests. Similarly, Hoekstra *et al.* (2005) attempted to determine which biomes should be given conservation priority. They found that temperate grasslands and savannas, and 'Mediterranean' forests, woodlands and scrub, are the least protected biomes (< 6% area protected). Tropical broadleaf forests received relatively high protection (16%). Unfortunately, many of the so-called 'protected areas' in the tropics are receiving only 'paper protection' and, in reality, continue to lose area and biodiversity (Curran *et al.* 2004; DeFries *et al.* 2005). Taking a somewhat different tack, Cardillo *et al.* (2006) identified priority conservation areas for mammals based on 'latent' extinction risk – areas that contain diverse mammalian assemblages but are at high risk from future, rather than present, land use and climate change. Despite their different methods, all of the approaches of strategizing conservation we have described above show a reasonable degree of congruence. For instance, 60% of Global 2000 eco-regions and 78% of EBAs overlap with the biodiversity hotspots (www.biodiversityhotspots.org), and the primary focus for conservation efforts is clearly and consistently identified as the tropical realm.

1.4   Summary

1 Tropical habitats are disappearing extensively and rapidly.
2 Drivers of this massive ongoing land conversion relate largely to the ever-burgeoning human population and its related activities such as agriculture, urbanization and logging.

3   Weak institutions and rampant corruption thwart efforts to curb tropical
    habitat loss.
4   Tropical habitats are critical for global biodiversity and should be adequately
    protected.

1.5   Further reading

Achard, F., Eva, H.D., Stibig, H.-J., Mayaux, P., Gallego, J., Richards, T. and Malingreau, J.-P. (2002)
    Determination of deforestation rates of the world's humid tropical forests. *Science* 297, 999–1002.
Laurance, W. F. (1999) Reflections on the tropical deforestation crisis. *Biological Conservation* 91,
    109–117.
Myers, N., Mittermeier, R. A., Mittermeier, C. G., da Fonseca, G. A. B. and Kent, J. (2000) Biodiversity
    hotspots for conservation priorities. *Nature* 403, 853–858.

2

Invaluable Losses

Nature's species and ecosystems provide many 'free' services to humanity (Figure 2.1). In this chapter, we discuss the role of nature in regulating climatic and hydrological processes, carbon sequestration, maintaining soil stability and in providing other ecosystem services, such as pollination, that otherwise would be costly and logistically challenging to replace artificially. We also highlight the cultural, utilitarian, economic, health and evolutionary values of nature that are being eroded by the degradation of natural habitats and the loss of biodiversity. Indeed, many of nature's ecosystem services are in decline as a result of unsustainable human activities (Figure 2.2), and much has been written on the biodiversity, economic and ecosystem aspects of tropical habitat loss, particularly in rain forests. In this chapter, we summarize what has been reported to date, and provide an explicit context to how these issues relate to tropical ecosystems.

2.1   Environmental filters

Our atmosphere works somewhat like a large, complex and self-regulating greenhouse. Short-wave radiation emitted by the sun passes through the atmosphere to the earth's surface, and some of the sun's energy is radiated back into the atmosphere in the form of long-wave infrared radiation. Certain trace (greenhouse) gases, such as carbon dioxide (CO_2) methane (CH_4) and nitrous oxide (N_2O), trap a portion of this infrared radiation within earth's atmosphere and warm the planet (Houghton 2004).

Over most of the past few hundred million years, terrestrial plants have helped to maintain the balance of carbon in the atmosphere. Forests convert carbon into cellulose and release oxygen through photosynthesis. Through photosynthesis, forests act as carbon sinks, absorbing and storing atmospheric carbon (Page *et al*. 2002). However, as detailed in Chapter 1, a sizeable fraction of the world's tropical (and temperate) forests have been, and continue to be, depleted by recent human activities. This trend, coupled with the release of greenhouse gases through the

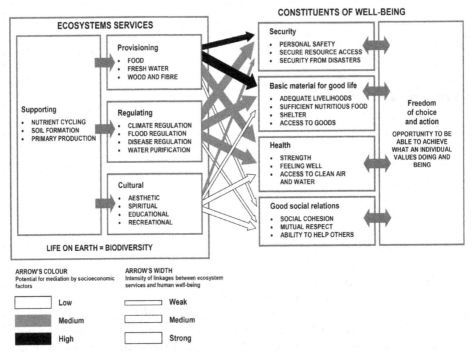

Figure 2.1 Commonly encountered linkages between the categories of ecosystem services and the components of human well-being, with the possibility of regulation of links by socioeconomic factors. The potential for mediation and the strength of the linkages differ among ecosystems and regions. Also, beyond the influence of ecosystem services, other environmental, economic, social, technological and cultural factors can influence human well-being, and ecosystems are in turn affected by these changes in human well-being. (After Millennium Ecosystem Assessment 2005. With permission.)

burning of fossil fuels for industry (releasing ancient accumulations of carbon), is jeopardizing the atmospheric carbon balance. After fossil fuel consumption, human modification of vegetation and soils are the next major causes of global carbon emissions (Flint 1994), with atmospheric CO_2 concentrations having now increased by an alarming 34% since 1750 (Millennium Ecosystem Assessment 2005). More worrying is that Global Carbon Project reports that CO_2 levels have increased by 2.5-fold since 2000 (news.bbc.co.uk/2/hi/science/nature/6189600. stm). This increase in atmospheric CO_2 is estimated to be responsible for about half of all global warming alone, and is unprecedented over at least the past few million years (Pataki 2002). Plants and soil harbour between 460 and 575 billion metric tonnes of carbon and, as a result, tropical deforestation contributes a large proportion of total CO_2 emissions (Flint 1994).

When a forest is felled and burnt for the establishment of cropland and pastures, the stored carbon is combined chemically with oxygen in the combustion process and liberated into the atmosphere. The large-scale release of CO_2 enhances the greenhouse effect, and almost certainly contributes to elevating global temperatures (IPCC 2001a). Tropical forests have an especially high potential for carbon storage (Malhi and Grace 2000). The carbon-sequestering capabilities of tropical forests have steadily decreased between 1999 and 2005 as a result of

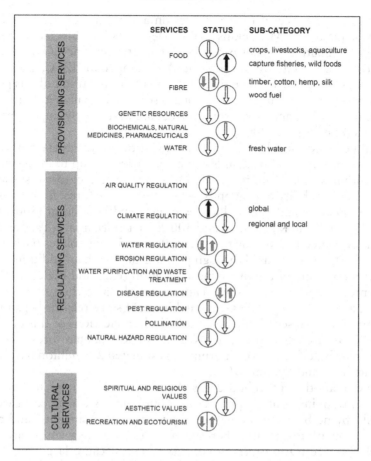

Figure 2.2 Global status of the state of provisioning (top), regulating (middle) and cultural (bottom) ecosystem services. The status illustrates whether the condition of a particular service has been enhanced (black arrows) or degraded (white arrows) by human activities, or a mix of both (grey arrows). (Redrawn from Millennium Ecosystem Assessment 2005. With permission.)

deforestation (FAO 2005). Mean carbon fluxes increased from 0.6 petagrams (Pg) per year (i.e. 6×10^{11} kg) to 0.9 Pg/year from the 1980s to the 1990s because of tropical deforestation (after taking forest regrowth into account) (DeFries *et al.* 2002), although the precise figures remain the subject of debate. Net carbon flux for 1 ha of tropical forest varies between 151 and 190 tonnes (Achard *et al.* 2002). Using area estimates of tropical forest coverage, it is believed that deforestation in Southeast Asia alone releases approximately 465 million tonnes/ year of carbon into the atmosphere (Phat *et al.* 2004). This represents 29% of the total global carbon release due to deforestation.

Carbon emissions resulting from deforestation are roughly equivalent in Brazil and Indonesia, and these releases have been exacerbated by recent forest fires in both areas (Santilli *et al.* 2005; Figure 2.3). Total CO_2 emissions might increase by 100–200 times by 2100, largely because of the continued burning of fossil fuels, the clearance of native vegetation and release of the carbon it had previously sequestered (M. Jenkins 2003). Deforestation thus facilitates global warming by eliminating a major potential sink of atmospheric CO_2 and releasing forests' stored carbon into the atmosphere. According to the IPCC, global surface temperatures will be 2–6°C higher in 2100 than today as a result of this pollution (M. Jenkins 2003). Thus, undisturbed forests serve as globally vital carbon sinks (W.F. Laurance 1999), and large tropical trees are particularly important carbon sequesterers (Bunker *et al.* 2005).

Tropical peatlands also lock away large quantities of terrestrial carbon (Page and Rieley 1998). Natural peatlands can be composed of up to 20-m-thick peat deposits within swamp forests, so drainage and forest clearing substantially increase the susceptibility of peatlands to fire. During the forest fires in Indonesia in 1997–98, an estimated 1.45 million ha of peatlands was burnt (300 000 ha in Sumatra, 750 000 ha in Kalimantan and 400 000 ha in Irian Jaya; Page and Rieley 1998). It is estimated that during this event, 0.48–0.56 gigatonnes (Gt) of carbon ($4.8-5.6 \times 10^{10}$ kg) was released through peat burning, with an additional 0.05 Gt released from burning of the overlying vegetation (Page and Rieley 1998). This release represents 13–40% of the mean annual global carbon emissions from fossil fuels, and is the largest single annual increase since records began in 1957 (see Chapter 4 for a discussion of the ecological implications of this event). The earlier 1982–83 fires that spread across the East Kalimantan forests of Borneo may have contributed even more, burning an estimated 2–3 million ha of tropical forest (Hartshorn and Bynum 1999).

Methane is another greenhouse gas that is over 100 times more effective than CO_2 in trapping heat (Keppler *et al.* 2006), and is produced in wetlands, paddyfields, by herbivorous animals (e.g. termites and ruminants) and through the combustion of fossil fuels. Forests in some areas (e.g. Asia) are cleared extensively for rice (*Oryza sativa*) cultivation, and because large land areas of Southeast Asian countries are now under rice cultivation (44 million ha) (FAO 2004), the volume of methane release represents a real concern. World methane emissions increased by 49% between 1940 and 1980, and it is estimated that paddy cultivation contributed 83% of this increase (Bolle *et al.* 1986). Although

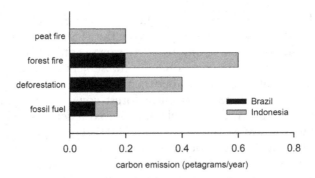

Figure 2.3 The estimated carbon emissions from peat fires, forest fires, deforestation and fossil fuels in Brazil and Indonesia. (Data derived from Santilli *et al.* 2005.)

the exact amount of methane emission through rice cultivation remains debatable, it is clear that the expansion of rice paddy cultivation is a major potential source of elevated methane emissions. Other forms of cultivation and livestock production release between 106 and 201 million tonnes of methane every year (Millennium Ecosystem Assessment 2005). Intact forest canopies also absorb another greenhouse gas, N_2O, and other atmospheric pollutants, and thus help maintain air quality (Chivian 2002) if they are retained rather than converted for agricultural production of greenhouse gas-emitting crops.

2.2   Precipitation and temperature regulation

Tropical forests are vital mediators of both local and regional climates. They act as heat pumps by distributing solar radiation from the equator to temperate zones, thus warming the temperate areas and cooling the tropics. Hence, deforestation may ultimately result in warmer and drier climates in the tropics (Berbet and Costa 2003). Tropical forests also affect local and regional precipitation, by releasing large amounts of water through evaporation and evapotranspiration, which generates clouds and precipitation. It is predicted that deforestation may result in an 8% decline in rainfall in Southeast Asia over the next decade, with a much steeper precipitation decline of 17% in Indonesia, which currently still contains substantial forest cover that is under threat (M. Hoffmann *et al.* 2003). Another study predicted that annual rainfall will be reduced by as much as 172 mm in Southeast Asia from deforestation (Zhang *et al.* 2001). Similarly, large-scale deforestation of the Amazon basin may reduce rainfall by as much as 20% in the region (IPCC 1996). Thus, the reduction in precipitation as a result of deforestation will result in elevated temperatures, reduced humidity and enhanced drought conditions in tropical regions, and will make forests and peatlands ever more susceptible to fires, creating a vicious positive feedback loop.

2.3   Water purification

Forest soils are also thought to assist in purifying water (Chivian 2002). With their high water retention capacity and ability to alter flow regimes, especially during flooding events (see below), the organic and inorganic material constituting the forest floor has remarkable properties that enable it to remove impurities from water (Fujii *et al.* 2001). It has been shown that when flood waters flow through alluvial forests in temperate systems, forest tracts remove a large component of the phosphate and nitrate (Sanchez-Perez *et al.* 1991a; Sanchez-Perez and Tremolieres 1997) and of ions such as chloride, sodium, calcium, magnesium and potassium (Sanchez-Perez *et al.* 1991b). Forest soils can also reduce the concentration of noxious bacteria in contaminated wastewater, with the result that more such bacteria make their way into catchments (Kermen and Janota-Bassalik 1987). Although the degree to which tropical forests participate in water purification is contested (Sanchez-Perez *et al.* 1991a,b; Sanchez-Perez and Tremolieres 1997;

Fujii *et al.* 2001), the high vegetation biomass of tropical forests coupled with the relatively high seasonal water flux typical of these regions suggests that water purification is highly compromised by deforestation.

2.4   Protecting catchments and soils

Forests can also assist in regulating the water flow to downstream areas. Thus, deforestation can alter the natural water flow, resulting in either flood or drought episodes, with such effects being most pronounced immediately after deforestation (Giambelluca 2002). Forest canopies reduce the force with which rain water hits the soil, thereby reducing its erosional impact. Roots bind soil, so that it is less likely to be washed away during flooding events (Chivian 2002). In Southeast Asia, an undisturbed forest can intercept 35% of rainfall, which is 15% more than regenerating logged forest, and 23% more than rubber or oil palm plantations (Myers 1997). Undisturbed forests are better at mitigating the impact of tropical downpours, and prevent the break-up of ground soil and flushing loss of nutrients. Landslides initiated by heavy rains and exacerbated by rampant illegal logging caused the death of 1400 people in the Philippines (Anonymous 2004a), illustrating the importance of intact rain forests for maintaining soil stability. It has been estimated that every year between 30 and 75 tonnes of topsoil is washed away from each hectare of deforested land in Nepal (Cool 1980). Such substantial soil loss may have economic repercussions, especially for agriculture (Myers 1997). Loss of topsoil due to deforestation can reduce rice output by 1.5 million tonnes per year, rice that could otherwise feed up to 15 million people annually (Magrath and Arens 1989). Deforestation-driven siltation may also reduce the life of dams, clog natural waterways and affect offshore fisheries (Myers 1997). In addition, soil erosion following deforestation can reduce soil water storage capabilities (Bruijnzeel 2004).

Forest ecosystems are probably responsible for the regulation of about half of the water drainage systems on the planet. Indeed, roughly 5 billion people rely on water supplies from forest ecosystems (Millennium Ecosystem Assessment 2005). Worldwide, at least 70% of this water is used for agricultural purposes, and water withdrawals by humans doubled between 1960 and 2000 (Millennium Ecosystem Assessment 2005). Such a high consumption of freshwater may result in future shortages, which would clearly affect human health, food production and economic stability.

2.5   Forests and floods

Surface runoff and river discharge usually increase when natural vegetation, especially forests, are removed (Bruijnzeel 2004). For example, a study in the Tocantis Tiver basin in Brazil recorded a 25% increase in the river discharge rate between 1960 and 1995, probably attributable to increased surface flow caused by large-scale conversion of natural vegetation into cropland (Costa *et al.* 2003). Water discharge that occurs independently of changes to precipitation can result in flooding. The frequency and impacts of floods have increased in the past 50 years, mainly as a result of human-mediated land use changes (Millennium Ecosystem Assessment 2005).

For centuries, forests have been purported to provide flood protection (Clark 1987). However, until recently, such claims have not been backed by broad-scale quantitative evidence that forests play an important role in modifying flood risk and severity (Kaimowitz 2004). Using data on flooding events collected between 1990 and 2000 from 56 developing countries in Africa, Asia and South America, Bradshaw *et al.* (2007a) demonstrated that after taking rainfall, average gradient, antecedent soil moisture regime and the proportion of degraded land area into account, the cover of intact natural forest did explain some of the excess variation in flood risk among countries. The amount of natural forest that had been lost (as opposed to the area remaining) also correlated with the excess variation in flood risk, indicating the important role of natural forest cover in reducing the average flood frequency experienced by any one country. Bradshaw *et al.* (2007a) also predicted that a mere 10% decrease in natural forest cover could result in a 4–28% increase in the risk of flooding, possibly because of reductions in the interception of rainfall and the evaporation of water from the tree canopy and reductions in the hydraulic conductivity of soils, all of which can cause increased runoff (Gentry and Lopez-Parodi 1980; Clark 1987).

Further, Bradshaw *et al.* (2007a) report that tropical countries with relatively higher natural forest cover may also experience, on average, less severe flooding events as measured by the total damage exacted by floods in economic terms. The flood events that they examined between 1990 and 2000 resulted in the deaths of nearly 100 000 people and the displacement of 320 million more, with total reported economic damages exceeding US$1.15 trillion. As such, the global-scale relationships found by Bradshaw *et al.* (2007a) suggest that catastrophic flood events may increase in frequency and magnitude in the future because of continuing heavy deforestation in developing countries. Clearly, these findings demand urgent actions by all stakeholders (e.g. governments and international organizations). Tangible actions to mitigate these impacts should include the protection of existing natural forests and massive reforestation activities. Moreover, the latter should use native trees, because exotic tree species have low conservation value for native biodiversity (Sodhi *et al.* 2005a) and may not provide as effective flood reduction characteristics (Bruijnzeel 2004).

2.6   Nitrogen flux

The global nitrogen cycle consists of a large, well-mixed pool of nitrogen (N_2) in the atmosphere (~78% by volume) and a smaller quantity of molecular nitrogen bound within plants, animals, soils, sediments and solutions (Schlesinger 1991). This nitrogen cycle has been modified extensively by human activities, such that now more nitrogen is fixed annually by human-driven than by natural processes (Vitousek 1994). Considerably more nitrogen is mobilized through biomass burning, land clearing and conversion and wetland drainage than is estimated to have occurred in pre-human times (Matson *et al.* 1987; Crutzen and Andreae 1990).

For example, a study of the effects of slash-and-burn fire on the biogeochemical cycle in the neotropical dry forests of Brazil demonstrated that up to 96% of the pre-fire above-ground nitrogen reserves were lost due to burning, equating to > 500 kg of nitrogen lost per hectare (Kauffman *et al.* 1993). Based on the measured losses of nutrients from single slash-burning events, Kauffman *et al.* (1993) predicted that reaccumulation of soil nutrients to pre-slash levels would probably require a century or more of natural sequestration. Similarly, Keller and Reiners (1994) found that the largest nitric oxide (NO) emissions occurred in formerly forested and recently abandoned pasture sites in Costa Rica, but that soil–atmosphere fluxes of nitrous oxide (N_2O) and NO could be restored to pre-disturbance rates during secondary succession. Neill *et al.* (1997) also determined that net nitrogen mineralization rates and net nitrification were higher in forest than in pasture soils in Brazil.

Human alteration of the nitrogen cycle has global consequences for climate by affecting nitrous oxide emissions and the carbon cycle, for biodiversity by affecting species diversity, and for a host of other ecological processes (Vitousek 1994). It has been suggested that in tropical systems natural riparian nitrogen fluxes are greater than those in the temperate zone, but higher general rates of deforestation may be contributing to these higher tropical fluxes. Regardless, projected increases in fertilizer use and atmospheric deposition of nitrogen in the tropics are likely to increase nitrogen loading in many tropical river systems (Howarth *et al.* 1996), leading to eutrophication.

2.7   Eutrophication

Eutrophication is defined as the enrichment of an ecosystem (although the term is generally restricted to aquatic ecosystems) with chemical nutrients – typically, phosphorus and nitrogen. Although eutrophic systems generally have higher biological production, excessive nutrient loading can result in extremely high levels of plant production, followed by high bacterial activity associated with the eventual decomposition of this production and deoxygenation of water. Deforestation leads to two main problems with respect to nutrient enrichment: (1) increased water flows can result in higher siltation and nutrient run-off into aquatic systems and (2) the change in land use from forests to agricultural activities is often accompanied by higher terrestrial inputs of fertilizers that further exacerbate the nutrient build-up in forest catchments (Novotny 1999).

Eutrophication has caused cascading ecological effects and exacerbated species extinctions in one of the world's largest and most biodiverse freshwater lakes – Lake Victoria in East Africa. Limnological changes suggestive of eutrophication have been documented since 1960: hypolimnetic anoxia (deoxygenation), increases in phytoplankton productivity and shifts from diatom to blue-green algae dominance (Gophen *et al.* 1995). These changes have occurred as a result of increased inputs of nitrogen and phosphorus, derived largely from agricultural and urban developments and increased runoff due to deforestation. Similar patterns of aquatic ecosystem degradation from eutrophication have been observed elsewhere in the subtropics and tropics (OgutuOhwayo *et al.* 1997; Nauta *et al.* 2003; Rosenmeier *et al.* 2004; Berrio *et al.* 2006).

In marine ecosystems, nutrient enrichment resulting from deforestation and topsoil erosion can also cause an imbalance in the exchange of nutrients between the symbiotic zooxanthellae (dinoflagellate algae) and the host coral living in coral reefs (Dubinsky and Stambler 1996). Such eutrophication, leading to excessive phytoplankton growth, can cause reduced light penetration within the water column and reduced photosynthetic activity in the coral symbionts, as well as encouraging the proliferation of weeds (Dubinsky and Stambler 1996). Excessive seaweed growth can smother, and eventually replace, the slower-growing coral reefs adapted to the typically low nutrient levels of tropical seas (Dubinsky and Stambler 1996; see Chaper 7).

2.8   Nature's pharmacy and goods

Tropical forests are a source of food, remedies, natural products and construction materials for many local communities. At least 25% of medicines patented by Western pharmaceutical companies are derived from medicinal plants identified and prepared through traditional indigenous techniques (Posey 1999). Forest products remain the only pharmaceutical option for some remote local communities. For example, 150 medicinal plants are regularly used by people native to the Mekong wetlands of Southeast Asia (Millennium Ecosystem Assessment 2005). In addition, tropical forests remain a major source for a variety of commercial products, ranging from food and construction material to latex and perfumes (W.F. Laurance 1999; see Chapter 6). For example, there has been a recent discovery that two species of *Calophyllum* tree from the rain forest of Sarawak may produce anti-HIV (human immunodeficiency virus) agents (Chung 1996). The therapeutic value of rain forests remains vastly underexplored, with myriads of new medicines and products potentially awaiting discovery (W.F. Laurance 1999). However, medicinal plant abundance has generally declined due to overharvesting (Millennium Ecosystem Assessment 2005). Eight of the 12 top-selling medicinal plants in Amazon Brazil are found only in those forests (Shanley and Luz 2003). These plants have no botanical substitute and, sadly, their abundance is rapidly declining because of timber overharvest (Shanley and Luz 2003).

2.9   Human health and nature

The destruction of rain forests can also facilitate the spread of human diseases. For example, malaria, a disease caused by the mosquito-borne protist *Plasmodium*, is one of the main causes of human sickness and death in the world, especially in developing tropics. Deforestation seems to lead to an increase in the distribution of mosquitoes and concomitant increases in the transmission of arthropod-borne pathogens (van der Kaay 1998; Kidson *et al.* 2000; D. R. Norris 2004). Deforestation improves the habitat of mosquitoes by compromising drainage (and thus creating small, isolated ponds of still water that lack predatory fish), increasing light and temperature to facilitate the growth of algae (the main food of mosquito larvae) and facilitating the deacidification of standing water (Chivian 2002). Further, several disease-carrying mosquitoes (*Anopheles* spp.) are increasing in the tropics, probably as a result of changes in land use (Aiken and Leigh 1992) and global warming (Epstein *et al.* 1998).

In formerly pristine areas, logging activities, slash-and-burn farming, road construction and transmigration projects attract immigrants, most of whom have little or no immunity to the human diseases in tropical forests (Aiken and Leigh 1992). Conversely, new colonists to the rain forest can also be major sources of malaria brought from more densely populated regions (Moran 1988). Human mortality due to malaria can be as high as 25% in some plantations (Ooi 1976). Increasing tropical deforestation coincides with an upsurge of malaria and its vectors, even after accounting for the effects of changes in human population density (Foley *et al.* 2005). In the Amazon, it is estimated that every 1% increase in deforestation boosts the number of malaria-carrying *Anopheles darlingi* by 8% (H. Pearson 2003). Similar increases in prevalence are expected intuitively for other insect vectors such as black flies (*Simulium* spp.), which carry the filarial disease 'river blindness' (onchocerciasis) (Dadzie *et al.* 1989). Increases in black rat (*Rattus rattus*) abundance in oil palm plantations can also make humans more vulnerable to scrub typhus transmitted by the rat's mites (Parshad 1999). The disease situation is exacerbated in new settlements owing to poor sanitation.

A major reason why disturbing wildlife habitats is damaging to human health may be the fact that approximately 75% of human diseases have origins in wild or domesticated animals ('zoonoses' – Foley *et al.* 2005). For example, a disease that emerged in Peninsular Malaysia between September 1998 and 1999 resulted in the death of 105 humans and the slaughter of over 1 million domestic pigs (*Sus scrofa*) (Yob *et al.* 2001). The culprit in that case was discovered to be a new disease, subsequently named Nipah virus after the town where it first appeared (Chua *et al.* 2000). The presence of Nipah virus was later discovered in four species of fruit bats (Megachiroptera) collected in various parts of Peninsular Malaysia (Yob *et al.* 2001), and the infection rates ranged from 4% to 31% for different species. In addition, an individual of an insectivorous bat species, the lesser Asian house bat (*Scotophilus kuhlii*), was also infected. It is hypothesized that the smoke from Southeast Asia's catastrophic forest fires in 1997–98 led to the failure of many forest trees to fruit, causing frugivorous bats to switch to fruit trees in Malaysia's large pig farms. These bats then probably passed on the deadly Nipah virus to the pigs, which then transmitted it to humans (Chivian 2002).

Severe acute respiratory syndrome (SARS) is another cause for concern from the perspective of increasing wildlife–human contact in the tropics. With over 8000 people infected and about 774 fatalities, SARS has proven to be a formidable hazard to human health (www.who.int). This disease first originated in Guangdong Province, China, in late 2002, and spread quickly to at least 30 countries by 2003, culminating in public concern worldwide. With the exception of Toronto (Canada), all SARS hotspots were in tropical or subtropical Asia (China, Hong Kong, Taiwan and Singapore). Caused by a coronavirus related to influenza, the origin of SARS remains unclear, but a zoonotic origin is suspected. The recent discovery of a SARS coronavirus in wild Chinese animals such as the masked palm civet (*Paguma larvata*) and raccoon dog (*Nyctereuteus procyonoides*) implicates these vectors in the outbreak (Guan *et al.* 2003). The animals tested were caught from the wild and were part of a booming market for wildlife meat in East Asia. Similarly, bush meat consumption of simians in Africa has been posited as a potential origin of HIV [causing acquired immune deficiency syndrome (AIDS)] origins, and this consumption will probably continue to expose humans to yet new viruses and resulting diseases (Wolfe *et al.* 2004). The emergence of tropical zoonotic diseases is of increasing concern as road building by logging companies facilitates access to formerly inaccessible regions and so increases bush meat hunting (see Chapter 6).

Mangrove forests are some of the most important types of coastal vegetation in terms of their benefits to human society; however, they are now highly threatened by human activities (see Chapter 1). It has been suggested that the catastrophic loss of life and property brought about by the Asian tsunami on 26 December 2004, would have been lessened had the mangrove forests not been cleared on the affected areas (Dahdouh-Guebas *et al.* 2005). Models show that as few as 30 mangrove trees per $100\,m^2$ in a 100-m-wide strip can reduce tsunami water flow by as much as 90%, and these natural barriers are considerably more effective than other artificial coastal structures such as buildings (Danielsen *et al.* 2005). As such, mangrove protection and restoration in coastal areas are urgently needed.

2.10   Ecosystem services from nature

Many mobile forest animals, including bees, butterflies, birds, bats and other mammals, help to maintain vital ecological processes such as pollination, seed dispersal, and pest control (Sekercioglu 2006; see Spotlight 2: Gretchen Daily). For example, at least 450 economic products, including fruits, timber, wood for fuel, medicines, tannins and dyes, are derived from plants whose pollen or seeds are dispersed by fruit bats (*Cynopterus* spp.) (Payne 1995). In assisting the reproduction and dispersal of these plants, fruit bats provide an enormously beneficial service to humanity that would be extremely difficult and costly to replace by artificial means. Nearby forested areas are essential for the pollination of some crops. Therefore, deforestation may adversely affect the agricultural production of many tropical areas. The loss of such 'free' ecosystem services in the pursuit of tangible economic gain can ironically cost human enterprise dearly (Myers 1996; Ricketts *et al.* 2004). The classical example of this is illustrated by the oil palm, which is native to the humid regions of Africa and was imported in the early 1900s into Southeast Asian countries, such as Malaysia and Indonesia, because of its commercial potential. However, the weevil (*Elaeidobius* spp.) responsible for pollinating the oil palm in its native habitat was not imported at the same time. This led the Malaysian and Indonesian planters to rely upon expensive and labour-intensive hand pollination techniques (Syed *et al.* 1982). In the early 1980s, the most abundant and efficacious weevil pollinator (*E. kamerunicus*) was released in Malaysia (Krantz and Poinar 2004), the establishment of which boosted fruit yield by up to 60% and generated savings in labour costs worth US$140 million per year (Kevan *et al.* 1986; Chivian 2002).

Deforestation has ramifications beyond those immediate to the loss of forests

Spotlight 2: Gretchen C. Daily

Biography

I am an ecologist working to develop a scientific basis and financial and institutional support for managing earth's life support systems. My primary efforts are focused on making conservation mainstream, economically attractive and commonplace. I came to environmental science at a young age. In the name of adventure, my family moved to West Germany when I was 12 years old. It was 1977, and I woke up to a turbulent world of street demonstrations against environmental devastation. Protesters said that acid was falling from the sky and that no one was doing anything about it. I could hardly fathom this. A few years later, I heard about a high-school science competition and, with the encouragement of a dream chemistry teacher, I signed up. He thought acid rain was a bit too ambitious a topic, but helped me to launch a study of the pollution in a nearby river, during which I discovered my passion for scientific research. Although I could spend lifetimes exploring the wonders of the universe for sheer pleasure, I find I cannot take my mind off of the big issues confronting society. It's clear that scientific understanding is but one of a complex of interacting factors shaping the future. Science *alone* will get us nowhere. So, after postgraduate school in ecology, I took up a wonderful research position designed to foster new, integrated approaches to environmental issues. I began building a background in economics, law and other key disciplines, and cultivating a network of people far removed from academia who shared a sense of urgency about the state of the environment. This diverse group helped me arrive at new ways of thinking about the environment, revealing the wealth of opportunities to make conservation both practical and profitable. While we still need a lot more scientific understanding to proceed effectively, our greatest challenge now lies in the social realm, including trying to scale up and replicate the small models of success to date.

Major publications

Chan, K. M. A., Shaw, R., Cameron, D., Underwood, E. C. and Daily, G. C. (2006) Conservation planning for ecosystem services. *PLoS Biology* **4**, e2138.

Daily, G. C. (ed.) (1997) *Nature's Services: Societal Dependence on Natural Ecosystems*. Island Press, Washington, DC.

Daily, G. C. and Ellison, K. (2002) *The New Economy of Nature: The Quest to Make Conservation Profitable*, Island Press, Washington, DC.

Pereira, H. M. and Daily, G. C. (2006) Modeling biodiversity dynamics in countryside landscapes. *Ecology* 87, 1877–1885.

Ricketts, T. H., Daily, G. C., Ehrlich, P. R. and Michener, C. (2004) Economic value of tropical forest to coffee production. *Proceedings of the National Academy of Sciences of the USA* 101, 12579–12582.

Questions and answers

How can the conceptual framework of 'countryside biogeography' be used to enhance biodiversity in human-dominated tropical landscapes?

'Countryside biogeography' is a new conceptual framework for elucidating the fates of populations, species and ecosystems in 'countryside' – the growing fraction of earth's unbuilt land surface whose ecosystem qualities are strongly influenced by humanity. With numerous collaborators, I have launched investigations of the countryside biogeography of a variety of strategically selected taxa, including birds, mammals, reptiles, amphibians, butterflies, moths, bees and plants. We have discovered that at least half of the native species (and the vast majority of families) in each group persist, over at least the short to medium term (50–100 years), in countryside typical of much of the tropics. We have also gained considerable insight into the role that different habitat elements and configurations play in driving patterns of distribution. Major questions are whether this high potential conservation value exists in countryside globally, and whether it can be sustained over the long term (centuries to millennia). Preliminary work suggests that the patterns we have found are predictable and general, and also that substantial conservation value can indeed be sustained for thousands of years (such as in India), in a variety of agricultural production systems.

Your research focuses on some fundamental conservation questions such as: which constituents of the earth's biota (and with which attributes) are likely to survive (assuming that human impacts on the environment intensify as projected)? How has your focus on this question changed over time?

When I began addressing this question, the prevailing view was that the clues to the future of biodiversity were to be found in remnants of native habitat – Noah's Arks floating in a hostile sea of development. The logic was that most organisms are highly adapted to their native habitats and that few, therefore, would be able to exploit areas heavily modified by human activities. In general, those few would not require or merit protection. Through my own work and recent research of others, I have since discovered that, to the contrary, human-dominated ecosystems can retain substantial conservation value, and that there is now a rapidly closing window of opportunity to sustain this value.

Why do we need to conserve tropical forests?

I am actively attempting to link projected changes in biodiversity and ecosystems to changes in 'services' to humanity. These services include production of goods (e.g. seafood and timber), life support processes (pollination and water purification), life-fulfilling conditions (beauty and serenity) and options (genetic diversity for future use). An example of this work demonstrated a high value of tropical forest as an input to coffee production via the supply of pollinators – showing that conservation investments can pay off even in prime farmland within existing economic and legal systems.

What strategies can make conservation attractive and commonplace worldwide?

I have worked with many colleagues to launch The Natural Capital Project, a partnership among The Nature Conservancy (TNC), World Wide Fund for Nature (WWF) and Stanford University to make conservation attractive and commonplace worldwide. The project aspires to provide maps of nature's services, assess their values in economic and other terms, and – for the first time on any significant scale – incorporate those values into resource decisions.

per se. For example, almost all flowering plants in tropical rain forests are pollinated by animals (Bawa 1990), and it has also been estimated that one-third of the human diet in tropical countries is derived from insect-pollinated plants (Crane and Walker 1983). Accordingly, a decline in forest-dwelling animal pollinators may impede plant reproduction not only in the forests, but also in neighbouring agricultural areas. Lowland coffee (*Coffea* spp.) is an important tropical cash crop, and it depends, at least partially, on bees for cross-pollination. A study in Costa Rica found that forest bees increased coffee yield by 20% in fields within 1 km of the forest edge. Between 2000 and 2003, the pollination services provided by forest bees were worth US$60 000 to a 1100-ha farm (Ricketts *et al.* 2004). A forest patch as small as 20 ha located near farms can increase coffee yield and thus incur large economic benefits to the farmers (Ricketts *et al.* 2004). Similarly, in central Sulawesi (Indonesia), the number of social bee species visiting coffee fields was found to decrease with increasing distance from the forest edge (Klein *et al.* 2003a; Figure 2.4) and species-rich bee communities increased the pollination success of coffee plants (Klein *et al.* 2003b). Thus, native forests need to be preserved close to agroforestry systems to ensure the availability of pollinating insects that depend on the forests for nesting and reliable year-round sources of pollen and nectar. For optimum economic pollination, a coffee field should be situated within 400 m of a forest patch (Olschewski *et al.* 2006).

Many predators are assumed to be important agents of biological control of pests in agricultural areas (Polis *et al.* 2000). Natural pest control is usually more efficient and environmentally benign than human methods such as poisoning, and may be critical for food security and sustaining rural livelihoods (Naylor and Ehrlich 1997); indeed, in the absence of control, herbivorous insects can destroy 25–50% of crops either before or after the harvest (Naylor and Ehrlich 1997). As such, it is estimated that natural enemies (predators, parasites and pathogens) of agricultural pests save humanity US$54 billion annually (Naylor and Ehrlich 1997). Up to 80% of arthropods (> 5 mm in length) may be taken in coffee plantations in Guatemala by insectivorous birds, resulting in a substantial reduction in leaf damage (Greenberg *et al.* 2000). Further, experimental studies have found that

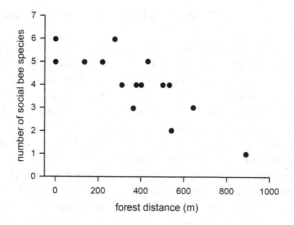

Figure 2.4 Negative relationship between the number of social bee species and the distance from the nearest forest. (After Klein *et al.* 2003a. Copyright, Blackwell Publishing Limited.)

insectivorous birds can prevent pest outbreaks in coffee plantations (Perfecto *et al.* 2004). The use of pesticides promotes the evolution of pest resistance and has other environmental repercussions (e.g. pollution); therefore, the natural enemies of agricultural pests should be encouraged to flourish in agricultural areas by keeping as many intact hedgerows and forest patches as possible to provide shelter, breeding sites and food for them (Naylor and Ehrlich 1997).

2.11 The direct economic value of nature

Just as ecosystem services are lost when natural habitats disappear, so too are the direct benefits they provide to local people. For instance, the collection of fruit, latex and nuts from a single hectare of Amazonian forest can provide native people with an annual income of between US$79 and $6300 (Myers 1997). Mangrove protection in Indonesia may help in maintaining viable fisheries stocks and provide benefits worth more than half a billion dollars per year (Ruitenbeek 1992).

However, the large-scale destruction of forests incurs costs beyond simply a loss of revenue and ecosystem services. In 1997–98, Southeast Asia experienced an extraordinary and widespread episode of forest fires when more than 5 million ha of Indonesian rain forest (in Sumatra and Kalimantan) was burnt (Schweithelm 1998; see Chapter 4). These fires, unprecedented in scale in modern times, were a result of the combination of El Niño-mediated drought conditions and poor land use practices. The resulting smoke and ash from these fires blanketed much of Indonesia, Malaysia, Singapore and northern Australia. This smoke not only jeopardized the health of approximately 20 million human inhabitants, but also disrupted the economies of these nations because of a decline in tourist numbers (Talbott and Brown 1998). The burning cost an estimated US$4.4 billion in terms of lost tourism revenues and increased health care costs (Tomich *et al.* 2004; Figure 2.5).

Intact nature areas also provide employment for many local people. Forests, for example, are important sources of employment and opportunities. In 2000 alone, the global forestry industry generated over US$350 billion – half of which was pulp and paper (Lebedys 2004). Furniture, which is counted separately, delivered another US$80 billion. The formal forestry sector employs 13 million people, more or less equally divided between the pulp and paper sector, wood-based industries and forestry activities in the field. The formal furniture sector employed an additional 3.5 million (Lebedys 2004). While forestry activities grew in Latin America and tropical Asia, they declined in Europe and Japan.

Africa accounted for only 2% of global forestry sector value-added exports in 2000 (Lebedys 2004). Nonetheless, logging and wood processing contributed more than 3% of the gross domestic product (GDP) in 21 African countries (Lebedys 2004) that have relatively poorly developed technical and industrial economies. Global forestry exports (adjusted for inflation) rose by 50% between 1990 and 2000 to US$144 billion. Developing countries account for only one-quarter of the value added by the global forestry sector; however, 70% of all forestry jobs in the sector are in developing countries. Between 1990 and 2000, the forestry industries of Latin America and the developing countries of Asia added 1 million new jobs (Lebedys 2004). From the above, it is clear that forests have a large impact on developing economies, fostering prosperity and reducing poverty. However, it is disconcerting that forest management practices are so poor in the tropics (Putz *et al.* 2000), making sustainability of this enterprise questionable. It should also be realized that many wild places and species generate money through ecotourism. For example, in 1999, 41% of US$1 billion earned by Costa Rica was generated by visiting bird watchers (Sekercioglu 2002).

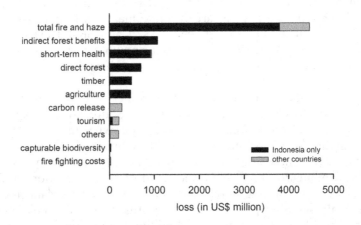

Figure 2.5 Estimates of damages to Indonesia and other countries resulting from fires and haze from 1993 to 1998. (Data derived from Tomich *et al.* 2004.)

2.12 The role of nature in human culture

It is clear from the preceding sections that nature provides many benefits to humanity. Indeed, overall, it has been estimated that the conservation of natural habitats can generate economic benefits (e.g. through fuel, fibre and pharmaceuticals) whose value is at least 100-fold higher than would be realized in the destruction and conversion of habitats for other land uses (Balmford *et al.* 2002). In addition, biodiversity provides strong aesthetic, moral and spiritual benefits. Humanity and biodiversity are inextricably linked, far beyond simple measures of monetary value (Gaston and Spicer 1998; Posey 1999). Human cultures, knowledge and religions are strongly influenced by nature. Many indigenous communities in the tropics protect refugial forested areas as 'worship forests' or 'sacred groves'. These worship forests are used during religious festivals and rituals, or as graveyards (Grove *et al.* 1998; Posey 1999), yet worship forests continue to be degraded by contemporary human activities, indicating cultural erosion that has its source in deforestation. Clearly, the benefit to humanity of conserving nature has many foundations: economic, ecological, moral and sociocultural.

2.13 Loss of knowledge

The biodiversity of tropical areas is much richer, yet, because of the predominance of formal Western science in Europe and North America, this diversity is also far less known than its temperate counterpart (Figure 2.6). With a few notable exceptions, all tropical regions remain poorly studied (Figure 2.7), and there is an alarming dearth of information on major taxonomic groups such as vascular plants, amphibians, reptiles and mammals. This information gap suggests that the documented biological impacts of habitat disturbance may represent just the tip of a grossly underestimated iceberg (Figure 2.8). Without doubt, many great, but as yet unrealized scientific discoveries are hidden in tropical habitats. The recent discoveries of a new, now extinct, dwarf human species from Flores (Indonesia)

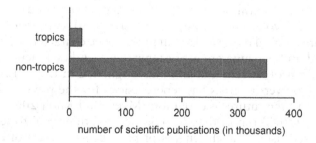

Figure 2.6 Total number of published articles on biodiversity-related issues in tropical and non-tropical regions. All comparisons were based on the number of internationally peer-reviewed research articles (excluding exclusively marine studies) published between 1986 and 2005, extracted from the database BIOSIS Previews.

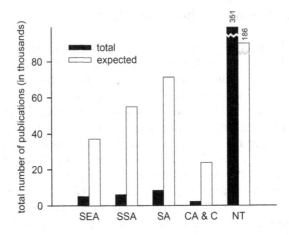

Figure 2.7 Total number of published (filled bars) and expected (unfilled bars) articles on biodiversity-related issues between tropical regions such as Southeast Asia (SEA), sub-Saharan Africa (SSA), South America (SA), Central America and Caribbean (CA & C), and non-tropical (NT) regions. All comparisons were based on the number of internationally peer-reviewed research articles (excluding marine studies) published between 1986 and 2005 extracted from the database BIOSIS Previews. The expected number of publications for each geographical region was calculated by dividing the number of publications evenly among geographical regions based on the total number of species per region. Taxonomic groups involved are vascular plants, freshwater fish, amphibians, reptiles, birds and mammals. Total species numbers were obtained from the World Resources Institute website (earthtrends.wri.org).

and several new species of palms, butterflies and frogs in the Foja mountains of New Guinea are strong testimonies to this statement (Morwood *et al.* 2004; Cyranoski 2006). Some species new to science are even sold in local markets. A new species of rodent was recently discovered being sold for food in local markets in Lao People's Democratic Republic (P.D. Jenkins *et al.* 2005). Some of these undiscovered wonders may aid humanity, such as the discovery of new medicines that could halt the spread of emerging diseases (see above).

Thus, extinctions through habitat loss and degradation of pristine areas will not only impede scientific knowledge, they may also profoundly affect human well-being and evolution. Owing to massive land conversion and invasive species, future tropical biodiversity will probably be characterized by fragmented, homogenized and depauperate ecosystems that lack key functional elements. A large proportion of current global biotas originated in tropical regions, and so the tropical biodiversity crisis has repercussions for the preservation of current biodiversity and the future of evolution (Myers and Knoll 2001). For instance, Sechrest *et al.* (2002) report that 70% of the evolutionary histories (343 million years; measured using branch lengths of phylogenetic trees) of carnivores and primates are harboured in 25 biodiversity hotspots covering just a few per cent of the earth's surface. Therefore, these biodiversity hotspots, of which most are tropical (see Chapter 1), are vital not only for endemic species, but also for supporting unique and threatened evolutionary trajectories.

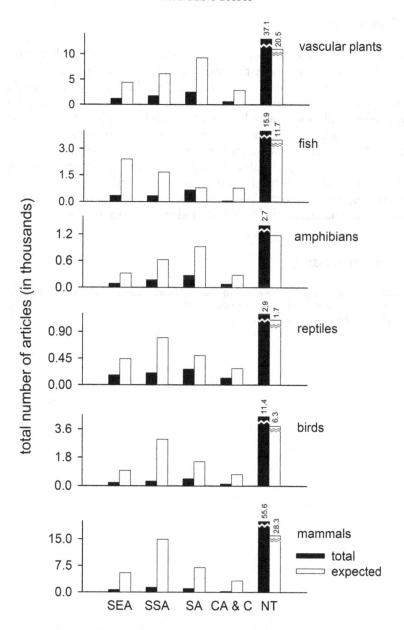

Figure 2.8 Total number of published (filled bars) and expected (unfilled bars) articles on biodiversity-related issues between tropical regions and non-tropical regions for each taxon group. All comparisons were based on the number of internationally peer-reviewed research articles (excluding marine studies) published between 1986 and 2005 extracted from the database BIOSIS Previews. For each taxonomic group, the expected number of publications for each geographical region was calculated by dividing the number of publications evenly among geographical regions, weighted by the total number of recorded species per region for that particular taxonomic group. Species numbers were obtained from the WRI website (earthtrends.wri.org). For abbreviations of different areas see Figure 2.7.

2.14 Summary

1 Habitat destruction in the tropics, in addition to being a direct determinant of biodiversity loss, may elevate the emission of greenhouse gases, alter regional and local weather conditions, enhance the spread of disease, increase the risk of flooding and, thus, ultimately, affect humanity in many undesirable ways.
2 Water safety and quality may be compromised by tropical deforestation.
3 Many economies, particularly in the developing tropical countries, are inextricably linked to and dependent on natural ecosystems.
4 With the loss of tropical habitats, not only is the composition of existing biotas compromised, so too is their future evolution.
5 Improved scientific knowledge in the tropical region is needed for effective conservation management of habitats and their associated biotas.

2.15 Further reading

Chivian, E. (2002) *Biodiversity: Its Importance to Human Health*. Centre for Health and the Global Environment, Harvard Medical School, Cambridge, MA.
Daily, G.C. (1997) *Nature's Services*. Island Press, Washington, DC.

3

Broken Homes: Tropical Biotas in Fragmented Landscapes

As a result of massive deforestation and forest degradation (Chapter 1), habitat fragmentation is becoming one of the major issues in tropical conservation biology. Tropical forest fragments (also called 'remnants' or 'patches') are scattered among urban areas, pastures, agricultural areas and other types of land uses. Millions of hectares of tropical forests currently exist as fragments (Wright 2005). In this chapter, we first discuss the theoretical premises upon which the understanding of the processes involved in habitat fragmentation are based. Next, we discuss abiotic and biotic effects of fragmentation, including a detailed discussion of 'edge effects'. Additionally, we report on the species and ecosystem processes vulnerable to fragmentation and the importance of surrounding non-forested habitats. We also discuss the conservation value of tropical fragments over the long term and how this can be enhanced through connecting corridors.

3.1   Theoretical premises of fragmentation

Habitat fragmentation occurs when a large expanse of a particular, broadly defined habitat type is reduced to smaller patches that are isolated by surrounding, but different habitats (generally termed 'matrix' habitats – Wilcove *et al.* 1986). To date, most fragmentation studies have been done in forested landscapes, and the general consensus is that fragmentation effects may be stronger in tropical than in temperate systems (Fahrig 2003). Forest fragmentation results in three prominent changes: (1) reduced forest area, (2) increased isolation among fragments and (3) the creation of edges where forests abut non-forested habitats. Each of these outcomes of fragmentation affects populations, communities and ecosystem processes (Kupfer *et al.* 2006; Figure 3.1). Kupfer *et al.* (2006) argue that the matrix (non-forest habitat surrounding forest fragments) can also exert strong influences on processes within the fragments themselves (see below). Reduction in forest area can result in losses in habitat diversity and resources, with ensuing declines in animal and plant populations that make them more vulnerable to

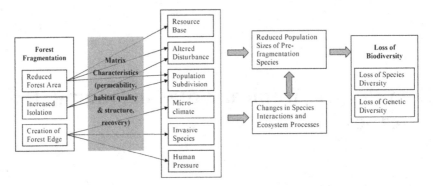

Figure 3.1 A conceptual model of the effects of forest fragmentation on biotas. The influence of matrix (habitat surrounding fragments) is also included. (After Kupfer *et al.* 2006. Copyright, Blackwell Publishing Limited.)

extinction (Brook *et al.* 2003a). Small populations are more vulnerable to extinction due to their heightened susceptibility to extreme environmental events (Shaffer 1981; Gilpin and Soulé 1986; see Chapter 9) and the loss of genetic variability (Spielman *et al.* 2004).

Increased isolation can hamper dispersal, and so gene flow among fragments may be compromised. Fragmentation can result in subpopulations within a 'metapopulation' (a set of local populations among which there may be gene flow, and extinction and colonization – Hanski 1991; see below) becoming genetically isolated from the remaining patches due to a disruption in dispersal. Fragmentation also alters the edge-to-interior ratio of forest fragments. Forest edges tend to have harsher microclimates for some forest biota due to the greater penetration of solar radiation and wind, higher ambient temperature and lower humidity (Murcia 1995). There may also be synergistic processes involved in the fragmentation process; for example, larger patches may be easier to locate by dispersing animals.

Upon isolation, a fragment is likely to harbour more species than it is capable of maintaining. Through the negative effects of fragmentation, some of these species are eventually lost – a process known as 'species relaxation'. Removal of surrounding vegetation can also result in the concentration of displaced individuals. This 'crowding effect' can alter intra- and inter-specific interactions through increased competition and other density-dependent effects. If predators are involved, crowding can also result in elevated rates of predation (D.A. Saunders *et al.* 1991). As the forest was felled around fragments in Brazilian Amazon (see below), many bird species found sanctuary in the remnant forests (Bierregaard *et al.* 1992). After isolation, bird capture rates at least doubled in newly isolated fragments. These elevated bird numbers persisted for about 200 days until population sizes fell below pre-isolation levels (Bierregaard *et al.* 1992), with the extent of faunal collapse being greater in smaller fragments.

Larger fragments tend to have a greater immunity to the negative influences of fragmentation because of their larger core area (i.e. the region unaffected by edge effects) and relatively smaller proportion of edge. Such fragments may also contain a higher diversity of habitats and more species and individuals than smaller ones, even after accounting for their larger area. Fragment shape is also an important determinant of the severity of negative edge effects. Narrow linear fragments have more edge than square or round fragments as a result of the higher edge-to-core ratio. However, the shape of a fragment is probably more important for smaller than for larger fragments (D.A. Saunders *et al.* 1991).

The initial debates about habitat fragmentation centred on whether the equilibrium theory of island biogeography (MacArthur and Wilson 1967) can be applied to fragments (Simberloff and Abele 1976a). MacArthur and Wilson (1967) proposed that the number of species on an island would reflect a balance between the rate at which populations of established species are lost and the rate at which new species colonize it. However, habitat fragments are not oceanic or land-bridge islands, and they are not surrounded by matrix as inhospitable to terrestrial species as vast expanses of water; thus, island biogeographic theory may not be directly applicable to the processes of habitat fragmentation (Gilpin and Diamond 1980). A subsequent debate raged on the concept of whether a single large or several small reserves (called the 'SLOSS' debate) was more effective for preserving species subject to fragmentation (Simberloff and Abele 1976b). Although the SLOSS debate spawned many interesting studies, it has been suggested that the scientific minutiae may have hampered adequate reserve/fragment planning and management (D.A. Saunders *et al.* 1991).

Two other major conceptual pathways have permeated fragmentation research. Populations in patchy habitats can function as metapopulations when and if patch extinctions can be offset by colonizations from patches retaining the capacity to produce dispersing individuals (so-called 'source' populations). At theoretical population equilibrium, the rate of extinctions is equal to the rate of colonizations (Levins 1970; E.O. Wilson 1975; Hanski 1989). However, owing to the poor dispersal capacity of many tropical taxa (Sodhi *et al.* 2004a), it is unlikely that many tropical forest fragments can function as metapopulations to maintain equilibrium.

As a subset of metapopulation theory, source-sink theory can also be applied in the fragmentation context. Source-sink theory assumes a consistent net flow of migrants from populations in superior habitats to those in poorer habitats (Pulliam 1988). Fragmented landscapes may be partitioned into population sources and sinks. 'Sources', owing to their superior habitat quality, may have reproductive surpluses that engender consistent net exportation of organisms. Large patches of continuous forests can serve this purpose in fragmented landscapes. In 'sinks', local reproductive success cannot offset local mortality, so net importation of individuals is required to avoid extinction. This immigration process is also known as the 'rescue effect' (immigration by unrelated individuals into isolated populations; J.H. Brown and Kodric-Brown 1977). Small patches in fragmented landscapes can act as sinks.

Despite these inspiring theoretical developments, Haila (2002) argued that the concept of habitat fragmentation has remained conceptually ambiguous for three reasons. First, habitat fragmentation is not a simple phenomenon affecting issues of connectivity and microhabitat quality – the process also results in the absolute reduction of available habitat and a change in spatial configuration. The corollary of these complex and interacting processes is that species composition and extinction risk can be difficult to predict. Second, many habitats are naturally fragmented by mechanisms such as fire, seasonal flooding and insect outbreaks. Finally, different organisms respond differently to fragmentation, again rendering various predictions uncertain.

3.2   Abiotic and geometric components of fragmentation

There is increased solar radiation within a fragment following isolation (D.A. Saunders *et al.* 1991). In addition, air temperatures are higher in fragment edges and surrounding matrix areas than in the interior (Kapos 1989). Increased soil heating may affect soil microorganisms and leaf litter invertebrates and thus reduce litter decomposition and soil moisture retention. Fragmentation may also expose fragments to more wind, resulting in elevated vegetation mortality directly through physical damage or by increased evapotranspiration and desiccation following reduced humidity (D.A. Saunders *et al.* 1991). Elevated tree damage results in more litter fall and thus alters the habitat for ground-dwelling biota. Fragmentation can also alter hydrological cycles. Loss of native vegetation may reduce rainfall inception, thus increasing water runoff and soil erosion.

Forest fragments are susceptible to invasion by weedy species and other negative processes (Gascon *et al.* 2000). Fragmentation can also increase vulnerability to extreme weather conditions such as El Niño-induced drought (W.F. Laurance *et al.* 2000; W.F. Laurance and Williamson 2001). These droughts can elevate tree mortality, increase leaf litter and change plant phenology. Owing to their dry conditions, fragments remain vulnerable to fire, especially when surrounding matrix such as pastures are burnt repeatedly (Gascon *et al.* 2000; W.F. Laurance and Williamson 2001). Fragments may also provide easy access to hunters and poachers. Further, harvested populations in fragments may not be replenished and the prey base may be poorer as a result of faunal depauperation. Peres (2001) argued that in heavily fragmented areas exposed to hunting, most mid-sized to large vertebrates may be driven to local extinction in the Brazilian Amazon. Chemical compounds such as fertilizers and pesticides can also move from surrounding agricultural areas into the fragments and thus affect biodiversity negatively (Murcia 1995).

3.3   Biotic effects of fragmentation

Most of our empirical understanding of tropical fragmentation has been derived from the Biological Dynamics of Tropical Rain Forest Fragments Project (BDFFP). In 1979, the World Wide Fund for Nature (WWF) and Brazil's National Institute for Research in Amazonia launched the BDFFP near the city of Manaus, Brazil to determine how fragmentation impacts tropical biotas (see Spotlight 3: William Laurance). This massive project tested a host of fragmentation hypotheses within a cattle ranching area by exploiting existing forest fragments and creating others experimentally. The data collected from these fragments were also compared with

Spotlight 3: William F. Laurance

Biography

I am a conservation biologist, and I am especially interested in assessing the impacts of intensive land uses, such as habitat fragmentation, fires and logging, on tropical ecosystems. My team also studies global change phenomena, such as the effects of global warming on tropical biotas, and I'm broadly interested in conservation policy. Over the past two decades I've worked in the Amazon, central Africa, tropical Australia and Central America. Why do I work and live in the tropics? I have long been interested in nature conservation, and tropical forests are among the most biologically diverse and imperilled ecosystems on earth. I was raised far from any rain forest – in the western USA – but I worked in several zoos in my youth and in that way became intrigued by tropical species and communities. When it came time to do my PhD at the University of

California, Berkeley, I decided to study the impacts of forest fragmentation on tropical mammals. Later, I started working on tropical trees, and also used remote sensing and geographic information systems to study deforestation and land use change. Today, I pretty much work on anything – trees, vines, mammals, birds, amphibians – but the one common theme is that my research has a strong conservation focus. I am very much a believer that conservation biologists have to be active conservationists as well. This is especially so in the tropics, where biologists have led international efforts for nature conservation. As president of the Association for Tropical Biology and Conservation, the world's largest scientific society devoted to the study and protection of tropical ecosystems, I've tried to ensure that our organization plays a leading role in promoting conservation. Our main weapon is our scientific credibility, and the fact that we have a lot of expertise among our members. We've fought a number of important conservation battles, and won some of them. Sometimes we feel like the little Dutch boy with his finger in the dike, but if we biologists don't strive to slow rampant forest destruction, who will?

Major publications

Laurance, W. F. and Bierregaard, R. O. (1997) *Tropical Forest Remnants: Ecology, Management and Conservation of Fragmented Communities.* University of Chicago Press, Chicago.
Laurance, W. F. and Peres, C. A. (2006) *Emerging Threats to Tropical Forests.* University of Chicago Press, Chicago.

Laurance, W. F., Laurance, S. G., Ferreira, L. V., RankindeMerona, J. M., Gascon, C. and Lovejoy, T. E. (1997) Biomass collapse in Amazonian forest fragments. *Science* **278**, 1117–1118.

Laurance, W. F., Cochrane, M. A., Bergen, S., Fearnside, P. M., Delamonica, P., Barber, C., D'Angelo, S. and Fernandes, T. (2001) The future of the Brazilian Amazon. *Science* **291**, 438–439.

Laurance, W. F., Lovejoy, T. E., Vasconcelos, H. L., Bruna, E. M., Didham, R. K., Stouffer, P. C., Gascon, C., Bierregaard, R. O., Laurance, S. G. and Sampaio, E. (2002) Ecosystem decay of Amazonian forest fragments: A 22-year investigation. *Conservation Biology* **16**, 605–618.

Questions and answers

Is fragmentation always bad for tropical biodiversity?

It depends what you mean by biodiversity. Whenever there is an environmental change, there are winners and losers. If you fragment a forest, edge-adapted, generalist and exotic species proliferate, whereas old-growth specialists and area-demanding species (such as predators and large-bodied species that are vulnerable to hunting) decline. The reason we worry so much about habitat fragmentation is that the world has plenty of generalist and exotic species; we don't need to conserve them. The old-growth and area-demanding species, however, are a different story. In a fragmented landscape, their populations often collapse and vanish. So if our goal is to maximize the long-term survival of species – especially those that are most vulnerable to extinction – then habitat fragmentation is almost universally a bad thing. A final consideration is that fragmented landscapes tend to be far more vulnerable to fires, logging, and overhunting than are intact forests. In the Amazon, for example, fire frequency increases drastically within a few kilometres of forest edges relative to forest interiors. In times-series imagery from satellites, you can see the fragments imploding over time, because rain forests just can't survive this withering recurrence of destructive fires. Thus, fragmentation is bad from lots of different perspectives.

Do you think timber logging, agricultural expansion or global change phenomena pose the biggest threat to Amazonia this century?

I'd have to say agricultural expansion is the biggest threat, especially cattle ranching and industrial soy farming, simply because it's so apparent that it's devastating vast expanses of forest. Cattle ranching has exploded – the number of cattle in Brazilian Amazonia has risen from about 20 million to 60 million head over the last decade – while soy farms have also grown exponentially. Soy farmers not only clear forest themselves, they also buy up a lot of recently cleared land, and thereby force ranchers and slash-and-burn farmers to push ever further into the frontier and destroy even more forest. The soy farmers are also a powerful political lobby that is pushing for a massive expansion of highways, roads, and other transportation infrastructure in the Amazon. These new projects are criss-crossing the Amazon and are greatly increasing the pace of forest loss and fragmentation. It's far harder to predict the effects of global change phenomena. Some models suggest that increasing deforestation (which reduces evapotranspiration and hence rainfall) and global warming could both have major impacts on the Amazonian climate. But the different models vary a lot, and the bottom line is that there is still much we don't understand. The threat from global change might be relatively limited, or it might be massive.

Has the BDFFP delivered practical conservation outcomes, and can its principles be applied to other tropical regions?

Yes, I think so. We've demonstrated, for example, that even remarkably small clearings, such as a powerline corridor or highway, can dramatically inhibit the movements of many rain forest species. We've shown that smaller (< 100 ha) forest fragments rapidly lose many species and exhibit striking changes in their ecology. We've also found that edge effects drive many changes in fragmented rain forests, and this has implications for reserve and buffer zone management, and for the design of wildlife corridors. These are all quite practical conservation outcomes. In general, I think that many of these principles can be applied to other tropical regions, though of course that's not to suggest that all forests behave similarly. For example, the importance of edge effects may vary quite a lot among different tropical regions.

data from control sites within an adjacent continuous forest of several hundreds of hectares in area. In total, fragments of 1, 10, 100 and 200 hectares (ha) and control plots in continuous forest of 1, 10, 100 and 1000 ha were measured and surveyed (Bierregaard *et al.* 1992). This study, the only long-term fragmentation research ever done in the tropics, provided a wealth of information (some of which is summarized below), especially considering that most of the fragmentation studies in the tropics have been done on birds (I.M. Turner 1996). However, it is still unclear if a heavy reliance on one locality (BDFFP) is sufficient to generalize the effects of tropical forest fragmentation to other biotas, even considering the added information from other (principally) bird studies (I.M. Turner 1996).

Nonetheless, studies show area to be a good predictor of species richness in the fragments. For example, area was the strongest predictor of bird species richness in forest fragments in Singapore (Castelletta *et al.* 2005). However, there was a possibility that habitat type interacted with patch size when predicting species richness. Twenty bird species were restricted to forest patches containing vegetation that was > 50 years old (Castelletta *et al.* 2005). Similar results have been found in other regions. Richness of forest-dependent avifauna was found to increase with fragment size in the Cerrado region of central Brazil (Marini 2001). Likewise, J.E.M. Watson *et al.* (2004) found that remnant area was the main predictor of forest bird species richness in Madagascar (Figure 3.2). Van Balen (1999) sampled birds of 19 lowland forests (6–50 000 ha) on the island of Java and found that 30 forest species (i.e. those found in forest edges and interior habitats) occurred only in forests > 1000 ha in area, and that 13 of these were restricted to forests > 10 000 ha. This result suggests that many forest species in Java require large patches of intact habitat to survive. Additionally, many of these species occurred in only a few locations, suggesting a heightened vulnerability to environmental stochasticity and forest disturbance.

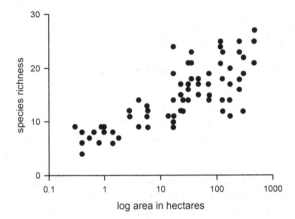

Figure 3.2 Mean total bird species richness per point count per remnant versus area in littoral forests of Madagascar. (After J.E.M. Watson *et al.* 2004. Copyright, Blackwell Publishing Limited.)

Similar area-related trends have been found for other tropical taxa as well. Fragment area and isolation explained 82% of the variation in species richness of arboreal marsupials in tropical rain forests in northern Australia (W.F. Laurance 1990). The most vulnerable species was the lemuroid ringtail possum (*Hemibelideus lemuroids*) with > 97% decline in abundance in forest fragments (1.4–590 ha) compared with a > 3000 ha selectively logged primary forest. Richness and abundance of dung and carrion beetles varied positively with fragment size and negatively with fragment isolation in Mexico (Estrada *et al.* 1998). In experimentally isolated forest fragments in the BDFFP, species loss for leaf litter beetles was 50% for 1-ha fragments, 30% for 10-ha fragments and 14% for 100-ha fragments (Didham *et al.* 1998).

As efficient decomposers and drivers of nutrient recycling, dung beetles are essential ecosystem engineers that act as a cornerstone for tropical biodiversity. Further, they act as vectors for seed dispersal, and they may control the spread of parasites to vertebrates by removing dung. Andresen (2003) found that dung beetle richness in a 1-ha fragment was half that found in continuous forest and in 10-ha fragments in the BDFFP. Dung and seed burial rates were also higher in continuous forest than in fragments, suggesting that some vital ecosystem processes may be compromised by fragmentation.

The BDFFP showed that, in addition to certain invertebrates, understorey birds and mammals were also sensitive to fragment area, with a number of the species disappearing even from fragments as large as 100 ha (W.F. Laurance *et al.* 2002). However, taxa respond differently to fragmentation. Frog richness in this area increased after isolation of fragments, apparently due to the resilience of this group to area and edge effects and the generation of higher habitat diversity at the edges exploited by edge or matrix specialists (Gascon 1993). However, whether this effect was temporary is still unknown. Similarly, there seemed to be little effect of fragmentation on small mammals in this area because of their capacity to exploit edge habitats and regenerating forests (Malcolm 1997). There is even evidence that congeneric species respond differently to fragmentation – the abundance of the frog *Eleutherodactylus chloronatus* increased with fragment size in Ecuador, but the sympatric *E. trepidotus* did not (D.M. Marsh and Pearman 1997). In the case of the latter species, reduced isolation from the nearest forest patches appeared to offset population decline.

The negative impacts of fragmentation have been also been documented for birds and mammals from two wildlife sanctuaries in Thailand (Pattanavibool and Dearden 2002). Both wildlife sanctuaries investigated (Om Koi and Mae Tuen) contained a mixture of montane and evergreen forests. Patches in Om Koi were larger (> 400 ha) and more connected whereas those in Mae Tuen were smaller (< 100 ha) and more isolated. Overall, Om Koi contained 119 bird species compared with only 89 in Mae Tuen. In the latter, almost half of the species were rare (< 1 individual/visit), but this was true for only one-third of the species at Om Koi. Additionally, large frugivores such as the brown hornbill (*Ptilolaemus tickelli*) and great hornbill (*Buceros bicornis*) were found only in Om Koi. This study illustrates the general rule of thumb in conservation ecology that protection of larger, connected patches is better for maintaining bird species richness. A similar conclusion was made regarding patch connectivity in Singapore after sampling birds from 17 patches of varying sizes (7–935 ha) (Castelletta *et al.* 2005).

A comparative study of the mammal species richness in seven protected areas (70–304 ha) and their adjoining logged forests in Peninsular Malaysia (states of Pahang and Selangor) (Laidlaw 2000) indicated that size of the natural forest area was the most important variable affecting mammal species richness. Indeed, sharp declines in richness were observed when natural forest fragments dropped below 164 ha in area. However, large herbivorous and carnivorous mammals in Peninsular Malaysia may require forest ≥ 10 000 ha in size. Laidlaw (2000) concluded that the preservation of undisturbed forest adjacent to disturbed areas can effectively enhance mammal richness. Large continuous forest is also required by large carnivores and herbivores such as the Asiatic elephant (*Elephas maximus*). Further, even within large areas of continuous forest (e.g. the 27 469 km² Taman Negara National Park in Peninsular Malaysia and contiguous forests in southern Thailand), some of the larger mammals, including the tiger (*Panthera tigris*), may avoid areas with high human traffic (Kawanishi and Sunquist 2004). Pattanavibool and Dearden (2002) found similar results for mammals in Thailand – larger fragment sizes and greater connectivity in Om Koi resulted in 10 more mammal species there than in the more fragmented Mae Tuen, and only Om Koi retained large mammals such as the Asiatic elephant.

Isolation can also affect species richness in fragments. For example, abundance and richness of understorey insectivorous birds declined after experimental isolation in the BDFFP (Stouffer and Bierregaard 1995a). As little as 80 m separating fragments was a strong barrier to movement by some insects and mammals and the vast majority of understorey birds they contained. Deep forest euglossine bees did not visit a 10-ha fragment that was 100 m away from the continuous forest (A.H. Powell and Powell 1987), which is particularly worrisome considering their importance as pollinators (see Chapter 2). Similarly, ecosystem-engineering dung and carrion-feeding beetles also seemed reluctant to negotiate the 100 m of open habitat isolating fragments (B.C. Klein 1989). Morphology, physiology and behaviour may constrain a species from crossing open areas. However, regenerating vegetation in the matrix can facilitate movements; for example, some insectivorous birds were able to move through regenerating forest in the matrix (Stouffer and Bierregaard 1995a).

Renjifo (1999) found that 30% of avifauna in sub-Andean Colombia perished due to forest fragmentation, although he did suggest that fragments may still serve as reservoirs for some globally endangered bird species and can be used to restore extirpated or perishing avifaunas. A similar argument has been made by I.M. Turner and Corlett (1996), who suggested that small fragments can serve as last refuges for some rain forest species and can thus provide resources to restore forests and their biological communities.

3.4 Long-term fate of fragments

Little is known regarding the conservation value of fragments over the long term because monitoring data span relatively short periods. Some taxa are long-lived (e.g. tropical trees) and may take decades or even centuries to perish following fragmentation. Brooks *et al.* (1999a) found that the half-life (the time taken to lose half of the species within a defined area) for avifaunal extinctions in fragmented forests in Kenya was approximately 50 years on average, but also that half-life decreased with decreasing fragment size. Indeed, 100-ha fragments may lose half of their bird species in less than 15 years (Ferraz *et al.* 2003; see Chapter 9). Likewise, local extinctions of butterflies, birds and primates in experimentally isolated fragments in the BDFFP occurred more rapidly in small (1–10 ha) than in large (100 ha) fragments (W.F. Laurance *et al.* 2002). Leck (1979) also reported a loss of 25 bird species from an 87-ha forest fragment in Ecuador after just 5 years of isolation.

Few fragments have been monitored for more than 50 years. The avifaunal changes over 100 years (1898–1998) in a 4-ha patch of rain forest in Singapore (Singapore Botanic Gardens) revealed that 49% of forest species were lost during this time, with the addition of invasive species such as the house crow (*Corvus splendens*) (Sodhi *et al.* 2005b; Chapter 5). By 1998, 20% of the birds observed were introduced species, with more native species expected to be extirpated from the site in the future through competition and predation. That study shows that small fragments decline in their conservation value for forest birds over time, and that they are vulnerable to competition from invasive species (Chapter 5). From the same fragment, similar effects have been observed for plants – 51% of plant species have been extirpated in this fragment since 1890, while an additional 80 introduced plant species have been recorded (I.M. Turner *et al.* 1996).

Sodhi *et al.* (2006a) reported avifaunal turnover in a 86-ha tropical woodlot (Bogor Botanical Garden, Indonesia) containing 54% native and 46% introduced plant species and receiving a mean of 83 649 human visitors per month. Since its isolation in 1936, subsequent surveys have shown a gradual reduction in avifaunal species richness. By 2004, the original richness of this woodlot had declined by 59% (from 97 to 40 species), and its forest-dependent avifauna had declined by 60% (from 30 to 12 species). All seven forest-dependent bird species that attempted to colonize this woodlot by 1987 perished thereafter, thus demonstrating that fragments may have poor long-term sustainability for most forest avifauna.

3.5   Edge effects

Fragmentation can reduce species diversity across a broad spectrum of taxa; however, mid-sized herbivores can increase in number following the demise of their carnivore predators (Terborgh 1992). These herbivore gains can then lead to an increase in seed predation and seedling damage, and an increase in competition among vulnerable herbivorous species. Exceptionally high densities of native wild pigs (*Sus scrofa*) (27–47 pigs/km^2) were found in the Pasoh Forest Reserve (Peninsular Malaysia) (Ickes *et al.* 2001). Ickes *et al.* (2001) attributed this to the local extinction of natural predators (e.g. tigers) and the year-round food supply of African oil palm fruits (*Elaeis guineenis*). Here, pigs reduced plant recruitment threefold, but seemed to increase plant growth by 52% for trees between 1 and 7 m tall (Ickes 2001).

Ruderal (weedy) plant species can be good indicators of forest disturbance because of their tendency to be recruited along forest edges (Sizer and Tanner 1999). For example, the fragmentation of Brazilian Atlantic forest fragments resulted in a boom in ruderal plant species (Tabarelli *et al.* 1999). These are shade-intolerant species and typically found only in forest edges and gaps, so their richness is generally higher in smaller fragments (Tabarelli *et al.* 1999; Figure 3.3).

Soulé *et al.* (1988) also suggested that the loss of predators results in small omnivorous species becoming abundant in fragments. Known as 'mesopredator release', the expansion of such species may elevate predation rates on small vertebrate species and promote an influx of omnivore predators and parasites

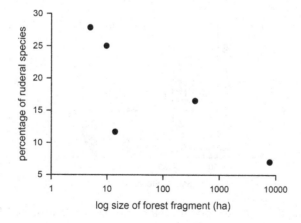

Figure 3.3   Percentage of ruderal floral species in montane Atlantic forests. (After Tabarelli *et al.* 1999. Copyright, Elsevier.)

from the matrix (see also Chapter 9). Superabundant omnivore populations can cause the reduction of some native species in fragments through these 'negative edge effects'.

Likewise, fragmentation can elevate predation by making bird nests more accessible to 'forest-avoiding' generalist predators, such as the house crow, that typify disturbed areas. In fact, high nest predation may have been one the drivers of the extirpation of forest birds from Singapore (Wong *et al.* 1998). Because predation events are difficult to observe, artificial nest experiments have been used to compare predation pressure among sites. In Singapore, 80% ($n = 328$ nests) of artificial ground nests were depredated, with primary forest experiencing at least 12% less predation than other fragments (Wong *et al.* 1998). Sixty-two per cent of arboreal nests ($n = 110$) were also depredated (Sodhi *et al.* 2003). However, there was no difference in predation pressure among the forest types, and other studies have reported similar predation rates. Burkey (1993) reported higher predation on chicken eggs in rain forest edges than in forest interiors in Belize and Mexico. Artificial nest predation has also been reported to be higher in fragments than in continuous forests (Loiselle and Hoppes 1983; Sieving 1992). However, artificial nest experiments may mimic natural predation events poorly because of factors such as lack of parental defence and differences in the characteristics of native and 'exotic' eggs that are used in most experiments.

Negative edge effects can also harm plants. The rates of tree mortality in experimentally isolated forest fragments in the BDFFP increased with decrease in distance to the forest edge (W.F. Laurance *et al.* 1998a). Tree mortality and damage was three times higher 0–60 m from the forest edge than in the forest interior (W.F. Laurance *et al.* 1998a, 2000; Figure 3.4). This may be due to harsher microhabitat conditions, such as high temperature and humidity, near edges and the increased exposure to strong winds. However, such effects

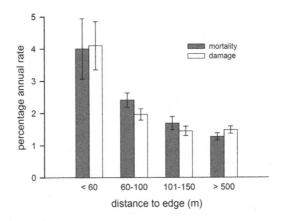

Figure 3.4 Annual rates of tree mortality and damage in Amazonian rain forests as a function of distance to the nearest forest edge. Only large trees (≥ 10 cm diameter at breast height) are included. Error bars represent one standard error. (After W.F. Laurance *et al.* 2000. Copyright, Cambridge University Press.)

penetrate deeper (60–100 m) into a fragment when the matrix comprises a more hostile habitat. For example, rain forest patches in the BDFFP had more extreme edge effects when the surrounding matrix was pasture instead of regenerating forest. Increased tree mortality due to fragmentation can also elevate carbon dioxide (CO_2) emissions. W.F. Laurance *et al.* (1998b) projected that carbon emissions from forest fragmentation range between 3 and 16 million tonnes/ year in the Brazilian Amazon alone, and between 22 and 149 million tonnes/ year from tropical forests globally. The other negative aspect of this story is that seedling density of shade-tolerant species was higher in continuous forest than in fragments (Benitez-Malvido 1998). In 100-ha fragments, seedling density was higher in forest interior than edges. Benitez-Malvido (1998) argued that this was an outcome of low seed rain following elevated tree mortality in fragments and their edges. The species richness of recruited herbs, lianas (structurally parasitic woody vines), palms and trees was also lower in the understorey of fragments than in continuous forest (Benitez-Malvido and Martínez-Ramos 2003), suggesting that forest regeneration may have low potential in fragments. However, lianas become more abundant near edges following fragmentation and may lead to higher rates of tree infestation and mortality (W.F. Laurance *et al.* 2001b). In addition, competition for light and nutrients with lianas may also increase tree mortality in fragments.

What factors affect species along the fragment edges? The species composition of leaf litter beetles in the Brazilian Amazon changed in relation to the distance from forest edge (Figure 3.5) and area, mediated primarily by air temperature, canopy height, per cent of twig ground cover, litter biomass and litter moisture content (Didham *et al.* 1998). The severity of edge effects also varies seasonally in the tropics. For example, *Norops* spp. lizards are abundant in forest edges

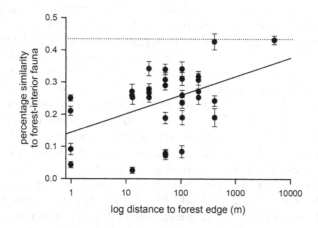

Figure 3.5 Variation in composition of leaf litter beetle assemblages as a function of distance from forest edges. Mean percentage similarity (± standard error) to corresponding forest interior is shown for each sample. The dotted line represents average background degree of similarity between different forest interior samples. (After Didham *et al.* 1998. Copyright, Blackwell Publishing Limited.)

during the dry season, but are found only in the forest interior during the wet season (Schlaepfer and Gavin 2001).

As mentioned previously, certain species can respond positively to the creation of edges. Birds that use treefall gaps, such as some habitat generalists, arboreal insectivores and hummingbirds, became abundant along edges in isolated fragments in the BDFFP (Stouffer and Bierregaard 1995a,b). Some frugivorous bats, marsupials and shade-intolerant plants also increased in abundance at edges in these areas (W.F. Laurance *et al.* 2002). Non-rain forest species can also invade fragments. For example, lianas, exotic palms, generalist fruit flies, light-loving butterflies, generalist frogs and non-rain forest birds invaded experimentally isolated fragments in the Brazilian Amazon (W.F. Laurance *et al.* 2002). In Singapore, the density of introduced bird species is lower in older secondary and primary forest fragments than in younger secondary and plantation fragments (Castelletta *et al.* 2005), indicating the usefulness of invasion statistics as a measure fragment habitat quality.

Although the evidence for various types of edge effects is vast, the results of the investigations carried out to date are not in most cases amenable to generalizations because of (1) poor study design (e.g. lack of replicates), (2) inconsistency of methods and (3) oversimplification of the perception of edge and edge effects because of factors such as differing species responses (Murcia 1995). Nonetheless, the general consensus among tropical ecologists is that edge effects are negative on the whole, and highly dependent on the structure of the composition, shape and form of the fragment matrix.

3.6   Vulnerability to fragmentation

Species belonging to a particular guild have similar ecological requirements and, therefore, similar roles in the community. It follows logically then that some guilds, such as species or taxonomic groups, are more vulnerable to fragmentation than others. Two avian guilds were particularly negatively impacted by fragmentation in the BDFFP: army ant-following birds (i.e. those that feed on insects disturbed by swarming army ants) and mixed-species flocking birds. Both groups disappeared from the 1- and 10-ha fragments (Harper 1987; Bierregaard and Lovejoy 1989) because they require large home ranges and are reluctant to cross even narrow (30–80 m wide) clearings (W.F. Laurance *et al.* 2000; see also Chapter 9).

In the Australian tropics, small forest specialist mammals declined while edge specialists and those exploiting the matrix increased following fragmentation (Harrington *et al.* 2001). Species with large home ranges were more prone to fragmentation than their more restricted counterparts (Harrington *et al.* 2001). Likewise, J.E.M. Watson *et al.* (2004) found that canopy insectivorous and frugivorous bird species in Madagascar were vulnerable to fragmentation through changes in microclimate that reduced insect and fruit abundance. Among tropical birds, understorey insectivores are particularly sensitive to fragmentation (see Chapter 9). Sekercioglu *et al.* (2002) postulated that high dispersal ability is the best determinant of persistence of understorey birds in Costa Rica.

Tabarelli *et al.* (1999) reported that, in the Atlantic forest fragments in Brazil, woody, shade-tolerant plant species occupying the upper strata of the forest canopy and which depend on animals for seed dispersal are more vulnerable to fragmentation. There was about a 9% decline in species richness in the families Myrtaceae, Lauraceae, Sapotaceae and Rubiaceae, all of which produce fleshy fruits eaten by vertebrate frugivores. Of course, species decline in these plant families may also be due to the loss of vertebrate frugivores themselves (Cordeiro and Howe 2003), implying a negative feedback loop or vortex. Seeds in forest fragments are also less likely to germinate than in continuous forest (Bruna 1999) because inclement microhabitat characteristics of small fragments (high temperature, drier conditions and increased light penetration) and accumulated leaf litter may also compromise seed germination (Bruna 1999). Therefore, even if vertebrates disperse seeds in forest fragments, many or most of these are unlikely to germinate.

Further, depauperate frugivore and pollinator communities in fragments compromise dispersal and result in loss of heterozygosity in seedlings due to reduced gene flow (Aldrich and Hamrick 1998; Dayanandan *et al.* 1999; Fuchs *et al.* 2003). The ensuing inbreeding depression resulting from this effect can cause reproductive success to decline for some tree species in the fragments (see also Chapter 2). For example, pollination success and seed production was found to be lower in smaller fragments in Costa Rica (Ghazoul and McLeish 2001), suggesting that some bees may not move widely among fragments, thus compromising gene flow among some plants. Likewise, native flower-visiting bees declined with fragment size in Argentina (Aizen and Feinsinger 1994), and 50% fewer flowers produced seeds in fragmented compared with continuous populations in the dry forests of Costa Rica, probably as a result of reduced pollinator activity (Fuchs *et al.* 2003). Some trees attempt to counter this process by producing more flowers (Fuchs *et al.* 2003); and other plants that rely on wind dispersal may be buffered against fragmentation by maintaining high pollen flow among fragments (White *et al.* 2002).

Sometimes an exotic species can negate the 'mutualistic vacuum' created by the loss of native species (Chapter 5). For example, exotic African bees (*Apis mellifera scutellata*) regularly visit the flowers of canopy-emergent trees (*Dinizia excelsa*) in Amazonian forest fragments. Such is the pollination assistance of this exotic species that individuals of this plant produce up to three times more seeds in fragments than in continuous forest, all the while maintaining genetic diversity (Dick 2001).

Fragmentation can also alter ecosystem processes such as trophic interactions. Both herbivory and parasitism in plants decrease with increased fragment size in Argentina (Valladares *et al.* 2006; Figure 3.6). However, herbivory is higher in fragment cores than at edges, while parasitism shows the opposite trend. Valladares *et al.* (2006) argued that this may result from herbivores avoiding the harsh microclimates at edges, whereas parasitoids become more efficient along edges by using visual cues enhanced by the patchiness created during fragmentation. That study showed that fragmentation can alter insect–plant food webs and may thus affect long-term ecosystem sustainability.

Mobile species and those that can tolerate habitat disturbances are able to persist in the forested fragments of Kenya (Lens *et al.* 2002). In some cases,

(a)

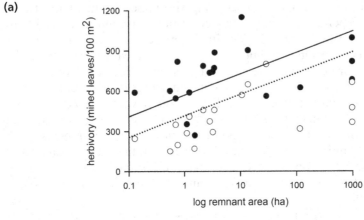

(b)

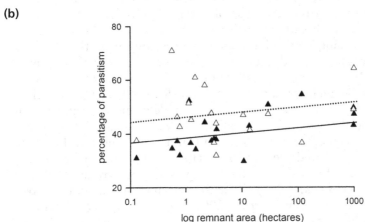

Figure 3.6 Relationships between trophic processes of insect–plant food webs and habitat fragmentation in Chaco Serrano, Argentina. (a) Herbivory by leaf miners (remnant edges = open circles; interior areas = filled circles) in relation to woodland area. (b) Parasitism on leaf miners (remnant edge = open triangles; interior areas = filled triangles) as a function of area. Least-squares lines of best fit (solid lines = interior; dotted = edge) based on analysis of covariance are shown. (After Valladares *et al.* 2006. Copyright, Blackwell Publishing Limited.)

nocturnal species such as bats and moths may be less sensitive to fragmentation than are diurnal species because of their heightened dispersing capability (Daily and Ehrlich 1996). Moths may also face similar ambient environmental conditions in fragments and in the surrounding matrix during the night because of the absence of the sun–shade transition zone. However, higher temperature, humidity and solar radiation in the matrix may act as movement barriers for some diurnal butterflies (Daily and Ehrlich 1996).

3.7   Importance of matrix

Different matrix types (logged, agricultural, pastures and urban) can differentially affect the movements of pollinators, seed dispersers and herbivores, thus altering the aforementioned processes and vegetation characteristics in the fragments they surround (Jules and Shahani 2003). Gascon *et al.* (1999) found that between 40% and 80% of primary forest frogs, birds and ants were detected in a matrix made either of pastures or of regenerating forests during the BDFFP. However, species that tend to avoid the matrix declined or disappeared from fragments following isolation, indicating that these species may not be able to cope with fragmentation and should be targeted for conservation attention. Matrix-tolerant species are generally better able to move among fragments and so are buffered against deleterious genetic and demographic effects (W.F. Laurance *et al.* 2000, 2002). Regenerating secondary forest in the matrix can also facilitate inter-patch movements (Castellón and Sieving 2006). Additionally, edge-related tree mortality can be lower in fragments surrounded by regenerating forest than in cattle pasture matrix (Mesquita *et al.* 1999).

3.8   Increasing fragment connectivity

The negative effects of fragmentation on biodiversity are generally reduced when patches are closer to one another. Species vulnerable to fragmentation do not cross open areas and also usually avoid forest edges (Woodroffe and Ginsberg 1998; Kinnaird *et al.* 2003). Camera-trap data from Sumatra (Indonesia) show that tigers and rhinoceros (*Dicerorhinus sumatrensis*) usually avoid areas within 2 km of forest boundaries, and elephants (*Elephas maximus*) usually avoid areas within 3 km of forest boundaries (Kinnaird *et al.* 2003; Figure 3.7). These results lend strong support to the importance of maintaining forest-like vegetation in the surrounding matrix so that species can use it as a stepping stone during inter-patch movements (Gascon *et al.* 2000). Forest corridors can be used to increase connectivity and thus facilitate movement between fragments (see Chapter 10).

Newmark (1991) recommended that forest corridors as wide as 200 m are needed to allow movements by East African forest birds between fragments. de Lima and Gascon (1999) surveyed species richness, composition and abundance of small mammals and leaf litter frogs in linear strips of primary rain forest (140–190 m in width) and adjacent continuous rain forest in Brazil. They found that species composition and abundance in both groups did not differ between strips and continuous forest. This suggests that linear primary forest strips can provide suitable habitat, at least for some groups, and this was reinforced by the fact that many small mammals and frogs were also breeding in these. de Lima and Gascon (1999) suggested that linear forest remnants can possibly serve as movement corridors and thus increase connectivity in fragmented tropical landscapes.

Similarly, S.G. Laurance and Laurance (1999) sampled arboreal mammals in linear rain forest remnants in tropical Queensland (Australia). They concluded that such floristically diverse remnants that were at least 30–40 m wide can function as habitat and possibly movement corridors for most arboreal mammals in this

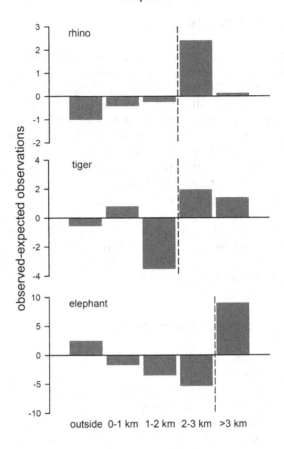

Figure 3.7 Three species of mammals avoiding areas near forest edges. Vertical dashed lines indicate natural breaks in distributions based on Jenk's optimization method. (After Kinnaird *et al.* 2003. Copyright, Blackwell Publishing Limited.)

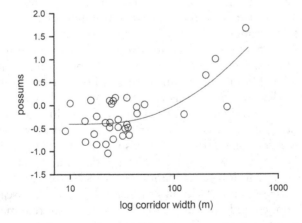

Figure 3.8 Two species of possums were found to be more abundant in wider corridors. (After S.G. Laurance and Laurance 1999. Copyright, Elsevier.)

region. However, the abundance of more sensitive lemuroid ringtail possums and Herbert River ringtail possums (*Pseudochirulus herbertensis*) increased with increasing corridor width (Figure 3.8). Therefore, S.G. Laurance and Laurance (1999) concluded that corridors in this region should be at least 200 m wide to cater for the species more sensitive to habitat fragmentation.

3.9   Summary

1   Theories of island biogeography, metapopulation and source-sink dynamics have been used to understand the effects of fragmentation on biodiversity loss and persistence.
2   Fragmentation alters the abiotic profile of tropical forests and may make them more vulnerable to fire, hunters and poachers.
3   Reduced area and increased isolation due to fragmentation affect biotic richness and ecosystem processes negatively.
4   Some species and groups (guilds) are more vulnerable to fragmentation than others.
5   Corridors and encouraging forest regeneration in the matrix can reduce the negative effects of fragmentation.

3.10   Further reading

Haila, Y. (2002) A conceptual genealogy of fragmentation research: From island biogeography to landscape ecology. *Ecological Applications* **12**, 321–334.

Laurance, W. F., Lovejoy, T. E., Vasconcelos, H. L., Bruna, E. M., Didham, R. K., Stouffer, P. C., Gascon, C., Bierregaard, R.O., Laurance, S. G. and Sampaio, E. (2002) Ecosystem decay of Amazonian forest fragments: A 22-year investigation. *Conservation Biology* **16**, 605–618.

Saunders, D. A., Hobbs, R. J. and Margules, C. R. (1991) Biological consequences of ecosystem fragmentation: A review. *Conservation Biology* **5**, 18–32.

Turner, I. M. (1996) Species loss in fragments of tropical rainforest: A review of the evidence. *Journal of Applied Ecology* **33**, 200–209.

4

Burning Down the House

Fire is a natural process that shapes many tropical forest systems. However, prolonged drought and poor land use decisions have made many tropical landscapes vulnerable to a higher frequency and intensity of fires, creating a negative feedback cycle that is damaging at both local (e.g. habitat loss) and global scales (e.g. atmospheric pollution). In this chapter, we report on the effects on tropical biotas of fire regimes modified by humans.

4.1   Forest fires

The periodic climatic phenomenon known as El Niño Southern Oscillation can induce severe and extended dry conditions in tropical areas. The combination of El Niño-mediated drought conditions and poor land use practices has facilitated the penetration of forest fires deep into tropical forests that were previously thought to be fire resistant (Uhl 1998; Figure 4.1). Transmigration of humans into forests and an increase in accessibility due to the creation of roads can further exacerbate the spread of forest fires (Stolle *et al.* 2003). Forest fires have always occurred in tropical forested landscapes, but factors such as rapid human population growth, changes in land use and apparently increasingly severe cycles of El Niño (see Chapter 8) are now working synergistically to increase the frequency and intensity of catastrophic fires (Kinnaird and O'Brien 1998; Taylor *et al.* 1999; Cochrane 2003). For example, large-scale fire cycles in the Amazon currently range between 7 and 14 years compared with natural fire rotation of at least hundreds of years (Cochrane *et al.* 1999; see Spotlight 4: Mark Cochrane).

The development of road networks for selective logging improves access for people engaged in both logging and non-logging-related activities, which could increase the chances of deliberate fires. Further, damage by logging activities in these areas leaves woody debris, making forests more vulnerable to fires due to an increase in available fuel. This vulnerability may persist for decades after

Figure 4.1   Land that has recently been cleared by fire. (Photo by Arvin Diesmos.)

cessation of logging activities (Holdsworth and Uhl 1997). Tropical deforestation can result in regional climate changes, such as increases in ambient temperature and wind, and decreases in precipitation and relative humidity (see Chapter 2) – all potentially contributing to increases in forest fires (W.A. Hoffmann *et al.* 2003). Indeed, a prehistoric loss of vegetation cover due to burning by people has even been implicated in a major weakening in the southward penetration of the summer monsoon in Australia (G.H. Miller *et al.* 2005).

During the intense El Niño event of 1997–98, Southeast Asia and Latin America experienced a remarkable and widespread period of forest fires. Approximately 20 million ha was burnt (Schweithelm 1998; Cochrane 2003), with countries such as Indonesia (Sumatra and Kalimantan) and Brazil each losing about 5 million ha of rain forest (Schweithelm 1998) and Bolivia losing 3 million ha (Cochrane 2003).

4.1.1   Effects of fires on flora and forest regeneration

Burning alters forest structure and composition. In Sabah (Malaysian Borneo), tree mortality after drought and fire ranged from 38% to 94% in logged forests and from 19% to 71% in unlogged (primary) forests (Woods 1989). Logged forests suffered more severe canopy loss than burnt unlogged forests in this area (Figure 4.2). Similarly, tree morality after fires was 37% in unlogged forest, but 55% in logged forest in Côte d'Ivoire (Ivory Coast; Swaine 1992). High tree mortality in logged areas following fire may diminish the prospects for

Spotlight 4: Mark A. Cochrane

Biography

I am a Professor at the Geographic Information Science Center of Excellence (GIScCE) at South Dakota State University and am jointly appointed with the Department of Biology and Microbiology and the Department of Geography. I conduct interdisciplinary work that combines remote sensing, ecology and other fields of study to provide a landscape perspective of the dynamic processes involved in land cover change. I first became interested in ecology through coursework while I was completing my baccalaureate in Environmental Engineering at the Massachusetts Institute of Technology. I then spent a year working on a variety of research projects in Antarctica, furthering my interest in science. Having had enough of cold weather environments, I chose to

do my postgraduate research in the Brazilian Amazon and, in 1998, received a doctoral degree in ecology from the Pennsylvania State University. I am among the world's leading experts on wildfire in tropical ecosystems. I am renowned for documenting wildfire characteristics, behaviour and severe effects in tropical forests, as well as how current systems of human land use foster wildfires. My research focuses on understanding spatial patterns, interactions and synergisms between the multiple physical and biological factors that affect ecosystems. Recently published work emphasizes the human dimensions of land cover change and the potential for sustainable development. My collaborative research with the Brazilian NGO IMAZON (The Amazon Institute of Man and the Environment) has been instrumental in the Brazilian government's recent (2003) programme to expand its national forest system in the Amazon to 50 million hectares. In my ongoing research programmes, I continue to investigate the drivers and effects of disturbance regime changes resulting from various forms of forest degradation, including fire, fragmentation and logging.

Major publications

Cochrane, M. A. (2001) Synergistic interactions between habitat fragmentation and fire in evergreen tropical forests. *Conservation Biology* 15, 1515–1521.

Cochrane, M. A. (United Nations Environment Program) (2002) Spreading like wildfire – tropical forest fires. In: *Latin America And The Caribbean: Prevention, Assessment And Early Warning*. United Nations Environment Program, Regional Office for Latin America and the Caribbean.

Cochrane, M. A. (2003) Fire science for rainforests. *Nature* 421, 913–919.

Cochrane, M. A., Alencar, A., Schulze, M. D., Souza, C. M., Nepstad, D. C., Lefebvre, P. and Davidson, E. A. (1999) Positive feedbacks in the fire dynamic of closed canopy tropical forests. *Science* **284**, 1832–1835.

Cochrane, M. A. and Laurance, W. F. (2002) Fire as a large-scale edge effect in Amazonian forests. *Journal of Tropical Ecology* **18**, 311–325.

Questions and answers

Are fires in tropical landscapes more frequent now than in the past?

Yes. Fires are much more prevalent in the tropics than in the past. The issues with fire in tropical landscapes centre not on the presence of fire in these landscapes, but on the frequency with which they are burning. In many regions of the tropics, current land use practices result in surviving forest fragments being subjected to fire so often that they are rapidly being degraded by a fire regime that they cannot withstand. Fire is the primary tool for clearing and maintaining agricultural land. Forests are slashed and allowed to dry before being burned to release their nutrients to the soils. Many lands are subsequently turned into pastures, which also are burnt frequently in order to keep trees from regrowing. Most land use is fire dependent. Forests are also subjected to selective logging, which removes only the valuable trees and leaves the remaining forests susceptible to escaped fires. Having fire-dependent agriculture embedded in fire-susceptible forests quickly leads to forest fires. Once this occurs, the landscape is quickly converted from one of a few flammable islands (e.g. pastures) within a near fire-immune sea of vegetation (i.e. rain forests), to one where a fire at any location can permeate an entire region, forests and agricultural lands alike.

Will global climate change be important in altering future fire regimes?

Yes. Global climate change will probably result in important changes in regional fire regimes. Global climate models (GCMs) do not all agree on what the future climates will be like, but it is fairly certain that temperatures will increase substantially, especially in interior regions such as the upper Amazon. Most models show a concomitant increase in rainfall for these regions, but it is uncertain whether or not these increases will be enough to offset the drought stress from rising temperatures. Rainfall reductions from ongoing deforestation, however, will probably be greater than any projected increases due to global climate change, leading to a net increase in drought stress and fire behaviour, further stressing these important ecosystems.

How important are the adverse effects of fire on tropical biotas relative to other drivers such as habitat loss and land use change?

Fire is what integrates tropical landscapes. Habitat loss is primarily seen as a function of deforestation, but forest degradation through logging and fire causes substantial changes in forest structure and composition that affect habitability. The key aspect of wildfire is that once it has damaged a region's forests it effectively changes the 'rules of the game' for land management. What was once a highly fire-resistant ecosystem becomes a highly fire-susceptible forest. A positive feedback of increasing fire frequency, fire intensity and fire severity can become established. Instead of burning once every thousand years or more, these forests may burn once every 10 years or less. These forests cannot withstand such a fire regime, and several fires can effectively deforest an area. As the landscape becomes more permeable to fire, even diligent landowners have difficulty protecting their lands from fires. Uncontrolled fires destroy large-investment, high-return perennial crops such as rubber tree, pepper and fruit plantations, making cattle ranching the most viable land use. Pasture grass is the most flammable land cover possible and only exacerbates a region's fire problems.

Is there a theory in fire ecology, and if yes, what is it?

The short answer is 'no'. To my knowledge there is no accepted theory underlying fire ecology.

vegetation recovery because the growth of grasses (which provide understorey fuel for future fires) will proceed unabated. In contrast, lower fire-related tree mortality in unlogged areas elevates the prospects for vegetation recovery (i.e. pristine forest is more resilient), although species diversity may be lower in the post-fire forest (Woods 1989). Similarly in the Amazon, canopy cover and live non-pioneer stem and sapling density have be found to decrease as the intensity of burning increases (Cochrane and Schulze 1999; Figure 4.3). In heavily burnt stands, live tree density was 10% of that in untouched stands (Cochrane and

(a)

(b)

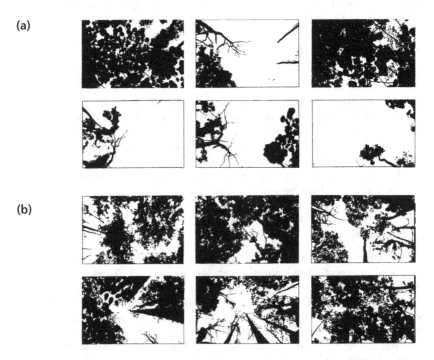

Figure. 4.2 A skyward view of the canopy from (a) logged and (b) unlogged burnt forests. (After Woods 1989. Copyright, Blackwell Publishing Limited.)

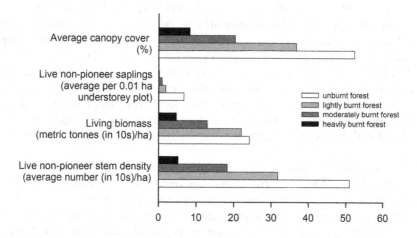

Figure 4.3 Comparisons of canopy cover, densities of non-pioneering saplings and stems and total living biomass between unburnt forest and three classes of burnt forest. (Data derived from Cochrane and Schulze 1999.)

Schulze 1999). Heavy burning also reduced live vegetation biomass by 83% in the Brazilian Amazon (Gerwing 2002).

Additional support for the reported losses of native vegetation through burning is provided by Slik *et al.* (2002). They compared tree (diameter at breast height ≥10 cm) species diversity of primary forests and burnt forests in Kalimantan (Indonesia). Even 15 years after fire, there were 22 more tree species recorded in the primary forest than in a forest burnt 1 year previously. Pioneer genera, such as *Macaranga*, dominated the canopy of regenerating forest 15 years after burning. Despite these observations, Slik *et al.* (2002) concluded that burnt forests, especially if only lightly burnt, still show signs of gradual recovery in tree diversity, and therefore retain some conservation value. In the same area, surface fires reduced seed availability by 85% in the litter layer, and by 60% in the upper 1.5 cm of soil (van Nieuwstadt *et al.* 2001). However, because further reductions in an already depleted seed bank can seriously reduce the sprouting potential of plants in areas that are burnt repeatedly, the conservation potential of such areas is diminished (van Nieuwstadt *et al.* 2001).

It is estimated that the 1997–98 fires in Sumatra killed 4.6% of the canopy trees, 70–100% of seedlings and 25–70% of saplings (Kinnaird and O'Brien 1998). The fires also seem to have facilitated the influx of exotic plant species such as *Chromolaena odorata* to the area. However, the full extent of damage caused by such opportunistic exotic plant species on the native flora could not be fully documented. Fires also reduced the number of flowering and fruiting trees in and near burnt areas (Kinnaird and O'Brien 1998), further reducing the prospects for recovery. This is particularly disconcerting from a conservation perspective because large, isolated, fruiting trees can attract frugivorous birds that disperse seeds that hasten forest recovery (Guevera *et al.* 1986).

The 1997–98 fires in the Brazilian Amazon resulted in 36% mortality of trees (≥ 10 cm diameter at breast height) and, as in Sumatra, an almost complete annihilation of saplings (Haugaasen *et al.* 2003). In the burnt areas, tree mortality seemed to increase (Figure 4.4), and thick-barked trees survived fires better than did thin-barked trees (Barlow *et al.* 2003). Buttressed trees were also particularly vulnerable to fires, probably because they easily accumulated flammable leaf litter and had thinner bark than non-buttressed trees (Barlow *et al.* 2003). In the dry tropical forests of Ghana, fire-induced vegetation mortality was inversely correlated with tree size, with smaller trees being more susceptible to fire (Swaine 1992).

Light-demanding, wind-borne pioneer species flourish in burnt forests (Cochrane 2003). Within burnt areas, unburnt forest fragments can serve as sources of seeds for post-fire vegetation recovery. For instance, in Sabah 1 year following fire, surviving adult dipterocarp trees provided enough seeds to cover the forest floor (Swaine 1992). However, repeated fires may also deplete this seed source (Cochrane and Schulze 1999) and so create a source-sink dynamic (see Chapter 3) that may compromise the viability of forests over large areas. Vines and grasses invade burnt areas and, given their high flammability, may slow forest regeneration and increase further vulnerability to fire (Woods 1989; Cochrane *et al.* 1999; Gerwing 2002). Further, a positive feedback exists between forest fires and future fire susceptibility and severity (Cochrane *et al.* 1999). In previously burnt areas, fire cycles may be as low as one every 5 years, and, based on estimated solar radiation intensities, 50% of burnt forest could burn again within 16 rainless days, compared with just to 4% in unburnt forest (Cochrane and Schulze 1999). Cochrane *et al.* (1999) postulated that severe fire regimes could eventually convert a large part of the Amazonian rain forest into scrub or grassland.

The majority of forest fires in Thailand are caused by humans, and they are occurring with increasing frequency. Marod *et al.* (2002) studied the effects of forest fires on seed and seedling dynamics in seasonal forest in western Thailand (Mae Klong Watershed Research Station), and found that many species, such as

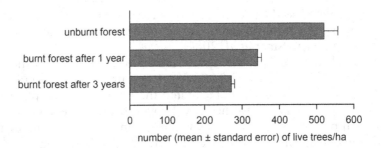

Figure 4.4 General decline in the density of live trees in burnt forests over time compared with an unburnt forest. (After Barlow *et al.* 2003. Copyright, Blackwell Publishing Limited.)

Pterocarpus spp., showed higher germination rates, a high level of resprouting, and the emergence of larger seedlings during fire years. Marod *et al.* (2002) hypothesized that in seasonal forests, many plant species may bet hedge against fire by producing seed banks and evolving life history strategies that permit rapid recovery from damage (e.g. resprouting, seeds that germinate after fire). However, this may not be the case for tropical rain forest plants that have evolved to cope with only sporadic fire episodes over time. Conversely, the effects of recent forest fires in seasonal forests may be underestimated because they occur at least annually and are deliberately started by people (Corlett 2004). Frequent fires in these drier forests probably lead to a reduction in the complexity of vegetation and suppress species diversity (Corlett 2004).

4.1.2   Effects of forest fires on fauna

Forest fires can affect native faunas in many ways. After the 1997–98 fires in Sumatra, the density of helmeted hornbills (*Buceros vigil*) declined by 50%, although two other hornbill species, the bushy-crested (*Anorrhinus galerius*) and wreathed (*Aceros undulates*) hornbill, seemed to be largely unaffected (Kinnaird and O'Brien 1998). In burnt areas, the loss of flowering and fruiting trees seemed to cause the precipitous decline of many frugivorous species (e.g. barbets, *Megalaima* spp.). Siamangs (the largest gibbons, *Holobates syndactylus*) disappeared from recently burnt areas, and groups of banded leaf monkeys (*Presbytis melalophus*) were 20% less abundant in burnt areas than in a nearby unburnt forest. Similarly, the horse-tailed squirrel (*Sundasciurus hippurus*), a primary forest specialist, also abandoned burnt areas (Kinnaird and O'Brien 1998). Factors such as the availability of resources (e.g. fruits and tree cavities for nesting) influenced this species' persistence in, and subsequent recolonization of, burnt areas. Similarly in Amazonia, Barlow and Peres (2006) found a 29% loss in fruiting tree basal area in forests that had suffered a single fire, and a 67% loss in those burnt twice. These reductions in tree biomass seem to have contributed to the decline of frugivorous vertebrates in burnt forests (Figure 4.5).

About a year after the catastrophic 1997–98 fires in the Amazon, habitat-specialist bird species were among the worst affected by fire, presumably because of their inability to withstand fire-mediated changes to their habitat (Barlow *et al.* 2002; Figure 4.6a). Dead-leaf gleaners, terrestrial gleaners and arboreal sallying insectivores suffered the most pronounced fire impacts (Figure 4.6b). Even 1 year after fire, there was little evidence that locally extirpated species had returned to the burnt areas (Haugaasen *et al.* 2003). Firelines seemed to act as dispersal barriers, preventing understorey primary forest birds from recolonizing burnt areas (Haugaasen *et al.* 2003). Even modest fires can alter avian communities in unlogged forests in a similarly insidious way as selective logging (Barlow *et al.* 2006). In Bolivia, there has been a report of higher species richness and abundance of amphibians, reptiles and small mammals in logged and burnt areas than in undisturbed forest (Fredericksen and Fredericksen 2002). This is as expected, because, even though disturbed areas are generally high in species richness, they almost exclusively contain more adaptive generalist species, rather than sensitive specialist species (e.g. those that exclusively inhabit primary forests)

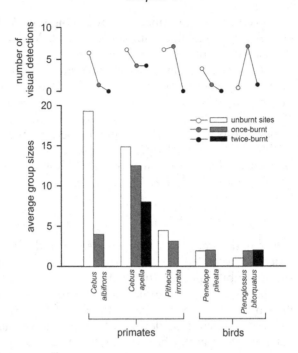

Figure 4.5 Vertebrate decline in the visual counts (top) and average group sizes (bottom) of frugivorous primates and birds in unburnt sites (unfilled bars and circles) and areas that were burnt either once (grey bars and circles) or twice (black bars and circles). (Data derived from Barlow and Peres 2006.)

of high conservation concern (Sodhi *et al.* 2005c). Similar to the Amazon, forest fires in Sumatra were particularly detrimental to the abundance of understorey insectivorous bird species (Adeney *et al.* 2006).

Fredericksen and Fredericksen (2002) hypothesized that small mammal richness is higher in disturbed areas because of an increased abundance of herbs, grasses and seeds. A higher abundance of amphibians in disturbed areas may be due to larger numbers of species that do not depend heavily on ambient moisture (e.g. the two-crested toad, *Bufo typhonia*).

In Sumatra, O'Brien *et al.* (2003) conducted a more detailed study to determine the longer-term effects of fires on siamangs; group sizes were smaller in burnt than in unburnt areas (average = 3.2 versus 4.0), and the chance of infants surviving to adulthood was lower in burnt areas (Figure 4.7). One of the reasons for this observation could be that strangler figs (*Ficus* spp.), the flowers of which are the siamang's main food, were reduced by 48% in burnt areas. In addition to the direct effects on siamang, the O'Brien *et al.* (2003) study implies that the decline in the number of seed dispersers, such as siamang, may also retard vegetation recovery in burnt areas.

Fires can also depress the reproductive success of other taxonomic groups. For instance, Cahill and Walker (2000) studied the impact of forest fires on the endemic red-knobbed hornbill (*Aceros cassidix*) in Sulawesi. Success of nesting attempts in years of fire was lower (62%) than in previous non-fire years (around 80%).

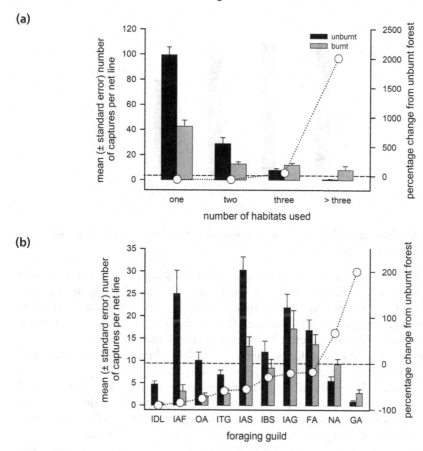

Figure 4.6 General decline in bird captures in unburnt forest (black bars) and burnt forest (grey bars), and the percentage change in abundance (open circles) from unburnt to burnt forest. All bird captures are grouped by (a) number of habitats used and (b) their foraging and dietary guilds. Guild classification: IDL, dead leaf-searching arboreal insectivore; IAF, gleaning terrestrial insectivore; OA, arboreal omnivore; ITG, gleaning terrestrial insectivore; IAS, arboreal sallying insectivore; IBS, bark-searching insectivore, feeding superficially; IAG, arboreal gleaning insectivore; FA, arboreal frugivore; NA, nectarivore; GA, arboreal granivore. (After Barlow *et al.* 2002. Copyright, Elsevier.)

Normal recruitment for this species has been estimated at 0.32 fledglings/female/ year (Kinnaird and O'Brien 1999), but this was almost halved to 0.17 fledglings/ female in the year of a fire. These results suggest that fire can compromise this species' reproductive success, which is particularly disconcerting considering that the species has also suffered from intense habitat loss in this part of Sulawesi.

Invertebrates are also severely affected by fires. In Bolivia, terrestrial invertebrate abundance was found to be lower in logged and burnt areas than in undisturbed forested areas (Fredericksen and Fredericksen 2002). Cleary and

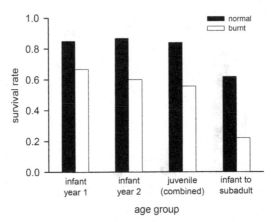

Figure 4.7 Survival rates of siamangs in unburnt (filled) and burnt (empty) forests. (Data derived from O'Brien *et al.* 2003.)

Genner (2004) found that El Niño-induced forest fires altered the structure of a butterfly community in Kalimantan. Two years after the fire, the community was dominated by large-winged generalist species and did not contain any endemic species that were present prior to the fire. This study shows that, at least over the short term, forest fires can selectively affect specialized butterflies.

In addition to facilitating forest fires, El Niño-induced drought can also disrupt vital ecological processes. Harrison (2000) found that in Sarawak such an episode reduced the production of inflorescences by dioecious figs (*Ficus* spp.) and may have led to the local extinction of up to eight species of dependent fig wasps (Agaonidae). Figs and fig wasps are a fascinating example of co-evolution – fig wasps raise their offspring in the secure environment of fig inflorescence, and in return provide the fig plant with an efficient pollination vector. Thus, the extirpation of fig wasps can disrupt the pollination of figs, and this may have cascading effects on frugivorous vertebrate species that depend on these for food (Harrison 2000; Koh *et al.* 2004a).

4.2   Burning savannas

Although natural ignition sources such as lightning have caused fires in tropical savannas for millions of years, 99% of the fires today are initiated by humans (Saarnak 2001). Most of these fires are deliberately lit and left to burn uncontrolled, with emissions from African fires transported to the mid-Atlantic, South Pacific and Indian Oceans (Cahoon *et al.* 1992). Africa is referred to as a 'fire continent' because of widespread fires caused by humans in savannas (Levine *et al.* 1995), and northern Australia is similarly aflame (Bowman 2000). Every year, burning of tropical savannas destroys an estimated three times as much dry matter as the burning of tropical forests (Andreae 1991). Savanna plants seem to be better adapted to fires than are forest species; bark thickness of savanna woody species is three times that of rain forest species, thus reducing their chances of mortality during fire (W.A. Hoffmann *et al.* 2003). Presumably in response to frequent fires, savanna species also reach maturity faster than congeneric forest species (W.A. Hoffmann *et al.* 2003).

However, the effects of fires on the biotas of savannas are poorly understood (Pardon *et al.* 2003). Frequent savanna fires can temporarily reduce shrub cover and promote herbaceous species (Sheuyange *et al.* 2005). This can lead to a positive feedback in which high herbaceous cover can attract herbivores, with their browsing preventing recruitment of trees in savannas (A.J. Mills and Fey 2005). In certain cases, however, high levels of grazing can prevent the spread of fire by breaking up the grass layer (Archibald *et al.* 2005). To counter high herbivory following fires, some resprouting plants, especially the *Acacia* of tropical Africa, allocate more resources to spine production (Gowda and Raffaele 2004). Fires may also cause shifts in the species composition of savanna grasses by favouring the persistence of annual flammable grass species that support higher dry fuel loads than perennial grasses, and can readily regenerate from seed following fire without exhausting their store of carbohydrates (Bowman *et al.* 2007). Fires in the savannas of northern Australia can influence flower and, ultimately, fruit production in woody species (Vigilante and Bowman 2004; Figure 4.8). Indeed, it is likely that Aboriginal burning in this area over the past 50 000 years has been used deliberately to manage fruit resources and facilitate hunting (Bowman 1998).

As with forest fires, savanna fires can also disrupt vital ecological interactions. For example, in burnt savannas there can be a reduced frequency of colonization by mycrorrhizal fungi ('fungus root'; Hartnett *et al.* 2004). These fungi are symbiotic with trees, in that the host plant photosynthesizes and provides carbon to the fungus and, in return, the plant benefits from increased nutrient supply by extension of the volume of accessible soil and the deterrence of parasitic fungi and nematodes. Nutrients are often rapidly depleted in areas directly around plant roots and the fungal hyphae are able to grow out beyond low-nutrient zones into

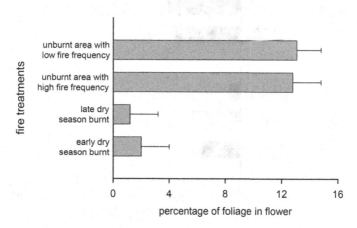

Figure 4.8 The flowering response of the green plum tree *Buchanania obovata* (large-leafed form) to four fire treatments (error bars indicate standard errors). [After Vigilante and Bowman 2004. Copyright, Commonwealth Scientific and Industrial Research Organisation (CSIRO) Publishing.]

otherwise inaccessible nutrient-rich places. Mycorrhizal symbiosis, because of its influence on plant allometric growth and population and community dynamics, is critical in mediating savanna structure and function.

Animals are also affected by burning tropical savannas. Granivorous and ground-feeding bird species become less common in severely burnt savannas, presumably as a result of depletion of seeds and ground cover (M.S.L. Mills 2004). In the Cerrado savannas of Brazil, different rodent species prefer different seral stages following burning, indicating that the habitat mosaics created by fires are important for maintaining high species diversity (Briani *et al.* 2004). However, the marsupial mouse opossum (*Gracilinanus agilis*) prefers only the later seral stages (> 23 years old) of post-fire savannas. Similarly, most rain forest plant and vertebrate species (amphibians, reptiles, birds and mammals) in Australia are more abundant in unburnt than in burnt savannas (Woinarski *et al.* 2004; Figure 4.9). Burning may therefore be detrimental to some rain forest species inhabiting the forest patches in savannas. Even those mammal species that have evolved to thrive in the fire-prone tropical savannas, such as the northern brown bandicoot

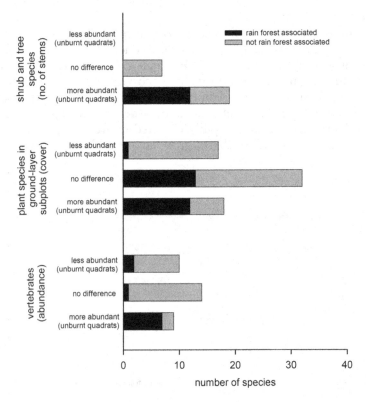

Figure 4.9 Total frequencies, based on relative abundance, of plants [shrub and tree species (top), ground layer plants (middle) and animals (vertebrates; bottom)] in unburnt and annually burnt sampling quadrats with respect to rain forest (black bars) and non-rain forest (grey bars) species. (Data derived from Woinarski *et al.* 2004).

(*Isoodon macrourus*), suffer unsustainable mortality when fires are too frequent or intense (Pardon *et al.* 2003).

Moreover, invasive species can exacerbate fires in the savannas and, conversely, altered fire regimes can facilitate their spread and enhance their competitive advantage. For instance, gamba grass (*Andropogon gayanus*), a perennial grass from Africa, has invaded northern Australia after being introduced as a pasture grass for grazing stock. Because gamba grass supports extremely high fuel loads, and is tall and well aerated when dry, the intensity of fires is eight times higher in areas invaded by this grass than in those that have not been invaded (Rossiter *et al.* 2003). Thus, this invasive grass is considered a serious threat to Australian tropical savannas and an engineer of changed fire regimes. Similarly, the exotic molasses grass (*Melinis minutiflora*), also originating in Africa, has increased the frequency of fire in Brazilian savannas (Mistry and Berardi 2005).

4.3   Tropical fires in the global context

Unprecedented in scale in modern times, the 1997–98 Southeast Asian fires seem to have been exacerbated by poor logging practices and both small- and large-scale land clearing for agriculture and tree plantations. The resulting smoke and ash from these fires blanketed much of Indonesia, Malaysia, Singapore and northern Australia. This smoke not only jeopardized the health of approximately 20 million of the region's inhabitants, it also induced a precipitous decline in tourist numbers, which disrupted the economies of these nations (Talbott and Brown 1998). It is estimated that in the worst-hit areas, the effect of these fires on humans was equivalent to each inhabitant smoking four packs of cigarettes each day (Talbott and Brown 1998). Owing to cardiovascular and respiratory complications arising as a result of elevated air pollution levels, these fires also seem to have increased the mortality of older (aged 65–74) people, based on an epidemiological study conducted in Kuala Lumpur (Sastry 2002). In Indonesia alone, 32 000 people suffered respiratory ailments that were most likely triggered by these fires (Kinnaird and O'Brien 1998). The total economic loss to the region from this environmental disaster was estimated at US$4.4 billion (Kinnaird and O'Brien 1998). Similarly, the 1997–98 forest fires in Brazil resulted in the death of thousands of people from fire-related illnesses (Cochrane 2003). These mortality figures do not include deaths attributable to smoke-related accidents. Savanna fires are also reported to have increased the number of cases of asthma in northern Australia (Johnston *et al.* 2002).

Depending on previous fire and land use history, net annual carbon emissions from the burning of tropical forests are estimated to be between 8 and 70 Mg (megagram)/ha (Cochrane 2003). It has been estimated that the total carbon emissions from human-caused burning of peat swamps in Kalimantan in 1997–98 were higher than those caused by the forest fires and probably exceeded 0.23 gigatonnes (Page *et al.* 2002). This release represents about 13–40% of annual carbon emissions resulting from the burning of fossil fuels and is the highest annual increase since recording of atmosphere carbon dioxide (CO_2) began in 1957. It is estimated that African savanna fires also regularly release large amounts

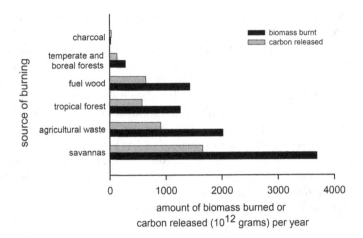

Figure 4.10 Global estimates of annual amounts of biomass burning and of the resulting release of carbon into the atmosphere. (Data derived from Andreae 1991.)

of greenhouse gases such as CO_2, carbon monoxide (CO) and methane (CH_4) (Koronti 2005). In fact, savanna burning is estimated to release considerably more carbon per annum than the burning of tropical forests (Levine *et al.* 1995; Figure 4.10).

4.4   Fire modelling and mitigation

Based on solar radiation intensities, Cochrane and Schulze (1999) estimate that burnt Amazonian forests are highly flammable (see above). The relationship between climate, fuel moisture and hotspots of fire occurrence is currently used to predict fires in Malaysia and Indonesia (Dymond *et al.* 2005). However, models to predict the fire susceptibility of tropical humid forests are certainly still in their infancy (Cochrane 2003). Current models based on the time since previous rain and leaf shedding may have poor predictive power due to the lack of a comprehensive accounting of moisture dynamics. The deep roots of tropical trees and tight hydrological cycles make the development of predictive fire susceptibility models challenging (Cochrane 2003). Improved knowledge of landscape-level occurrences of fires will be necessary if more robust predictions of fire occurrence are to be achieved in the future (Cochrane 2003).

Fire prevention is one of the primary concerns of managers of forested land, and regular and prescribed burning in savannas is used as a tool to avoid wildfires that might be otherwise caused by lightning. A recent study, however, has shown that rainfall patterns, not management actions, are the main determinant of fire in African savannas (Van Wilgen *et al.* 2004).

Fire management has probably played an integral role in the traditional lifestyle of Aboriginal people in northern Australia over the last 50000 years (Anderson *et al.* 2004), such that it is assumed that burning by Aborigines has also transformed Australian biotas (Bowman 1998). Currently, prescribed burning during the early part of the dry season (May-June) is undertaken to limit the extent and severity of fires in Australian savannas. This management practice, however, may not replicate the traditional fire management by Aboriginal people (Bowman *et al.* 2004). Further, the frequency and scale of prescribed Aboriginal burning was lower than by current European fire regimes; European fire regimes may have unwittingly triggered more fires (Bowman *et al.* 2004). One of the impediments in correctly mimicking fires initiated by Aborigines is the uncertainty about the background rate of fires caused by lightning (Bowman 2005). This deficiency is critical because a lack of understanding of indigenous fire management can result in policies that exacerbate fires, undermine the maintenance of habitat heterogeneity, are unfavourable to the biota adapted to an alternative pattern of burning, and may result in conflict between locals and managers (Maxwell 2004; I. Rodriguez 2004). For example, the impetus for continuing the emphasis on early dry-season fires has been challenged by recent work demonstrating that fires occurring earlier in the dry season tend to decrease juvenile tree growth and survival, probably as a result of these trees being physiologically more active during that time (Prior *et al.* 2006). Indigenous people in the Cerrado region of Brazil, as in Australia, have been practising prescribed burning during most of the dry season as well (Mistry *et al.* 2005). This practice has resulted in mosaics of burnt and unburnt savannas.

In West Africa, the education of local farmers is considered to be one of the critical management options for controlling fire given that these people are the main instigators of ignition (Swaine 1992). In Thailand, a participatory community fire management programme has been initiated in cooperation with local communities to prevent and extinguish fires (Hoare 2004). Another management option is to leave 'green' firebreaks to prevent the spread of fires. Maintaining intact vegetation as a guard against future fires should also be a major focus of effort for managers because such barriers can also assist in forest vegetation recovery. Planting trees in burnt areas can also be attempted to assist vegetation recovery.

The burning of peat swamps in Kalimantan provides a noteworthy example not only of flawed land use policy, but also of successful remedial management. In 1995, the Indonesian dictator Suharto announced that 1 million ha of land in Kalimantan would be converted for rice production. This resulted in the clearance and draining of extensive areas of peat swamp (Aldhous 2004). The peatlands, however, proved to be too acidic for growing rice, resulting in massive failure of this 'Mega Rice Project'. In 1997, the severe El Niño event resulted in the deliberate or accidental burning of these dried peat areas, and concomitant heavy carbon emissions (see above). Efforts are now being undertaken to ensure that the above scenario is not repeated. The key to success is to restore the water table and plant saplings on remoistened peat swamps. The group Climate Change, Forests and Peatlands in Indonesia (CCFPI), which has the backing of the Wetlands International and Wildlife Habitat of Canada and the European Union, is providing funds to build dams to rehydrate these peatlands (Aldhous 2004). CCFPI is also providing loans for the impoverished locals to buy livestock, and, if they plant and subsequently protect saplings, their loans are forgiven (Aldhous 2004).

4.5   Summary

1   Increasingly severe El Niño events, coupled with poor land use practices, are driving an increase in the frequency and severity of catastrophic forest fires in the tropics. Fires, especially when they occur repeatedly in rain forests, can diminish the chances of vegetation recovery and reduce the productivity of the habitats upon which tropical faunas depend.
2   In general, the vegetation of savannas is better adapted to fires than that of rain forests. However, overly frequent fires can disrupt community structure and invasive grasses can exacerbate fire intensity.
3   Fires in tropical landscapes result in heavy emissions of CO_2 and other greenhouse gases, potentially enhancing global warming.
4   Heavy smoke from catastrophic tropical fires can be detrimental to human health.
5   It may be difficult to predict the occurrence of forest fires. Fires need to be managed in conjunction, where possible, with indigenous and local knowledge.

4.6   Further reading

Bowman, D. M. J. S. and Franklin, D. C. (2005) Fire ecology. *Progress in Physical Geography* 29, 248–255.
Cochrane, M. A. (2003) Fire science for rainforests. *Nature* 421, 913–919.

5

Alien Invaders

Biotic invasions are emerging as a key driver of global biodiversity change. In this chapter, we report on the documented impacts of invasive species (also termed 'invasives') on tropical biotas, first by defining 'invasive' species, and then by providing a broad range of tropical examples and methods to manage existing problems.

5.1   What are invasive species?

Invasives are defined as those species that have spread from the point of introduction into an area they have not previously occupied and become abundant, threatening the native biota (Kolar and Lodge 2001; see Spotlight 5: Daniel Simberloff). Invasive species are also referred to as introduced species, non-indigenous species, exotics or aliens. Such species can be introduced deliberately through the pet trade, for agricultural purposes, as garden plants, or to support recreational, subsistence or commercial fishing and hunting. Invasive species may also be a by-product of commerce, e.g. stowaways in shipping material, ballast water or ballast soil. Some invasive species can also result from range expansion facilitated by human land use. For example, the expansion of common mynas (*Acridotheres tristis*) in Peninsular Malaysia and Singapore has most likely been facilitated by heavy deforestation and accompanying habitat modifications (Sodhi and Sharp 2006).

The negative impacts of biological invasions on native biodiversity are perhaps as severe as other high-profile threats such as global climate change (Chapter 8) and habitat loss (Chapter 1; Vitousek *et al.* 1996). Modern trade and transport have also facilitated the spread of biological invaders. Indeed, it is estimated that the current annual economic damage caused by biological invaders worldwide is US\$336 billion (Pimentel *et al.* 2000). This figure includes economic losses and control costs, but does not take into account the unmeasurable costs associated with biodiversity losses (e.g. loss of ecosystem services, Chapter 2). With increasing

Spotlight 5: Daniel Simberloff

Biography

As a child in rural Pennsylvania, I was fascinated by nature, collecting insects from age 5 and keeping insects, turtles and fish as pets. This idyll crashed to a halt at age 11 when we moved to an industrial suburb of New York City. I attended Harvard College, where I became excited by maths and majored in it. My junior year, while enjoying a non-majors biology course and realizing that I wasn't so enthusiastic about a maths career, I consulted a biology department advisor about postgraduate work. Frank Carpenter, an insect palaeontologist, astounded me by saying I could go to graduate school in biology with a little coursework in my final year. He also directed me to Ed Wilson as a potential graduate advisor. Ed introduced me to ecology and argued that maths is crucial to the maturation of ecology. He taught me  an enormous amount of biology and encouraged me in a great doctoral dissertation project, testing the theory of island biogeography that he had recently propounded with Robert MacArthur. We collaborated in fumigating small Florida mangrove islands and studying their recolonization by insects. Bill Bossert taught me about computers well before everyone knew about them. Beginning with my doctorate, my main interest has been how different species fit (or do not fit) together in communities, and this interest led to research in conservation issues, most notably on refuge design and impacts and management of introduced species, as well as more academic aspects of ecology, like the role of inter-specific competition. As a faculty member at Florida State University and now the University of Tennessee, I have learnt an enormous amount from excellent colleagues, postgraduate students and post-doctoral fellows. I also quickly interacted with policy makers and non-governmental organizations on conservation issues, first at the local and state levels, then nationally. My most important advice to prospective conservation biology students is to interact with challenging people doing interesting research and to engage in local conservation issues.

Major publications

Simberloff, D. (1997) *Strangers in Paradise: Impact and Management of Nonindigenous Species in Florida*. Island Press, Washington, DC.

Simberloff, D. (2003) How much information on population biology is needed to manage introduced species? *Conservation Biology* **17**, 83–92.

Simberloff, D. (2002) Managing existing populations of alien species. In: *Alien Invaders in Canada's Waters, Wetlands, and Forests* (eds. R. Claudi, P. Nantel and E. Muckle-Jeffs). Natural Resources Canada, Canadian Forest Service, Ottawa.

Simberloff, D. and Holle, B. V. (1999) Positive interactions of nonindigenous species: Invasional meltdown? *Biological Invasions* **1**, 21–32.

Simberloff, D., Parker, I. M. and Windle, P. N. (2005) Introduced species policy, management, and future research needs. *Frontiers in Ecology and the Environment* **3**, 12–20.

Questions and answers

What are the defining characteristics of an alien organism that make it 'invasive'?

An invasive introduced species is one that spreads into more or less natural ecosystems and thrives there, affecting native species.

Is there merit in the idea that tropical ecosystems are more resistant to invasion due to their species richness and complexity?

I don't believe either premise, that tropical ecosystems are particularly resistant to invasion or that complex ecosystems with many species are particularly resistant to invasion. There are many invasions into tropical ecosystems, including species-rich ones, and extensive research fails to support Elton's hypothesis that biological invasions are greatly facilitated by low native species richness. Any ecosystem is invasable by the right invader.

How common are 'invasional meltdowns', and what are the best tropical examples?

It is too early to know just how common invasional meltdowns are, but every year more cases are suggested, and some are buttressed by careful study. An important tropical example is the near destruction of the forest ecosystems of Christmas Island (Indian Ocean) by the introduced yellow crazy ant, whose population explosion was facilitated by later invasions of scale insects that produced a honeydew fed on by the ants. The ants devastated populations of the keystone species, the famed red land crab, and their impact on the forest was exacerbated by a plant pathogen (a sooty mould). See O'Dowd *et al.* (2003).

Biological, chemical, mechanical, genetic or ecosystem control: any to recommend?

The best defence against invasive introduced species is to avoid introducing them in the first place, by having more stringent regulations on movement of goods and by better inspections of luggage and cargo. The second line of defence is an effective early warning/rapid response system, which requires both ongoing monitoring and the institutional and legal mechanisms to act quickly when an invasion is discovered. Many introduced species have been eradicated before they spread too far, and many more could have been eradicated if a good early warning/rapid response system was in place. Once a species is established, biological, chemical and mechanical control, genetic intervention and ecosystem management all have roles to play in particular cases in maintaining invader populations at low levels. However, ecosystem management and genetic intervention, though widely discussed, have so far rarely been used to deal with introduced species, particularly in natural environments. By contrast, there are many successful uses of biological, chemical, and mechanical control. It is nonetheless important to recognize that the great majority of biological control projects do not control the target pest, that species introduced for biological control can and sometimes do attack non-target native species and that, once a biological control agent is well established, it is difficult if not impossible to eradicate it, even if it turns out to be problematic. The last feature contrasts with mechanical and chemical control, which can simply be terminated if they are not working as planned. For this reason, I feel that it is important to consider all possible means of dealing with introduced pests, and to take account of the irreversibility and historically low success rate of biological control.

Do large-scale, coordinated efforts like the United Nations Global Invasive Species Programme (GISP) offer the best means of addressing the problems of alien species?

Large international efforts, such as GISP, are important tools in dealing with introduced species. They can particularly aid in slowing down the rate at which non-native species arrive and in publicizing the problem and educating people about how to deal with particular invasions (as GISP has done). However, the most important response to introduced species will always be the efforts that each nation will mount against invaders that breach its borders, in terms of early warning/rapid response, attempted eradication and effective maintenance management.

commerce and habitat loss, invasive species pose a serious potential threat to the future of tropical biodiversity and human welfare.

Biotic invasions also have an important sociopolitical dimension. Some critics from fields outside biology, such as history, philosophy and sociology, have labelled attempts to control invasive species as being influenced by nativism, racism and xenophobia (Simberloff 2003a). Although earlier attempts in the nineteenth century to control invasive species focused on aesthetics, today the aim of such actions is to prevent ecological and economic damage (Simberloff 2003a). In 1997, the Global Invasive Species Programme (GISP) was developed as a corollary of the Convention on Biological Diversity (CBD). The Convention calls on participating parties to 'prevent the introduction of, control or eradicate those alien species which threaten ecosystems, habitats or species'. The GISP is coordinated by the Scientific Committee on Problems of the Environment (SCOPE) in collaboration with the World Conservation Union (IUCN) and CABI (Centre for Agriculture and Biosciences International). The GISP has identified the three key phases in the typical timetable for any species invasion: (1) introduction – the species overcomes a major geographic barrier such as crossing the sea from one island to another, usually with the help of humans; (2) establishment – the species overcomes local environmental obstacles, survives and begins to reproduce in its new location; and (3) dispersal – the species' distribution expands beyond the original site of introduction, spreading widely into new areas. In short, to qualify as an effective (and potentially harmful) invasive, an introduced species must arrive, survive and be able to spread (Allendorf and Lundquist 2003; Figure 5.1). The progression of an immigrant to invader usually involves a delay or lag time that is followed by rapid population increase and spread (Kowarik 1995; Mack *et al.* 2000; Sakai *et al.* 2001; Figure 5.2).

What are the factors facilitating some introduced populations to become invasive, while others, even close relatives, do not? Establishment success is

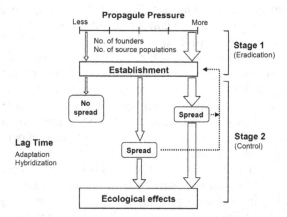

Figure 5.1   The two stages of invasion that generally coincide with different management responses. Propagule pressure is a continuum, with greater pressure leading to increased chance of establishment and spread with shorter lag times. If spread involves small groups of dispersing individuals, each group must be able to establish itself in a different area. Establishment or subsequent spread may be inhibited where groups reach the limits of particular environmental conditions. (After Allendorf and Lundquist 2003. Copyright, Blackwell Publishing Limited.)

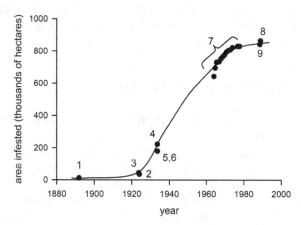

Figure 5.2   Many invaders, such as terrestrial plant invaders, occupy new ranges at an accelerating rate with a pronounced lag phase, which is more than 30 years in the figure. (From Mack *et al.* 2000. Copyright, Ecological Society of America.)

strongly affected by the number of individuals originally introduced ('founders'; Figure 5.1), the number of introduction attempts (i.e. the number of source populations; Figure 5.1), the introduced population's region of origin, and its dietary breadth, dispersal capability and life history traits that facilitate rapid population growth such as short generation time and high reproductive rate (Newsome and Noble 1986; Sakai *et al.* 2001). The first two components dictate propagule pressure because of their ability to persist through a genetic bottleneck

(Allendorf and Lunfquist 2003; Figure 5.1). High fecundity, dispersal capability and the similarity of original to new climatic conditions are among the primary factors thought to promote invasive spread (K.A. Lee and Klasing 2004). Escape from natural predators and pathogens in the newly colonized habitat (known as 'enemy release') may also facilitate invasion success. In contrast, many invasive populations end up collapsing before successful establishment and spread as a result of their inability to overcome novel parasites and diseases (Simberloff and Gibbons 2004). As such, variation in immune defence might explain invasiveness success, with successful invaders relying heavily on less expensive antibody-mediated immunity (K.A. Lee and Klasing 2004).

Invasive species can also be quintessential models for rapid evolution (C.E. Lee 2002; Phillips *et al.* 2006). Invasive success is influenced by genetic composition and its role in facilitating rapid adaptations is shaped by natural selection in the new environment. Most invasive species are thought to have overcome genetic bottlenecks through a number of release events, each expanding the genetic heterozygosity of the founding population (Allendorf and Lundquist 2003). Despite this, there are many studies showing low heterozygosity in invasive populations relative to native populations (Stockwell *et al.* 1996; Tsutsui *et al.* 2000; DeYoung *et al.* 2003; G.A. Wilson *et al.* 2005), thus indicating that some invasive populations may flourish despite low heterozygosity.

Other drivers of biodiversity change, land use and climate change (see Chapters 1 and 8) may actually facilitate the spread of invasive species (Dukes and Mooney 1999; Table 5.1). For example, many plant invasives are responding positively to elevated carbon dioxide and nitrogen deposition levels (Dukes and Mooney 1999) in their new exotic localities.

5.2   Invasive species in tropical realms

Of the world's 100 worst invasives, more than half (56) are found in the tropics (www.issg.org). Invasive species can have four major effects: (1) extinction of native biota, (2) alteration of the abiotic environment such as nutrition cycles and fire regimes, (3) damage to agriculture as pests and (4) harm to humans, native flora and fauna through the introduction or emergence of virulent diseases (Mack *et al.* 2000). Of 680 reported animal species extinctions, approximately 20% were probably caused by invasive species (Clavero and García-Berthou 2005).

Table 5.1 Possible general impacts of global change elements on the prevalence of invasive alien species. (From Dukes and Mooney 1999. Copyright, Elsevier.)

Element of global change	Prevalence of invaders
Increased atmospheric CO_2 concentration	Might increase prevalence of some invaders and reduce prevalence of others
Climate change	
Increased nitrogen deposition	Expected to increase prevalence of invaders in many affected regions
Altered disturbance regimes	
Increased habitat fragmentation	

In addition, there is a ubiquitous, positive correlation between the dominance of invasive species and native species decline (Didham *et al.* 2005). Although invasive species may not always be exclusively responsible for the negative consequences on biodiversity (Didham *et al.* 2005), there are a number of examples showing how invasive species generally create ecological havoc in the tropics. However, the field of invasion ecology is a still in infancy in this region, so it is likely that more evidence for the negative effects of invasives in the tropics will emerge over time. Furthermore, processes such as habitat modification and loss (e.g. logging) may promote the establishment of non-native species, thus acting synergistically to threaten native biodiversity. Below, we review some key examples of tropical invasions and their reported impacts on the environment.

5.2.1　Native extinctions by invasives

The probability of local species extinction on islands is elevated by the introduction of novel predators (Blackburn *et al.* 2004; also see Chapter 9). Several ecological and life history attributes of island species, such as their relatively small range and population sizes and the absence of co-evolved traits to thwart the invading species' arsenal, make island biotas vulnerable to displacement by invading species (Simberloff 1995). For example, the introduction of the brown tree snake (*Boiga irregularis*) shortly after the First World War wreaked biodiversity havoc on the island of Guam in the South Pacific. Tree snakes were probably directly responsible for the loss of 12 of 18 native bird species, and they also reduced the populations of other vertebrates such as flying foxes (*Pteropus mariannus*) (Fritts and Rodda 1998; Wiles *et al.* 2003). While the evidence for a direct link between invasion and extinction is still inconclusive in the case of some extirpated species, alternative explanations for this biotic catastrophe, such as disease, habitat alteration and overhunting, have all been discounted. Despite an annual expenditure of US$44.6 million, tree snakes on Guam are still not under control (Simberloff 2001), largely because of their ability to penetrate artificial snake barriers such as fences.

On many islands in French Polynesia, monarch birds (*Pomarea* spp.) are rarely found on islands invaded by the brown rat (*Rattus rattus*) (Seitre and Seitre 1992), suggesting either that they avoid this potential nest predator or that brown rats have contributed to local extirpations (Thibault *et al.* 2002). Although declines are caused primarily by nest predation, population reductions are exacerbated by competition with another invader – the red-vented bulbul (*Pycnonotus cafer*).

The mosquito (*Culex quinquefasciatus*) that carries the avian malaria parasite (*Plasmodium relictum*) was inadvertently introduced to Hawaii in 1826, with the parasite arriving thereafter (Mack *et al.* 2000; Woodworth *et al.* 2005; see also Chapter 8). Since then, avian malaria has been responsible for the decline and extinction of some 60 species of endemic forest birds on the Hawaiian islands (van Riper *et al.* 1986; Atkinson *et al.* 2005). Without having evolved in the presence of the disease, Hawaiian bird populations were generally unable to cope with the novel parasite. However, since the establishment of the disease 100 years ago, some native thrushes (*Myadestes* spp.) are now showing resistance to the disease, although many other species, such as endangered honeycreepers (e.g.

Hawaii creeper, *Oreomystis mana*), are still vulnerable (Atkinson *et al.* 2005). The most feasible method of reducing transmission of malaria seems to be to reduce or eliminate vector mosquito populations through chemical treatments and elimination of larval habitats (Atkinson *et al.* 2005).

Perhaps one of the best-known examples of an invasion catastrophe in the world occurred in the largest tropical lake – Lake Victoria in East Africa. Celebrated for its amazing collection of over 600 endemic haplochromine (i.e. formerly of the genus *Haplochromis*) cichlid fishes (family Cichlidae) (Nagl *et al.* 2000), the Lake Victoria cichlid community is perhaps one of the most rapid, extensive and recent vertebrate radiations known to biology (L.S. Kaufman *et al.* 1997; Balirwa *et al.* 2003). A rich community of non-cichlid fishes also inhabits the lake (Balirwa *et al.* 2003). In addition to the threats posed to this unique biota by a rapid rise in fisheries exploitation, human density, deforestation and agriculture during the past century (see Chapter 7), without doubt the most devastating effect was the introduction of the predatory Nile perch (*Lates niloticus*) in the 1950s (Witte *et al.* 1992; Goldschmidt 1996; Lodge 2001). This voracious predator, which can grow to more than 2 m in length, was introduced from Lakes Albert and Turkana (Uganda and Kenya, respectively) to compensate for depleting commercial fisheries (Witte *et al.* 1999; Balirwa *et al.* 2003) in Lake Victoria. Although the Nile perch population remained relatively low for several decades after its introduction, a rapid population expansion in the 1980s caused the direct or indirect extinction of between 200 and 400 of the cichlid species endemic to the lake as well as several non-cichlids (Witte *et al.* 1992; Goldschmidt 1996; Lodge 2001; Balirwa *et al.* 2003). Although many other threats probably contributed to the observed extinctions, such as direct overexploitation and eutrophication from agriculture and deforestation leading to a change in the algal plankton community (see Chapter 7), there are few other examples of such rapid and mass extinction of a single endemic taxon anywhere worldwide.

Hybridization between native and invasive species can result in the production of sterile offspring and thus waste reproductive effort. The population of endemic and endangered Hawaiian duck (*Anas wyvilliana*) was initially threatened by habitat loss and degradation, but now its major threat is hybridization with the introduced mallard duck (*A. platyrhynchos*) (Rhymer and Simberloff 1996).

5.2.2 Agricultural pests

Some invasive species can also become agricultural pests. The freshwater apple snail (*Pomacea canaliculata*) has become a major crop pest in Asia and Hawaii. It was deliberately introduced as a potential human food, but now, in many parts of its new range, this snail has become a serious agricultural pest that may have caused the decline of some native freshwater snails (Halwart 1994). A voracious herbivore, the apple snail can decrease rice yields in the Philippines by 90% at densities of a mere eight snails per m^2 (Naylor 1996). Climate modelling suggests that major rice-producing countries such as China, India and Bangladesh are under threat of invasion by the apple snail (Lach *et al.* 2000) because the tropical climate is conducive to its rapid reproduction and growth, with individuals reaching sexual maturity in 2 months compared with the 2 years it normally takes them in their native range in South America (Lach *et al.* 2000). The control of this species includes handpicking, applying molluscicides and raising predatory fishes in paddyfields that target the snails as prey (Naylor and Ehlrich 1997).

5.2.3 Ecosystem disruption by alien plants

Alien plants threaten conservation areas on every continent excluding Antarctica (although climate change may not spare Antarctica for much longer), and all ecosystems are vulnerable to them (Cronk and Fuller 1995). The impact of invasive plants includes displacement and even extinction of native species, as well as the alteration of soil chemistry, fire regimes and hydrology (Cronk and Fuller 1995; Mack *et al.* 2000). The destruction of riparian habitats in southern Africa is probably due to invasive alien plant species (Richardson *et al.* 1997). For example, one possible means of introduction of alien species in Serengeti National Park area is through the importation of construction materials (Foxcroft *et al.* 2006).

Invasive plants can also be a major competitor with native vegetation and, in extreme cases, they can convert large areas to exotic monocultures. Tropical rain forests are the most speciose of all terrestrial ecosystems and it is traditionally believed that their high species richness provides a good barrier to biological invasions (Pimm *et al.* 1991). Therefore, the number of invasive species in a tropical forest can be a good indication of its level of deterioration. For example, despite Singapore being heavily deforested with a matrix (areas surrounding forest patches – see Chapter 3) rich in cultivated and naturalized exotic species (Corlett 1988), no introduced were species found in mangroves, and only a single tropical American plant species – a bird-dispersed shrub called Koster's curse (*Clidemia hirta*) – was found in primary and tall (15–25 m) secondary forests (Teo *et al.* 2003). Teo and colleagues' study supports the hypothesis of lower invasability of relatively intact tropical rain forests. One of the factors governing this low susceptibility to invasion is probably the presence of closed canopy because the number of exotic species is reported to increase as the canopy cover becomes less complete (Teo *et al.* 2003; Figure 5.3).

As in Singapore, Koster's curse was found in an undisturbed forest in Pasoh Forest Reserve (Peninsular Malaysia), being first discovered there in the early 1990s (Peters 2001). A survey in 1997 of a 50-ha plot in Pasoh recorded 1002 individuals, with 7% in reproductive condition. All but eight of these individual plants were recorded in forest gaps and gap edges. The location of Koster's curse in gaps was facilitated by disturbance caused by wild pigs (*Sus scrofa*), an animal now superabundant in the reserve. No reproductively active Koster's curse individuals were found in the forest understorey. Peters (2001) postulated that Koster's curse outcompetes native plants in forest gaps and so alters the process of forest regeneration following events such as natural treefalls or selective logging. That study showed that the alteration of biotic interactions (e.g. an abundance of pigs) in forests can have unanticipated incidental effects (i.e. changes in plant community composition).

In Bangladesh, the exotic shrubs, grasses and vines typical of open habitats can invade regenerating forest and replace regenerating primary forest species such

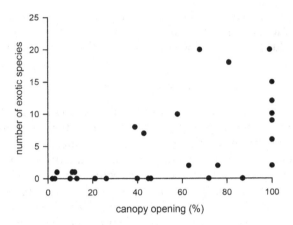

Figure 5.3 Relationship between the number of exotic species recorded per plot and percentage of canopy opening. (After Teo *et al.* 2003. Copyright, Blackwell Publishing Limited.)

as dipterocarps (Islam *et al.* 2001). Similarly, sites once logged in Madagascar do not recover in native plant species diversity owing to the dominance and persistence of invasive species (K.A. Brown and Gurevitch 2004). Some invasive species can assist in the spread of human diseases. For example, the invasive neotropical shrub (*Lantana camara*) provided habitats for stream-dwelling tsetse flies (*Glossina* sp.) in East Africa and thus has increased the incidence of sleeping sickness (Greathead 1968).

5.2.4   Biodiversity loss and resilience to invasives

As with plants, it was found that closed-canopy forest patches in Singapore (primary and late secondary forests) supported fewer introduced bird species than abandoned, open-canopy plantations and young secondary forests (Castelletta *et al.* 2005). Similarly, a small 4-ha patch of residual forest in the Singapore Botanic Gardens became home to more introduced species (e.g. the house crow, *Corvus splendens*; Sodhi *et al.* 2005b) than other larger areas of primary forest over a 100-year period (1898 to 1998). In 1998, 20% of birds observed in this patch were exotics, suggesting a degradation in habitat quality over time and a greater exposure to invasions due to more pronounced edge effects (Chapter 3). Sodhi *et al.* (2005b) hypothesized that a high occurrence of introduced birds in this fragment may eventually spell doom (e.g. through higher nest predation) for its existing native birds.

Forest-stream fishes are also largely immune to competition from invasive species in Singapore. Thirty-seven invasive freshwater fishes, turtles, molluscs and prawns have been now established on the island (Ng *et al.* 1993), yet these exotics do not appear to have affected the biodiversity of the aquatic fauna in streams within relatively intact forest. One of the reasons may be that a large proportion of the native freshwater fauna of Singapore (i.e. 80% of 54 freshwater fishes) prefer acidic waters, whereas introduced aquatic fauna are largely restricted to neutral or alkaline waters. However, a recent invasion of species that can thrive in acidic waters (e.g. tiger barb, *Puntius tetrazona*) is cause for concern. Therefore, Ng *et al.* (1993) recommended that tightly enforced legislation and strict quarantine measures are needed to ensure that no other species preferring acidic water species become established in Singapore, and attempts should be made to eradicate existing acid-water exotics.

5.2.5   Pollination and seed dispersal

Invasive animal species can also sometimes disrupt mutualistic interactions such as seed dispersal and pollination (Traveset and Richardson 2006). Pollination and seed dispersal are particularly critical for forest regeneration. Alien pollinators can depress the reproductive success of native plants if they do not transfer pollen effectively. On the other hand, alien pollinators might enhance plant reproductive success by increasing pollen transfer. Honeybees (*Apis mellifera*), for example, have invaded ecosystems worldwide and they can be effective pollinators in some, but not all, systems (Traveset and Richardson 2006). In some cases, invasive pollinators can serve the pollinator role that would be otherwise lost with the decline of native pollinators. In Hawaii, a vine (*Freycinetia arborea*) that was originally pollinated by an extinct bird now survives thanks to an introduced Japanese silvereye (*Zosterops japonica*) (Cox and Elmqvist 2000). Invasive species can also affect pollination by reducing the population of native pollinators. An invasive ant (*Wasmannia auropunctata*) is threatening populations of several species of geckos that pollinate and disperse numerous plant species in New Caledonia (Traveset and Richardson 2006). Another less obvious effect of invasive plants on pollination is their attractiveness to native pollinators and the concomitant reduction of those pollinators' visits to native plants. *Chromolaena odorata*, an invasive plant in tropical dry forests of Thailand, reduces the frequency of flower visits by native insect pollinators to indigenous *Dipterocarpus obtusifolius* (Ghazoul 2004).

Vertebrates can disperse the propagules of up to 90% of woody plant taxa in tropical rain forests (Herrera and Pellmyr 2002), but the effects of biotic invasions on seed dispersal are poorly documented in the tropics (Traveset and Richardson 2006). As with pollinations, introduced frugivorous animals can be inefficient dispersers of native fruiting plants by depositing seeds in unsuitable sites, or they can outcompete native seed dispersers. For example, the Argentine ant (*Linepithema humile*) has displaced several native ant species in South African fynbos, subsequently reducing the densities of large-seeded Proteacea that depend on them for seed dispersal (Christian 2001). Likewise, the invasive African big-headed ant (*Pheidole megacephala*) eliminates native ants and many other invertebrates from rain forest sites in northern Australia (B.D. Hoffmann *et al.* 1999), with hypothesized negative implications for native flora and fauna. On the other hand, introduced frugivorous animals can increase plant reproductive success when seed dispersal has been limited by extinctions of native frugivores (Traveset and Richardson 2006). In some cases, native frugivores can disperse the seeds of invasive plants and thus facilitate their spread. Up to 40% of invasive plant species may be dispersed by frugivores (Cronk and Fuller 1995). As such, because fruit traits such as size, the presence of inedible peel, crop size and phenology influence frugivore choice, these can be manipulated to dampen the spread of frugivore-dispersed invasive plant species (Buckley *et al.* 2006).

5.2.6 Continental invasions

As outlined previously, many of the more direct and easily observed impacts of terrestrial species invasions in the tropics have been documented in relatively closed systems such as islands and freshwater lakes. However, there are many examples of widespread invasions of terrestrial species in the tropics that have not necessarily caused drastic local extinctions, but their effects on the indigenous biota can still be pronounced.

For example, tropical northern Australia has a long and well-documented history of invasive species and their ecosystem effects. The most conspicuous populations of exotic animals in this region derive from animals introduced to support early human settlements and in the subsequent development of agriculture. Early settlements had a diverse complement of stock including cattle (*Bos* spp.), Asian water buffalo (*Bubalus bubalis*), pigs, goats (*Capra hircus*), sheep (*Ovis aries*), banteng (*Bos javanicus*) and horses (*Equus caballus*). Some of these species have thrived in north Australian environments, achieving densities never seen in their native habitats (Freeland 1990).

Indeed, Asian swamp buffalo have achieved populations numbering in the hundreds of thousands in Australia (Bayliss and Yeomans 1989). Banteng, whose wild ancestors are now regarded as endangered in their Asian native habitats (Bradshaw *et al.* 2006a), are also numerous in certain areas of Australia. Interestingly, the banteng have also recently developed a novel mutualism with a native bird, the Torresian crow (*Corvus orru*), whereby crows glean food from banteng (Bradshaw and White 2006). However, pigs may be the most successful large exotic vertebrate species in Australia and perhaps one of the most difficult exotic animal problems to resolve (Hampton *et al.* 2004). Estimates of their abundance are difficult to obtain and the likely costs of achieving control are high, largely because survey methods that give acceptable accuracy and precision are expensive (Choquenot 1995), but also because pigs have extensive dispersal capacity and high fertility rates (Dexter 2003).

These invasive species inflict excessive damage to Australian ecosystems by virtue of their sheer biomass. Pigs damage the landscape by digging for food in soft soils (Tisdell 1982), and they incur population-level effects on the wide range of plant, invertebrate and vertebrate prey. Buffalo also cause severe degradation to lowland environments due mainly to the facilitation of saltwater intrusion into low-lying freshwater floodplains (Mulrennan and Woodroffe 1998). Buffalo are also large animals (up 1200 kg) that consume huge quantities of vegetation per day within relatively restricted home ranges (Tulloch 1974). It is through these habitual behaviours and high densities (at times up to 34 individuals/km²) that buffalo create extensive damage.

The introduction of the South American cane toad (*Bufo marinus*) to Australia and Hawaii is perhaps one of the best-known failures in agricultural biological control. Unfortunately, it also represents one of the more infamous conservation disasters. In Australia, cane toads were originally introduced to Queensland in 1936 to control beetle pests in cane fields, but it was soon realized that not only were they incapable of controlling the pests, they are voracious predators that flourish by eating many native insects. To add injury to insult, cane toads are highly toxic and they have impressive dispersal ability (Phillips *et al.* 2006). Today the population has spread throughout northern Australia and is rapidly approaching the Western Australia border, with population densities known to exceed 2000 individuals/ha in suitable habitats (Freeland 1986).

Because of its high toxicity, almost any predator that attempts to eat a cane toad succumbs to rapid poisoning and death. Although no species is known to have become extinct as a result of cane toads, there are intense concerns about the future of much of tropical Australia's biodiversity. For example, Phillips *et al.* (2003) predicted that cane toads potentially threaten up to 30% of Australian snake species. Other reptiles at high risk include freshwater crocodiles (*Crocodylus johnstoni*) and certain monitor species (Burnett 1997; Lever 2001; Griffiths *et al.* 2004), because these species are often found dead with morbid or dead cane toads in their mouth (Burnett 1997). A mammal species at particular risk is the northern quoll (*Dasyurus hallucatus*), which has already disappeared from much of its former habitat in response to the recent cane toad invasion in the Northern Territory (M. Watson and Woinarski 2003; Oakwood 2004). Other groups also at risk include native freshwater snails that prey on cane toad eggs and tadpoles (Crossland 2000), some birds, other amphibians and other invertebrates (Catling *et al.* 1999).

In addition to the direct impacts associated with poisoning, cane toads also affect the environment in three other ways: (1) predation on native fish and invertebrates, (2) increased competition for prey and habitat, possibly displacing native species, and (3) ecosystem change through modifications to the plant and animal community structure. Research into these ecosystem-level effects is only in its infancy, and the full implications of the cane toad invasion in Australia have only begun to be detected. There are also serious cultural and economic consequences of the cane toad invasion through reduced food for Aboriginal Australians (Griffiths *et al.* 2004) and a reduction in wildlife ecotourism opportunities.

5.2.7 Marine invasions

Successful invasive marine species do not receive the conservation attention they deserve, largely because of the clandestine aspect of marine invasions, and it is only now that we are beginning to understand the extent of the threat they pose. Ever since humans have travelled in ships, invasive marine species have been carried to non-native areas both in ballast water and attached to vessel hulls (Bax *et al.* 2003). It has been estimated that at any given time, over 10 000 different species are being transported between different biogeographical regions in ballast water alone (Carlton 1999). Although most do not establish outside of their native regions (Bax *et al.* 2003), many species have successfully established populations well outside their native ranges to the detriment of local biota. For example, 91 of approximately 400 marine species in Pearl Harbor, Hawaii, are alien (Coles *et al.* 1999).

The Caribbean black-striped mussel (*Mytilopsis sallei*) is native to Central and South America but has become an important invasive pest in many tropical waters, including much of Southeast Asia, Fiji, India and northern Australia. The species is so successful that it has become the dominant bivalve in Tapong Bay, Taiwan (Lin *et al.* 2006). With a rapid growth rate, early age at sexual maturity and few known predators (B. Morton 1989; Lin *et al.* 2006), the species is particularly successful in forming dense monocultures on native and cultured oysters that can lead to substantial reductions in native marine shellfish biodiversity (Lin *et al.* 2006). In some areas, the black-striped mussel has been removed after establishment when the appropriate control measures were implemented in time (see section 5.3). Their establishment in Darwin Harbour, Australia, provoked a strong and effective reaction from both local and national authorities, driven by the high costs that can be associated with such organisms (e.g. in the USA the presence of the invasive zebra mussel costs industries $100 million annually – Pimentel *et al.* 2000).

The contribution of shipping to the spread of non-native marine organisms is particularly noteworthy. Subba Rao (2005) reviewed the literature for studies pertaining to the appearance of non-native marine taxa in the seas around India and discovered that, since 1960, 205 taxa have been introduced from various seas around the world (Figure 5.4). For example, 75 species hailed from the coastal seas of China and Japan, 63 from the Indo-Malaysian region, 42 from the Mediterranean, 40 from the western Atlantic and 41 from Australia and New Zealand. Additionally, 34 taxa from the eastern Atlantic, 33 from the Caribbean region and 29 from the eastern Pacific were found. Species common to other

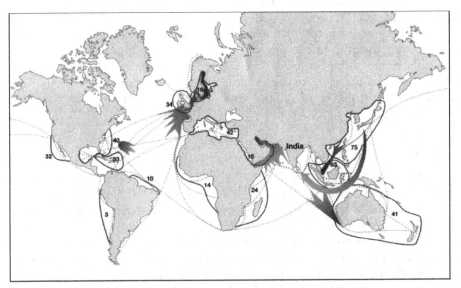

Figure 5.4 Number and origin of non-native marine taxa found in the seas around India since 1960. Dotted lines represent major shipping routes, solid lines delineate the areas sampled in various studies used in the review, and black arrows show the direction of spread of the biota. (After Subba Rao 2005. Copyright, John Wiley & Sons Limited.)

regions were comparatively rarer, with 24 from East Africa, 18 from the Baltic, 15 from the middle Arabian Gulf and Red Sea, 14 from the west coast of Africa and 10 from the Brazilian coast.

The taxonomic distribution of the non-native taxa was also amazingly wide: 21% were fish, < 11% were polychaetes, 10% were algae, 10% crustacean, 10% molluscs, 8% ciliates, 6% ascidians and the rest were minor invertebrates (Subba Rao 2005). Although the full effects of these invasions on local marine biodiversity are not yet fully understood, there also are important concerns regarding the introduction of pathogens or toxic bloom phytoplankton. Indeed, the pathogens *Vibrio* spp. that are toxic to humans have been found in fish and shellfish in western India and are thought to be derived from ballast discharges (Subba Roa 2002, 2005). Likewise, the global spread of toxigenic and paralytic shellfish poisoning algal blooms due to shipping ballast transport (Hedgpeth 1993) has reached India (Subba Roa 2005) and many other areas in the tropics in recent decades.

5.2.8 Freshwater invasions

Biotic homogenization of freshwater systems throughout the world is also increasing as a result of a replacement of native species with non-indigenous species that are more often than not introduced by humans (Rahel 2002). Moreover, as with terrestrial systems, degraded freshwater habitats are much more likely to succumb to successful invasion (Bunn and Arthington 2002; Dudgeon *et al.* 2006), so future prospects for improvement are bleak (Sala *et al.* 2000). This is particularly frightening considering that introduced species are considered to be the leading cause of species endangerment and extinction in freshwater ecosystems (Claudi and Leach 1999). Much of the existing literature in this field focuses on temperate and high-latitude examples (e.g. Rahel 2002); however, there is a growing body of evidence for important homogenization via invasion for many tropical freshwater systems. As mentioned in section 5.2.1, the deliberate introduction of non-indigenous fish can have catastrophic consequences for native fish biodiversity (e.g. Lake Victoria).

The tropical and subtropical tilapiine cichlids (genera *Oreochromis*, *Sarotheradon* and *Tilapia*) native to Africa and the Middle East are famed not only for their nearly worldwide distribution thanks largely to human activities (Canonico *et al.* 2005), but also for their legendary divine role in feeding the masses, as told in the Christian Bible (the so-called 'St Peter's fish'). Their popularity stems from their useful role as agents for the biological control of aquatic weeds and pest insects, as baitfish for some fisheries, as food in aquaculture and fisheries, and as aquarium specimens (Canonico *et al.* 2005). Given their ubiquitous distribution in aquaculture ponds and rice paddyfields, escapes into surrounding freshwater habitats are commonplace, thus leading to the rapid expansion of their range in recent decades (McCrary *et al.* 2001).

Although tilapias are generally thought to be mainly herbivores, detrivores or planktivores, they do consume the eggs and larvae of other fish, and they can even consume small adult fish. This omnivorous behavioural trait is thought to have caused the disappearance from Lake Apoyo in Nicaragua of native *Chara* spp.

water plants that act as an important habitat for native cichlids (McCrary *et al.* 2001). Tilapia have also been implicated in facilitating freshwater eutrophication. For example, high tilapia biomass in several tropical reservoirs in Brazil has resulted in increases in total phosphorus, chlorophyll *a* and cyanobacteria (Starling *et al.* 2002; Figueredo and Giani 2005). Non-native tilapias can also erode the genetic stock of native species, causing hybridization and reducing genetic diversity (Fitzsimmons 2001). Mosquitofish (*Gambusia affinis*) and guppies (*Poecilia reticulata*) have also been distributed extensively for various purposes, such as mosquito control, with possible negative implications for local biodiversity (Allan and Flecker 1993).

Many other examples of invasions into tropical freshwater environments abound. Exotic species invasion is also a serious problem for many of the world's tropical wetlands. Exotic neophytes have dispersed widely through the upper Amazon River floodplain (Seidenschwarz 1986 in Junk 2002) causing potentially important changes to the native flora and fauna. The Australian rice industry spends approximately AU$14 million annually on herbicides to control invasive weeds such as the South American *Echinochloa* sp., and many other neotropical aquatic weeds such as *Salvinia molesta*, *Eichhornia crassipes*, *Pistia stratiotes*, *Parkinsonia aculeata*, *Mimosa pigra* and *Urochloa mutica* have been spreading throughout much of northern Australia, threatening the sensitive wetlands of areas such as the World Heritage Area, Kakadu National Park (Cowie and Werner 1993; Douglas *et al.* 1998). As with the introduction of exotic marine shellfish such as the black-striped mussel (section 5.2.7), invasive freshwater mussels can develop large biomasses and impede hydroelectric and water treatment plants, although their effects on local biodiversity are currently unknown (Darrigran and Ezcurra de Drago 2000; Junk 2002).

5.2.9   Invasions and climate change

In addition to being facilitated by climate change (see Chapter 8), some invasive species can exacerbate the ecosystem changes brought about by climate change. For example, the deliberate sowing of palatable African grasses and their fire-aided spread has led to their proliferation in the Amazon Basin through their positive response to fire damage (positive feedback loop). It is predicted that the continued conversion of forests into grasslands will enhance carbon build-up in the atmosphere and will also alter regional climate through increases in air temperatures (Mack *et al.* 2000). Further, these exotic grasses prevent the recolonization of native species in degraded areas, thus maintaining the microclimatic differences that arise from fragmentation (see Chapter 3).

5.3   Managing and controlling invasive species

Detailed knowledge of invasive species' population biology is required for the effective control of widespread or well-established invaders (Simberloff 2003b). Indeed, any chance of control requires the early detection of invaders and an attempt to eradicate or contain them prior to any extensive spread. According to Simberloff (2003c), a successful eradication programme requires five main components: (1) adequate resources for the entire duration of the project, (2) clear lines of authority, (3) adequate biological knowledge, (4) a high detection probability of the target species at low population densities (i.e. early detection, see Figure 5.1) and (5) habitat restoration or other management options that reduce the chances of re-invasion. Furthermore, eradication should ideally not cause overt harm to the invaded ecosystem (i.e. making the conditions conducive to a new invader species – Simberloff *et al.* 2005). For example, the removal of pigs (*Sus* sp.) and goats from Sarigan Island (Commonwealth of the Northern Mariana Islands) resulted in a population explosion of an exotic vine (*Operculina ventricosa*) (Simberloff 2001). It also must be noted that sometimes exotic species can provide habitat for endemic fauna, so their removal may threaten those species as well (Van Riel *et al.* 2000).

Management of invasives is more often than not hampered by inadequate funding, insufficient biological knowledge and inappropriate or counterproductive policy (Simberloff *et al.* 2005). Unfortunately, many regulatory agencies adopt a 'wait-and-see' policy and fail to address exotic species issues until they become major problems, with the usual result being that eradication becomes extremely difficult, if not impossible (Simberloff 1997). Accordingly, many studies show that the 'wait-and-see' attitude is not the best approach in invasive species management (Simberloff 2003c). For example, Koster's curse could have been relatively easily eradicated from Hawaii when it was found to be restricted to < 100 ha in 1950s. This species can be controlled effectively by thrips (*Liothrips urichi*), but by the time of their eventual release in 1954 control efforts failed to restrict the spread of this species in Hawaii (Pemberton 1957). Subsequent releases of other biological agents have also failed, with Koster's curse now covering > 100 000 ha and classified as one of the worst weed species in Hawaii (C.W.I. Smith 2000).

5.3.1 Control techniques

Three main technologies, applied individually or in unison, can be used to control an invasive species: (1) chemical, (2) mechanical and (3) biological control (Mack *et al.* 2000). Chemical control is widely used to target invasive species in agricultural areas, although chemicals may be hazardous to humans and other non-target species. Even when the chemicals applied do not present any overt public health hazards, there may still be public opposition to their use.

Mechanical control (e.g. hand picking, hunting, trapping) can succeed in some cases and is more acceptable to many people. For example, hunting and trapping were effective techniques used to eradicate the feral goat menace threatening the unique biota of the Galápagos Islands. Goats, pigs and other animals were brought to the islands by pirates and whalers in the eighteenth century to supply subsequent visitors with a ready source of food. However, goat populations started booming in the late 1980s, probably due to El Niño-driven changes to the vegetation (Guo 2006). The feral species denuded the vegetation and competed with giant tortoises (*Geochelone elephantopus*) for food plants, so a 6-year project was initiated to eradicate invasive species from the Galápagos Islands in

2000 (Guo 2006). After an US$18 million campaign and 6 years of hard work by sharpshooters and hunter–dog teams, goats were finally eradicated in 2005 by culling > 140 000 individuals. Even after only a short period since eradication, many native plant species are showing signs of rapid recovery (Guo 2006). There are now efforts to remove rats and cats from the Galápagos because of the threats they pose to species such as mangrove finches (*Camarhynchus heliobates*) and the enigmatic Galápagos marine iguana (*Amblyrhynchus cristatus*). The mechanical eradication of plants is usually more difficult than for vertebrates because the latter are more amenable to eradication through traps and chemicals (Simberloff 2001), although there have been some successes. For example, a 10-year project succeeded in almost complete elimination of an invasive annual grass that dominated 30% of the vegetation in Laysan Island, Hawaii (Simberloff 2001).

However, equipment and logistic expenses and difficulty in locating target organisms can limit the use of mechanical means to control many invasive species. In some cases, then, biological control can be an effective alternative, and there are some good examples of this approach succeeding in the tropics. Effective control of the South American cassava mealybug (*Phenacoccus manihoti*) in Africa was achieved through the introduction of another South American species – an encyrtid wasp (Odour 1996). In several Caribbean countries, the successful control of a number of crop pests, such as the citrus blackfly (*Aleurocanthus woglumi*), the sugarcane borer (*Diatraea saccharalis*) and the lima bean pod-borer (*Etiella zinckenella*), other nematodes, pathogens and weeds has also been achieved through the introduction of exotic predators (Cruz and Segarra 1990; P.S. Baker *et al.* 1992).

With successes in biological control also, unfortunately, come many failures, especially with respect to the potential negative effects on non-target species, as the Australian cane toad example clearly illustrated (section 5.2.6). There are other examples – the importation of a predatory snail (*Euglandina rosea*) to control the giant African snail (*Achatina fulica*) in the Hawaiian Islands not only failed to reduce the latter, it also caused the extinction of numerous endemic snail species (Civeyrel and Simberloff 1996). It is therefore extremely important that the appropriate ecological and biological information be collected prior to any planned introductions to reduce the probability of corollary detriments to native flora and fauna. In addition, biological control by itself should not be considered a panacea for pest eradication because it is seldom effective on its own. Effective control must, therefore, be combined with other pest control techniques in integrated management programmes (Greathead 1991).

5.3.2 Integrated pest management – the house crow case study

Biological knowledge is most often the critical first step in achieving eradication success. In this section, we describe how integrating science and habitat management was used to guarantee the long-term control of an invasive bird species, the house crow. This species is widespread and a notorious pest in Asia and Africa. This relatively large (245–371 g), aggressive and gregarious black bird lives in close association with people, with a 'native' range spanning southern Iran, Pakistan, India, Myanmar and Sri Lanka. Its diet is varied and includes invertebrates, fruit, offal, carrion, eggs, young birds and human refuse – the last being particularly important in urban areas. It is capable of breeding throughout the year, though it favours the hotter and drier months. These behavioural and life history characteristics predispose the house crow to successful invasions of new areas. Over the last two centuries, the house crow has spread throughout East and Southeast Asia (China, Malaysia, Singapore), Africa (South Africa, Kenya, Tanzania, Egypt), the Middle East (Oman, Yemen) and various oceanic islands, because of both inadvertent introductions associated with increased global sea traffic and trade and deliberate introductions. Introduced to Peninsular Malaysia in the late nineteenth century as a possible biological control agent for caterpillars (a move that did not, unfortunately, have the desired effects), the crow soon spread to neighbouring urban areas, and to the main island of Singapore in the late 1930s or early 1940s (Sodhi and Sharp 2006). From an estimated 200–400 individuals in the late 1960s, the population expanded to over 4500 by the mid-1980s despite moderate, but consistent, levels of culling since 1973. By 2000, the population had exploded to over 130 000, representing a 30-fold increase in 15 years (Brook *et al.* 2003b).

Densities in Singapore were sufficiently high for the house crow to be considered a major urban pest. Large, communal roosts of up to 3000 birds occurred close to residential areas, creating high levels of noise pollution and an accumulation of faeces on human property. Further, parent birds aggressively protect their nests, leading to several reported attacks on people. The house crow is also a possible vector for human pathogens such as *Cholera, Salmonella, Giardia, Enteroamoeba, Shigella* and *Escherichia*. In addition, their belligerent behaviour suggests that high densities could cause displacement of indigenous bird species through predation and/or competition (Brook *et al.* 2003b).

Past attempts to control the house crow through poisoning, shooting and nest destruction have largely failed – a common outcome in invasive pest control. The large-scale, but ultimately unsuccessful, attempt to eradicate house crows in Aden (Yemen) in the late 1980s provides a germane example. Over 240 000 birds were killed (poisoned) during a 2-year government-funded programme, but a lack of similar effort in neighbouring Lahej and Abiyan meant that migration quickly replaced those losses (G. Jennings 1992). A similarly frustrating situation persisted in Singapore, where the growth of the house crow population over 30 years largely defied efforts by the government to suppress the population through shooting (Brook *et al.* 2003b; Figure 5.5). Common to these examples and others is a consistent lack of informed scientific input, such that management actions are implemented *laissez-faire*, with a consequent reduction in the chance of success.

Recognizing these problems, the Singapore government funded an initiative to study the autecology and behaviour of the house crows with a view to providing scientifically rigorous guidelines for managers prior to the initiation of a large-scale control programme. To understand the population ecology and behaviour of the house crow in Singapore as an aid to successful control, regular population size and roost surveys, dissections of birds shot (to provide age structure and breeding status), detailed nest site observation, and monitoring of coastal dispersal were carried out. Using a discrete-time, density-dependent population model to

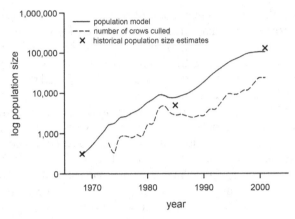

Figure 5.5 Logarithmic plot of the density dependent population model (complete line) for the house crow on Singapore Island in relation to the historical population size estimates (X) and past rates of culling (dashed line). (After Brook *et al.* 2003b. Copyright, Wildlife Society.)

synthesize this information, it was demonstrated that at least 41 000 crows needed to be culled in the first year of a control programme, and an equivalent effort maintained each year thereafter to be confident of suppressing the Singapore population from its 2001 density of 190 birds per km² to the management target of < 10 birds per km² within 10 years (Brook *et al.* 2003b). The culling target declines to 32 000 if culling is combined with other management strategies such as resource limitation (e.g. roost management) and nest destruction. Although complete eradication of the house crow from Singapore may be unrealistic due to the difficulties of detection at low population densities and an influx of migrants from neighbouring Malaysia, nest destruction combined with shooting reduced the crow population to 30 000 in 2003 (i.e. 40% of the original) (Sodhi and Sharp 2006), and it is probably much lower today, providing a notable success story.

5.3.3 Invasive species detection and action plans

Continued population suppression to low densities requires control efforts maintained over the long term, so effective prevention and control of invasive species requires a long-term and large-scale strategy (Mack *et al.* 2000). However, the point to note is that the precision and extent of the biological knowledge required for effective invasive species management depends on each situation. In some cases, basic natural history information, physical labour and basic tools such as guns for animals and root wrenches for plants can be used in successful eradications (Simberloff 2006). It is arguable that there have been too many 'scorch-the-earth' or 'brute-force' approaches to rid areas of invasive species. For example, removal of the Caribbean black-striped mussel (section 5.2.7) from Darwin Harbour in Australia was achieved through the application of 200 tonnes of bleach, which also killed numerous other native species (Bax *et al.* 2003).

The behaviour of species in invaded habitats can sometimes help facilitate their eradication (Simberloff 2003c). Introduced *Anopheles gambiae*, a disease vector for African malaria, was eradicated in the 1940s from north-eastern Brazil using chemicals targeting adults and larvae (J.R. Davis and Garcia 1989). The fact that this species of mosquito stayed near human habitation contributed to the success of its eradication.

The GISP is developing an 'early warning system' for the possible trajectory of any species once it is introduced into a new environment based on the species' known performance elsewhere. To prevent future invasions, better restrictions are needed for the importation of live soil, plants and animals. For example, quarantine legislation, which reduces the probability of exotic organisms entering new areas, has successfully prevented the spread of human parasites (McNeill 1976). In most cases, the costs of excluding exotic species through quarantine measures are trivial compared with what would be needed for their control after their establishment and consequent spread (Mack *et al.* 2000). The best approach in the management of most invasive species, therefore, is to prevent them from arriving in the first place (Windle and Chavarría 2005; see Spotlight 5: Daniel Simblerloff).

In addition to an early warning system, (1) any programme should be implemented through regular surveys, (2) people should be educated regarding the havoc created by invasives and (3) invasive species management should be integrated with other existing programmes such as fire management (Foxcroft *et al.* 2006). Additionally, more funding is needed to identify new invasions and to predict 'hotspots' for future ones. Long-term studies are also required to assist in a better understanding of the community and ecosystem-level effects of exotic species, resulting in better management where established and earlier eradication where exotics have recently colonized (Strayer *et al.* 2006).

5.3.4 Invasion meltdown

There often is a lag time when an introduced species is innocuous (Kowarik 1995; Sakai *et al.* 2001). To predict invasions and their potential effects more effectively, a better understanding of the mechanisms influencing these lag times is needed (Simberloff *et al.* 2005). More studies are also needed to understand the potential for 'invasion meltdown' (Simberloff and Von Holle 1999; see Spotlight 5: Daniel Simberloff) – that is, whether impacts by one invader are exacerbated by interactions with another exotic species (Simberloff and Von Holle 1999). Although there is a dearth of information on invasion meltdowns (Simberloff 2006), one classical example comes from Christmas Island in tropical Australia. There was a dramatic increase in the already established yellow crazy ant (*Anoplolepis gracilipes*) with the introduction of a scale insect and an outbreak of a native scale (O'Dowd *et al.* 2003; Abbott 2004). Because of the mutualistic interaction between ants and scale insects (ants protect scales from predators and parasites in return for scale honeydew), a greater food source initiated a rapid rise in the exotic ant population. The large ant population devastated the population of native red crab (*Gecarcoidea natalis*) and resulted in massive increases in forest undergrowth due to reduced levels of herbivory by crabs (O'Dowd *et al.* 2003).

The extirpation of red crabs from the forest floor could also make the island vulnerable to more plant invasions (P.T. Green *et al.* 2004).

Another meltdown example comes from the invasion of a small tree (*Myrica faya*) into the Hawaii Volcanoes National Park. This invasive tree fixes nitrogen faster than native plant species, thus increasing the nitrogen content of the nutrient-poor volcanic sands (Vitousek and Walker 1989). This enrichment has paved the way for other invasive plants that flourish in nitrogen-rich soils. *Myrica faya* also attracts seed-dispersing invasive birds such as the Japanese silvereye (*Zosterops japonica*), which competes with native bird species (Vitousek and Walker 1989). As such, issues pertaining to the associations between different invasive species beyond the more intuitive direct impacts of the target species alone need to be considered when designing management plans to avoid more serious ecosystem-wide impacts.

5.4   Summary

1   Invasive species generally affect native species and ecosystems negatively. Human economy also suffers because of biotic invasions.
2   Successful invasive species have characteristics such as high reproductive rate and they are good models to study rapid evolution.
3   Island biotas are particularly vulnerable to invasions. Invasions can also hamper some mutualistic associations between native species.
4   Chemical, mechanical and biological methods are used to control invasive species, with combinations of these providing the best effective prospects.
5   Eradication management plans must be carefully designed and implemented as early as possible after detection of exotic species. Adequate funding and long-term commitments are essential for eradication programmes to be successful.

5.5   Further reading

Mack, R. N., Simberloff, D., Lonsdale, W.M., Evans, H., Clout, M. and Bazzaz, F. A. (2000) Biotic invasions: Causes, epidemiology, global consequences, and control. *Ecological Applications* 10, 689–710.

Sakai, A. N., Allendorf, F. W., Holt, J. S., Lodge, D. M., Molofsky, J., With, K. A., Baughman, S., Cabin, R.J ., Cohen, J. E., Ellstrand, N. C., McCauley, D. E., O'Neil, P., Parker, I. M., Thompson, J. N. and Weller, S. G. (2001) The population biology of invasive species. *Annual Review of Ecology and Systematics* 32, 305–332.

Simberloff, D., Parker, I. M. and Windle, P. N. (2005) Introduced species policy, management and future research needs. *Frontiers in Ecology and the Environment* 3, 12–20.

6

Human Uses and Abuses of Tropical Biodiversity

Tropical biodiversity is under heavy threat from anthropogenic overexploitation (e.g. harvest for food or live specimens for the pet trade). For example, the hunting of 'bush meat' (or wild meat) is imperilling many tropical species as expanding human populations seek new sources of protein and potentially profitable new avenues for local and international trade. Here we highlight the effects of human exploitation on tropical biodiversity.

6.1   Bush meat crisis

Humans have been hunting wildlife since their origin in the tropics (Stanford 1999). Overhunting by prehistoric humans may have resulted in the loss of mammal megafauna (species > 45 kg; see Chapter 9). However, with the decline in forests and ever-increasing human densities, the pressure of hunting on wildlife has increased immensely in recent decades. The situation is exacerbated by factors such as the creation of roads into formerly inaccessible areas, technological advances in weaponry (e.g. better guns) and the dysfunctional nature of institutions meant to enforce quotas and hunting permits (Bulte and Horan 2002; Ling *et al.* 2002; R.J. Smith *et al.* 2003; see Spotlight 6: Bruce Campbell). There is a growing realization that wildlife overhunting in the tropics is not a hyperbole, and that the ensuing 'bush meat' crisis (overhunting of wildlife by humans for consumption) is one of the gravest threats to tropical animal biodiversity. The crisis is driven primarily by the following factors: (1) high market/subsistence demand for wild meat and associated animal products for traditional medicines and ornamentation, (2) poor enforcement of wildlife protection laws, including inadequate patrolling of protected areas, and (3) poor awareness in local communities of wildlife laws and the plight of wildlife. Unsustainable consumption of wildlife also affects local livelihoods – 60 million people living near and close to forests rely heavily on wildlife for protein (J.G. Robinson and Bennett 2004). Most of these people live at the fringes of market

Spotlight 6: Bruce M. Campbell

Biography

I was trained as an ecologist, going into ecology because of the enthusiasm of a mentor who really believed in the ability of individuals to make a difference in the world. But after moving to Zimbabwe and initiating work in the tropical savannas, where humans have had an impact for thousands of years, I found that a purely ecological perspective limited my ability to grapple with complex conservation issues. So I branched out into resource economics, and into institutional arrangements for common property management – I did this by reading basic texts, but more importantly by working closely with some world-class resource economists and sociologists – they were important in shaping my career. For about 20 years I focused mainly on African tropical woodlands and savannas, but I then joined the Centre for International Forestry Research (CIFOR) based in Indonesia, and started work in the humid tropics on three continents. CIFOR, with sites throughout the tropics, offers a wonderful environment for in-depth cases studies combined with synthesis based on a global perspective. Now at Charles Darwin University, I have started work on Aboriginal natural resource management, while still working with teams of researchers in some 20 developing countries. My work currently covers household economics (can natural resources lead to pathways out of poverty?), conservation and development dynamics (can there be win–win situations for forests and livelihoods?), and common property management (can collective action and community-based management lead to improved outcomes for forests and livelihoods?).

Major publications

Campbell, B. M., Costanza, R. and van den Belt, M. (2000) Special section: Land use options in dry tropical woodland ecosystems in Zimbabwe: Introduction, overview and synthesis. *Ecological Economics* **33**, 341–351.

Campbell, B., Mandondo, A., Nemarundwe, N., Sithole, B., De Jong, W., Luckert, M. and Matose, F. (2001) Challenges to proponents of common property resource systems: Despairing voices from the social forests of Zimbabwe. *World Development* **29**, 589–600.

du Toit, J. T., Walker, B. H. and Campbell, B. M. (2004) Conserving tropical nature: Current challenges for ecologists. *Trends in Ecology & Evolution* **19**, 12–17.

Cunningham, A. B., Belcher, B. and Campbell, B. M. (2005) *Carving Out a Future: Forests,*

Livelihoods and the International Woodcarving Trade. People and Plants Conservation Series. Earthscan, London.

Sayer, J. and Campbell, B. (2004) *The Science of Sustainable Development: Local Livelihoods and the Global Environment.* Cambridge University Press, Cambridge.

Questions and answers

Is big industry such as logging companies, or expanding human populations the more important threat to tropical ecosystems?

This depends on the context – but in general I believe that population, especially local population, is not a major driver. Colleagues and I work in the Brazilian Amazon, in the forests of central Africa and in Indonesian Borneo. In the Amazon site the most important threat relates to the expansion of commercial agriculture (soya beans and livestock), resulting in considerable forest loss. In central Africa forest is not really being lost – but the bush meat trade for expanding urban populations is negatively affecting biodiversity. At the Indonesian site forest destruction is driven by logging companies and those issuing the permits. Context, context, context!

Why is the bush meat trade, an ancient human activity, no longer sustainable in many tropical areas?

The scale of hunting is now much greater than before – to supply the expanding urban markets. But I should also point out that areas far from roads (and there are many of those) are not very severely affected. However, as roads penetrate further into the forest, bush meat exploitation will follow.

In what way can small-scale enterprises that rely on the exploitation of tropical forest products be beneficial to conservation?

Many conservation agencies will spend considerable resources to improve the livelihoods of those living in and around protected areas. Small-scale enterprises can be used to build good will among local people towards conservation areas. Where local people value a particular enterprise and want it to be maintained, they can themselves institute management measures to ensure sustainable harvest. Some of the best examples of this are in southern Africa, where wildlife hunting is proving a win–win situation for people and conservation (but, even there, context is important – some schemes will not work because of a number of factors, e.g. too many people, too few prized hunting animals, discord in communities).

How can cross-disciplinary cooperation and facilitation in conservation be made to work given the lack of incentives for such teamwork from bureaucracies?

I think there are a number of success factors for cross-disciplinarity. Questions posed by conservation biologists that require expertise from another discipline don't generally provide enough excitement to that other discipline. To engage other disciplines in any real sense, one must invite resource economists, sociologists, etc. to the initial meetings where the research questions are defined. Respect is key – with the different languages and approaches of different disciplines it is easy to find fault – there must be a high degree of mutual respect in a team. To build respect and derive common visions about the research requires time. There are bureaucratic hurdles but I think these can be broken down by committed teams.

How is the Centre for International Forestry Research (CIFOR) being effective in changing policy and having an on-ground impact?

CIFOR has a very conscious approach to achieving impact. This can be broken down into four elements: (1) understanding where impact is possible and getting focus on a few topics; (2) building impact strategies so that one clearly understands, for each programme of work, the kinds of research that will be produced, who the users of that research will be, what formats are needed for the research outputs and what processes need to be engaged with (e.g. if impact is sought at the international convention level, how do we interface with the United Nations Convention on Biological Diversity?); (3) building partnerships with agencies that are crucial for the uptake of the research; (4) making sure that there is time and budget for the necessary processes and products after the formal research is completed (e.g. media campaigns, policy briefs, key stakeholder meetings).

economy and rely on wild meat for sustenance and cash (Carpaneto and Fusari 2000; Figure 6.1). Sustained poverty drives locals to hunt in protected areas – close to 60000 people derive their primary income from illegal hunting in the protected areas of Tanzania (Loibooki *et al.* 2002). Thus, precipitous loss of wildlife is also jeopardizing local livelihoods, so the bush meat crisis demands biological, as well as sociological, solutions.

6.1.1 Harvest rate

The estimates of annual tropical wildlife harvest remain high but imprecise. Overall, it has been estimated that wildlife is extracted from tropical forests at approximately six times the sustainable rate (E.L. Bennett 2002). For example, an estimated 23500 tonnes of wildlife meat are consumed annually in Sarawak (Malaysian Borneo) alone, with an estimated 2.6 million animals shot there every year (E.L. Bennett *et al.* 2000; E.L. Bennett 2002). It is estimated that in neighbouring Sabah an incredible 108 million animals are shot every year (E.L. Bennett *et al.* 2000). Five million tonnes of meat is extracted from the Afrotropical and neotropical forests annually (Fa *et al.* 2002). Moreover, in the Brazilian Amazon, 23.5 million animals are consumed every year (Peres 2000). Such figures are clearly alarming, and it is perhaps not surprising that the Peres' (2000) study from the Amazon showed that the overall biomass of vertebrate communities declined with increasing levels of hunting pressure (Figure 6.2).

In other words, the quantity, and most likely the diversity, of human prey is diminishing. Using simulation models, Barnes (2002) showed that hunting can result in the population collapse of common human prey species in Africa: the spot-

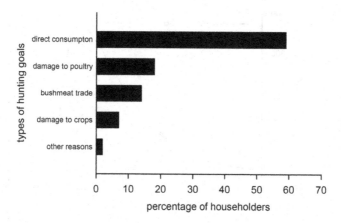

Figure 6.1   Percentage of householders with various reasons for hunting. (After Carpaneto and Fusari 2000. Copyright, Springer Science and Business Media.)

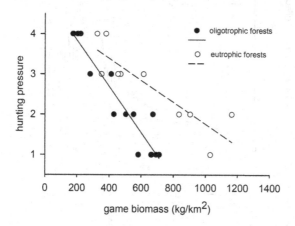

Figure 6.2   Graphs demonstrating the negative correlations of increasing hunting pressures with decrease in game biomass in oligotrophic (filled circles) and eutrophic (unfilled circles) forests. (Data derived from Peres 2000.)

nosed monkey (*Cercopithecus nictitans*), medium-sized bay duiker (*Cephalophus dorsalis*) and small blue duiker (*C. monticola*) (Figure 6.3). The particulars of individual species' life histories affect the steepness of the relationship between population decline and hunting. For example, slow-breeding monkey species experience more rapid population collapse than faster-breeding ungulate species (Barnes 2002). In general, the species' generation time (birth to reproductive age), is also a function of body mass (allometry), and larger, longer-lived and slow-reproducing animals are more susceptible to overharvest and extinction (Jerozolimski and Peres 2003; Brook and Bowman 2005).

There is evidence that wildlife hunting will expand in the future. Between

(a)

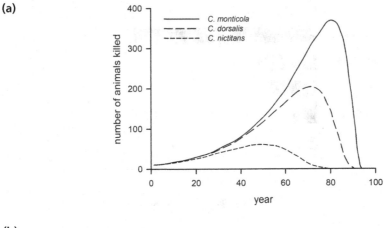

(b)

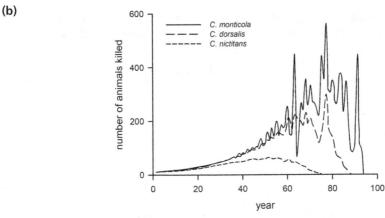

(c)

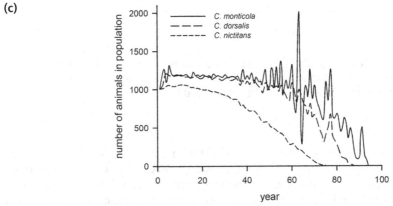

Figure 6.3 Effects of increasing hunting pressure on one monkey (*Cercopithecus nictitans*) and two duiker species (*Cephalophus dorsalis*, *C. monticola*) simulated by harvest models. Harvests are predicted by (a) the deterministic model, (b) the model that incorporates random variation and (c) the number of individuals in each population predicted by the harvest model incorporating random variation. Common names are in the text. (After Barnes 2002. Copyright, Cambridge University Press.)

1991 and 1996 in Bioko Island (Equatorial Guinea), there was a 13% increase in biomass and a 60% increase in the number of animals killed by humans for food (Fa *et al.* 2000). Increasing human populations across most of the tropics (Chapter 1) will probably lead to more wildlife hunting (Bulte and Horan 2002) given the positive correlation between human density and total biomass of wildlife harvested (J.G. Robinson and Bennett 2004; Figure 6.4). Human populations are expanding particularly quickly in urban centres (Figure 1.6), and the excessive hunting that results leads to a halo of defaunation around these areas, as has occurred in the Congo Basin (D.S. Wilkie and Carpenter 1999). In apparent contrast to the problems associated with overharvesting of bush meat sustained by poverty, the consumption rate of bush meat can also increase as average income rises. Indeed, there is evidence that increasing wealth among the growing urban population in Africa will increase the demand for bush meat (East *et al.* 2005).

Because slow-breeding large animals such as apes, large carnivores and elephants are particularly vulnerable to hunting (J.G. Robinson *et al.* 1999), the potential for population recovery in these animals over short time scales is extremely low (Brook and Bowman 2004). In support of this hypothesis is evidence that within the past 40 years 12 large vertebrate species have been extirpated from Vietnam primarily as a result of excessive hunting (Milner-Gulland and Bennett 2003). Some of these extinctions may also have implications for forest recovery – the loss of keystone species such as seed dispersers (e.g. hornbills) can be detrimental to forest regeneration (see Chapter 9).

Commercial logging and other industrial activities such as oil exploration (Thibault and Blaney 2003) are key facilitators of rampant overexploitation of forest-dwelling species by increasing access to remote areas (i.e. road development), by bringing more people into such regions and by altering local economies and resource consumption patterns (see Chapter 1). For example, bush meat forms up to two-thirds of the diet of highland people in Sarawak

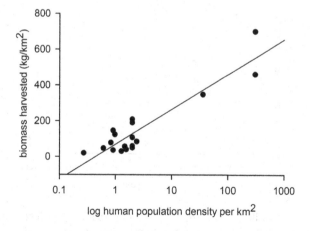

Figure 6.4 Correlation of biomass harvested with human population density. (Data derived from J.G. Robinson and Bennett 2004.)

(E.L. Bennett *et al.* 2000), with most hunting occurring along the roads opened up by logging (J.G. Robinson *et al.* 1999). Commercial logging companies can, however, help to alleviate wildlife hunting by enforcing or promoting control among staff. 'Green labelling' and independent-party certification can be used as positive incentives for forest managers to encourage logging companies to support environmentally friendly practices such as ensuring that their staff do not hunt or bring native wildlife meat into logging camps (J.G. Robinson *et al.* 1999; Wilkie and Carpenter 1999).

The overexploitation of wildlife jeopardizes not only the animal populations that fall prey to this pressure, but also the survival and cultural heritage of the people who rely on them (E.L. Bennett *et al.* 2000). The most vulnerable consumers remain traditional forest peoples. Fa *et al.* (2003) found that in the Congo Basin bush meat extraction is unsustainable because it is harvested at a higher rate than production (Figure 6.5a). They projected that bush meat protein supplies will decline in the region (Figure 6.5b), with insufficient non-bush meat proteins available to offset this decline.

6.1.2 Case studies

Below, we present some of the case studies reporting major impacts of bush meat hunting on tropical fauna. Galliforms (including grouse, pheasants, partridges

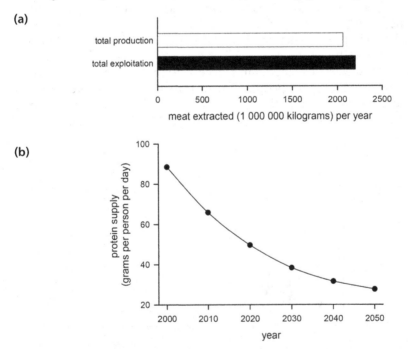

Figure 6.5 Unsustainability of bushmeat in the Congo Basin, with (a) total amount of wild meat extracted (filled bar) exceeding the total produced (unfilled bar) annually and (b) the general decline in protein supplements of local people. (Data derived from Fa *et al.* 2003. Copyright, Cambridge University Press.)

and quails) are particularly vulnerable to hunting because they are predominantly ground-nesting and can occur at high densities in certain locations (McGowan and Garson 2002). About one-quarter of all galliforms are now listed as threatened by IUCN (The World Conservation Union), and in 90% of cases overhunting is one of the listed threats (McGowan and Garson 2002). To highlight the plight of galliforms, we provide an example of the maleo (*Macrocephalon maleo*). Endemic to the islands of Sulawesi and Buton (Indonesia), the charismatic maleo unfortunately faces a high risk of imminent extinction (Gorog *et al.* 2005).

Weighing about 1.6 kg, the maleo is a terrestrial megapode (Figure 6.6) that has a unique nesting behaviour. As many as 100 pairs nest communally at traditional sites around geothermally heated soils in forests or in solar-heated sandy beaches and riverbeds. Using their powerful feet, adult birds excavate individual 60- to 70-cm-deep burrows. Each burrow contains one egg, buried a specific depth depending on the surrounding soil temperature. Eggs are covered with sand and are left to incubate via environmental sources of heat – adults do not provide any

(a)

(b)

Figure 6.6 (a) The maleo (*Macrocephalon maleo*), endemic to the islands of Sulawesi and Buton (Indonesia), and (b) the size of its egg (relative to a human hand). (Photos by Connie Bransilver, conniebransilver.com.)

parental care. Young maleo are able to dig their way to the surface and can feed independently almost immediately after emergence. Breeding between October and April, a pair of maleo can lay up to 12 eggs per year (Prawiradilaga 1997), and, because each egg weighs about 250g and has a high yolk content (Figure 6.6), this concentration of highly nutritious protein has long attracted humans to maleo nesting grounds.

Initially, local people used to collect maleo eggs for traditional ceremonies; however, this situation has now changed because maleo eggs are considered both delicious and nutritious throughout Indonesia. In some local markets, maleo eggs are worth between 2500 and 5000 rupiah each (< US$0.60). Despite being listed on the CITES (Convention on International Trade in Endangered Species of Wild Fauna and Flora) Appendix I (see below) and protected under Indonesian law, uncontrolled egg collecting by the burgeoning human population of Sulawesi has led to maleo population declines and abandonment of many former nesting sites. Of 131 known nesting sites, 42 have been abandoned in recent times, with another 38 severely threatened, 12 of unknown status and only five (4%) not yet considered threatened (Butchart and Baker 2000). The global population is currently estimated to be between 4000 and 7000 breeding pairs. However, this population is declining rapidly, with 90% declines in some areas since the 1950s (Argeloo and Dekker 1996).

In addition to heavy human exploitation, maleo are also losing foraging habitat as a result of extensive deforestation (A.J. Whitten *et al.* 1987). Clearly, the future persistence of this species is in extreme jeopardy. There is a dire need for a better understanding of the ecology (e.g. recruitment and site fidelity) of this unique bird, and the status of this species needs to be evaluated critically throughout its range with data collected over many years to identify precisely those areas experiencing population declines. To halt further population decline, immediate actions are needed. Although over 50% of the known maleo nesting sites are protected (Dekker 1990), such protection is inadequate and poorly enforced. The chance of a nesting area remaining active over time is 55% higher in protected than in unprotected areas (Gorog *et al.* 2005). There are numerous maleo captive breeding programmes in central Sulawesi (Christy 2002), but their success will be limited without the assurance that the released birds have a legitimate chance of surviving to reproduce in the wild. To achieve tangible conservation outcomes, social aspects need to be integrated fully into conservation plans. For instance, there is a need to develop and promote sustainable egg-harvesting regimes with the full involvement and endorsement of local residents. Local communities should be assisted in exploiting alternative food sources, as has been done for a community in Pakuli (central Sulawesi). Here, locals are encouraged to rely more on fish (*Osteochilus* sp.) to reduce their dependence on maleo eggs (Sodhi and Brook 2006). This community is a custodian of their local maleo nesting grounds. In our minds, similar approaches adopted across Sulawesi will work best to ensure the survival of maleos and other exploited species more generally. Transition to alternative prey, however, is difficult at least for some other local communities (see E.L. Bennett *et al.* 2000).

How likely is the maleo to be driven to extinction by human activities? This question can be answered, at least approximately, with the use of inferential population modelling. Brook *et al.* (2006b) derived a series of multivariate generalized linear mixed-effects models that statistically approximated the minimum viable population size (MVP) for 1198 well-studied species by relating extinction thresholds estimated through population viability analysis (PVA) to a suite of ecological 'correlates', including body size, generation time, niche breadth, reproductive rate, dispersal ability, range size and ecological flexibility. An MVP can be broadly defined as the minimum population size required for a species to have a predetermined and reasonable probability of persistence over a specified length of time (usually 100 years or 40 generations), given the inherent uncertainty in demographic, genetic and environmental processes that affect populations. Similar concepts are inherent in the IUCN categories of endangerment, e.g. following criterion E, a 'Least Concern' population has a greater than 90% probability of survival over 100 years (IUCN 2005). An inference based on the known life history and environmental attributes and the model of Brook *et al.* (2006b) suggests that for the maleo an MVP with a less than 10% probability of extinction over the next 100 years is at least 2300 individuals. These MVPs represent viable populations in the absence of external deterministic population pressures such as overexploitation by humans and habitat loss. Thus, the maleo with its current population size of 4000–7000 breeding pairs, is already close to its MVP, and will in all likelihood be driven to extinction within the next century if current anthropogenic pressures cannot be mitigated.

Birds are not the only heavily hunted group. The endemic Sulawesi crested black macaques (*Macaca nigra*), a type of monkey, declined substantially between 1978 and 1994 (Rosenbaum *et al.* 1998) (Figure 6.7). One of the primary reasons for this decline appears to have been overhunting – these macaques have been found caught in snare traps set for forest (wild) pigs (*Sus scrofa*). There are predictions that this macaque will be wiped out within the next 20 years if the current levels of hunting continue unabated (R.J. Lee *et al.* 1999). In the same area (Sulawesi), O'Brien and Kinnaird (1996) determined how hunting and habitat change were affecting the populations of a range of birds and mammals. In general, those species known to be hunted declined between 1979 and 1994, whereas those thought to be unaffected by hunting (e.g. Sulawesi tariatic hornbill, *Penelopides exarhatus*) did not (Table 6.1).

The latter group (unaffected by hunting) is considered less vulnerable because these species are not overly susceptible to the hunting techniques used predominantly in the study area (i.e. traps, snares, air-powered pellet rifles and dogs). The ground-nesting maleo, as described earlier, is a heavily exploited species on the verge of local extinction due to high rates of egg collection, with only seven pairs left. These observations suggest that excessive hunting has been responsible for population declines of some of the species in this area. O'Brien and Kinnaird (1996) believe that urgent efforts are needed to curtail hunting to mitigate this loss, both in the study area and across the region.

Does hunting by traditional indigenous people substantially impinge on wildlife populations? Hunting by Maya communities in Quintana Roo (Mexico) for over 4000 years seems to be have been sustainable (Jorgenson 2000). Alvard and Winarni (1999) studied the impact of traditional subsistence harvest by indigenous people called the Wana on birds in Sulawesi. Today, there are several thousand Wana living in and around the reserve. These people have little contact with the outside world, practise slash-and-burn horticulture and obtain their protein by hunting and trapping animals. Their animal prey include mice, birds, bats, reptiles, amphibians, ungulates and primates. Alvard and Winarni (1999) found that the more abundant a bird species was, the higher were its chances of being hunted by the Wana, a potentially sustainable tendency.

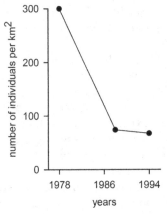

Figure 6.7   Decline in population density of the Sulawesi crested black macaque (*Macaca nigra*) from 1978 to 1994. (After Rosenbaum *et al.* 1998. Copyright, John Wiley and Sons. Photo by Cagan Sekercioglu, naturalphotos.com.)

Table 6.1 Change in vertebrate populations due to hunting in Sulawesi. (After O'Brien and Kinnaird 1996. Copyright, Cambridge University Press.)

Common name	Species	Observed change	Expected response to hunting
Anoa	*Bubalus depressicornis*	Decline	Decline
Sulawesi pig	*Sus celebensis*	No change	No change/ decline
Crested black macaque	*Macaca nigra*	Decline	Decline
Bear cuscus	*Phalanger ursinus*	Decline	Decline
Babirusa	*Babyrousa babyrussa*	Decline	Decline
Maleo	*Macrocephalon maleo*	Decline	Decline
Tabon scrubfowl	*Megapodius cumingii*	Increase	No change
Red junglefowl	*Gallus gallus*	Decline	No change/decline
Red-knobbed hornbill	*Aceros cassidix*	Increase	Decline
Sulawesi taritic hornbill	*Penelopides exarhatus*	Increase	Decline

However, hunting by humans does not seem to affect seriously the viability of bird populations in this area. Of the 46 prey species, 31 were more abundant at Wana-occupied sites than in regions not experiencing bird hunting. Although habitat heterogeneity and differential detectability can also influence such a result, this study does suggest that some traditional hunting practices may not cause heavy declines in prey (bird) populations. Alvard (2000) found that Sulawesi wild pig (*Sus celebensis*) hunting by Wana also seems to be sustainable. In a similar finding to the above, a low density of Aboriginal people using primitive hunting techniques (e.g. snares) with only limited access appeared to be having little impact on tigers and their prey in Peninsular Malaysia (Kawanishi and Sunquist 2004).

The above cases may nevertheless be anomalies rather than the norm for the region. For instance, a group of indigenous Penan people appear to have been responsible for the disappearance of an entire population of Bornean gibbons (*Hylobates muelleri*) from a primary forest in Sarawak (E.L. Bennett *et al.* 2000). Jerozolimski and Peres (2003) reported that subsistence hunters in neotropical forests deplete large mammals and then switch to smaller prey. Rao *et al.* (2005) also found that hunting by locals in Myanmar was indiscriminate and depended largely on the relative abundance of the prey (Figure 6.8). Even endangered species such as wild dog (*Cuon alpinus*), capped langur (*Trachypithecus pileatus*) and hoolock gibbon (*Bunopithecus hoolock*) were hunted. Rao *et al.* (2005) recommended a better legal framework and its application to curb the indiscriminate slaughter of wildlife.

Other studies report similar predicaments. In addition to habitat loss, the main direct threat to the mammals of Sangihe and Talaud Islands (Indonesia) has been hunting. Of 57 farmers interviewed, 77% admitted to hunting wildlife (Riley 2002). Hunting was the major threat to at least two mammalian groups: fruit bats and cuscus (Figure 6.9). Cuscus (arboreal possum-like animals) have a number of subspecies endemic to this region (e.g. bear cuscus, *Ailurops ursinus melanotis*) (Figure 6.9), and 5 of the 11 fruit bats are listed as globally threatened (e.g.

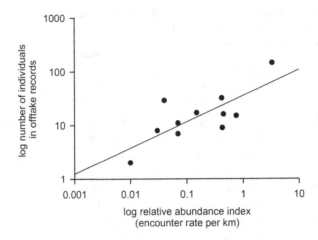

Figure 6.8 Relationship between the number of harvested individuals against relative abundance (encounter rates of tracks and signs per km of transect). (Data derived from Rao *et al.* 2005.)

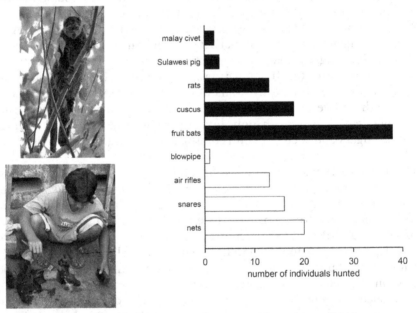

Figure. 6.9 Cuscus (top left) and fruit bats (bottom left) are the predominant mammal species hunted (filled bars) in the Sangihe and Talaud Islands (Indonesia). Number of animals captured using different hunting methods (unfilled bars). [Data derived from Riley 2002. Photos (cuscus) by Cagan Sekercioglu, naturalphotos.com, and (fruit bats) Tien Ming Lee.]

Talaud Islands flying fox, *Acerodon humilis*; Riley 2002). Farmers predominantly used mist nets and snares to catch prey (Figure 6.9). The bats are regularly sold in some markets in Indonesia (Figure 6.9). Similar results have been reported on hunting patterns by locals from other regions as well (e.g. Carpaneto and Fusari 2000).

Bush meat statistics have been compiled for other systems under threat. Up to 90 000 mammals are sold annually in a single market in north Sulawesi (Clayton and Milner-Gulland 2000), and it has been estimated that 86 198 kg of wild prey were consumed by people living around Manembonembo and Gunung Ambang Nature Reserves in Sulawesi alone (R.J. Lee *et al.* 1999). In nine markets in north Sulawesi, 24 864 individuals of 27 mammal species were sold (R.J. Lee *et al.* 1999) (Figure 6.10). Of these, approximately 1% of individuals harvested were protected species (e.g. dwarf cuscus). This apparently low proportion still represents many hundreds of individual animals, which for species with small, already threatened populations may be sufficient to lead eventually to their extinction (R.J. Lee *et al.* 1999). Creation of highways in Sulawesi facilitates this wild meat trade by allowing quick transport to markets.

6.1.3   War and bush meat hunting

Civil wars break social order and can facilitate environmental destruction (Talbott and Brown 1998; Draulans and van Krunkelsven 2002). For example, in Rwanda, selling of bush meat by poachers is more common than it was before the civil war (Plumptre *et al.* 1997). This trend has led to calls for peace parks to protect biodiversity in war zones (Draulans and van Krunkelsven 2002).

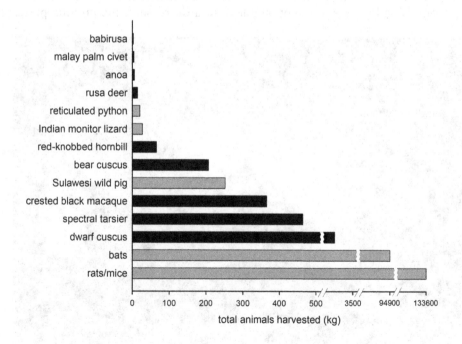

Figure. 6.10   Total mass of protected (black bars) and non-protected (grey bars) harvested animals from nine markets in northern Sulawesi. (Data derived from R.J. Lee *et al.* 1999.)

6.1.4   Sustainability of harvesting bush meat

J.G. Robinson and Bennett (2004) estimated that because of the presence of ungulates and rodents, the supply of wildlife is relatively higher in dry forests and savannas than in moist forests. They also concluded that demand for wild meat exceeds supply in moist forests. This disparity between supply and demand can be offset by a reliance on secondary and disturbed forests. Fa *et al.* (2005) reported that ungulates formed 73% of hunted animals from Afrotropical forests. Overall, an annual average of 2000 carcasses (16 000 kg in biomass) were extracted from 36 sites in seven countries – rates considered to be unsustainable.

Clayton *et al.* (1997) developed a spatial model to describe the hunting of two endemic pig species, the babirusa (*Babyrousa babyrussa*) and the Sulawesi wild pig, in Sulawesi (Figure 6.11). They found that hunting will have a greater effect on the former species than on the latter because babirusa are restricted to the rain forest, are suffering from rapid range contraction due to loss of habitat, and have a total wild population of only 5000 (Clayton *et al.* 1997). The Sulawesi wild pig is not protected by law, but the babirusa is. The penalty for killing a babirusa is considerable, with a maximum fine of US$45 454 or 5 years in prison. However, chances of dealers actually being fined are minimal because they routinely bribe the relevant officials with meat (Clayton *et al.* 1997). Clayton *et al.* (1997) recommended that the most realistic means of curtailing the hunting of babirusa will be consistently to impose fines on illegal traders attempting to sell babirusa meat at the markets. They also predicted that future increases in economic conditions may eventually lead to a decrease in the hunting pressure

Figure. 6.11   A successful hunt on one of the endemic Sulawesi wild pig. (Photo by Cagan Sekercioglu, naturalphotos.com.)

– a hypothesis not supported by a study from Borneo showing that even affluent people continued to hunt wildlife (E.L. Bennett *et al.* 2000).

Refisch and Koné (2005) estimated that 249 tonnes of primate meat was extracted near Taï National Park (Côte d'Ivoire) in 1999. Models showed that harvesting of only red colobus monkey (*Procolobus badius*) was sustainable, while in the case of other species (the black and white colobus; *Colobus polykomos*; sooty mangabey, *Cercocebus atys*; Diana monkey, *Cercopithecus diana*; and Campbell's monkey *C. campbelli*), harvesting exceeded sustainability by up to three times (Figure 6.12). Refisch and Koné (2005) recommended that farming of domestic animals and pond fish be increased in the area to decrease harvest of primates in this region.

6.1.5   Possible solutions

Although the 'bush meat crisis' has galvanized conservation biologists, the solutions may be difficult (but not impossible) to implement (J.G. Robinson and Bennett 2002). In a study from Peninsular Malaysia, it was found that a 76-ha patch of lowland rain forest amidst oil palm plantations that was protected from hunting retained an extremely rich primary forest bird and mammal fauna (E.L. Bennett and Caldecott 1981). However, a large-scale total ban of wildlife hunting in the tropics is neither possible nor feasible (Rowcliffe 2002), although there have been calls to regulate harvest to ensure sustainability (Ling *et al.* 2002). As hunting in the tropical forests is largely illegal and not capitalized or industrialized, implementing regulatory mechanisms will be difficult. Further, most tropical countries lack governmental institutions or the political will to manage hunting activities. Therefore, imposing hunting quotas, regulating hunting (e.g. specifying the ages and sexes of animals that can be hunted) and educating hunters will be difficult (J.G. Robinson *et al.* 1999; but see below).

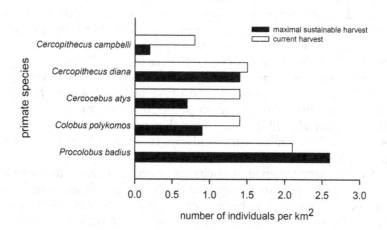

Figure 6.12   Commercial hunting of primates in Côte d'Ivoire with reference to current (white bars) and maximum sustainable (black bars) number of individuals per square kilometre. (Data derived from Refisch and Koné 2005.)

Nonetheless, to control the supply and demand loop, more regional rather than local actions are needed (Corlett 2007).

There is broad consensus that scientific understanding needs to underpin policies for bush meat hunting (E.L. Bennett *et al.* 2000; McGowan and Garson 2002; Rowcliffe 2002). However, studies on the biological and social components of the bush meat crisis are rare (E.L. Bennett *et al.* 2000). There is a need to develop more and better models of sustainable harvesting (i.e. ensuring that the combination of hunting plus natural mortality does not exceed population recruitment) for key species, even in situations where knowledge of a given species' ecology and demography is imperfect. In addition, broad-scale analyses that pinpoint local and regional 'hotspots' of the bush meat crisis are needed so that immediate actions can be taken where they are most urgently required (Fa *et al.* 2002).

Development agencies need to take the bush meat crisis into account so that poverty alleviation programmes are not mutually exclusive of the health and long-term well-being of wildlife populations that many indigenous people utilize (G. Davis 2002). It should also be borne in mind that bush meat can sometimes pose a serious health risk to humans if precautions in food preparation and animal handling are inappropriate. Bush meat consumption of simians in Africa has been posited as a potential source of HIV (Wolfe *et al.* 2004). Similarly, severe acute respiratory syndrome (SARS) originated in Asia from wild animals caught for meat (Bell *et al.* 2004). Bush meat will probably continue to expose humans to more new viruses and resulting zoonotic diseases (Wolfe *et al.* 2005).

Alternative protein sources should also be explored when and where possible. In Bolivia, the price of fish and meat from livestock is positively correlated with the consumption of wildlife (Apaza *et al.* 2002). This trend suggests that policy makers can alleviate bush meat hunting by reducing the price of meat from domestic animals. Brashares *et al.* (2004) found that between 1976 and 1992 in Ghana the annual counts of hunters were inversely correlated with regional fish supply (Figure 6.13). As the number of hunters was positively correlated with wildlife declines, it appeared that hunters venture into the reserves more frequently for bush meat when fish supplies are low. Similarly, in Gabon, as the price of bush meat increases, its consumption falls and results in increased consumption of fish (Wilkie *et al.* 2005). Brashares *et al.* (2004) also supported actions such as building up agriculture and fish stocks to ease hunting pressure on wildlife. This is critical given that 16 of 41 mammal species have become locally extinct in the surveyed reserves in Ghana, most likely due to unsustainable harvest (Brashares *et al.* 2004). However, policy makers have to be mindful that increases in household wealth will not always result in a decrease in bush meat consumption (Wilkie *et al.* 2005).

A multifaceted management approach is most needed to alleviate the bush meat crisis, with long-term solutions requiring the joint consideration of economic, social and biological issues (E.L. Bennett *et al.* 2000; Pratt *et al.* 2004). A tangible approach should include efforts to (1) provide alternative income and protein sources for people who rely on wildlife for their everyday subsistence needs, (2) curtail or at least mitigate the wildlife trade through better regulation, (3) more effectively protect 'protected areas' and (4) educate hunters and buyers

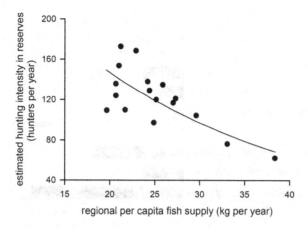

Figure 6.13 Negative relationship between hunting intensity and regional fish supply. [After Brashares *et al.* 2004. Reprinted with permission. Copyright American Association for the Advancement of Science (AAAS)].

on the risks associated with overexploitation of wildlife resources (Riley 2002; J.G. Robinson and Bennett 2002; Milner-Gulland and Bennett 2003; Corlett 2007). The farming of some wildlife can be an option to alleviate hunting of wild populations (Njiforti 1996). Sarawak provides a heartening example where some of these measures are currently being implemented to alleviate bush meat hunting (E.L. Bennett *et al.* 2000). However, complicated social issues (e.g. finding alternative protein sources) need to be taken into account when attempting to achieve tangible conservation goals such as limiting wildlife harvest for food.

6.2 Captivity trade

The global trade in wildlife destined for captivity is enormous, involving millions of individual plants and animals and an estimated annual net worth of billions of dollars (www.traffic.org). As large as these figures appear, they exclude domestic wildlife trade and illicit (illegal and unregulated) trade, which has been estimated to account for over one-quarter of the total trade (Beissinger 2001). Illicit wildlife trade can flourish in certain regions and situations, especially during an economic crisis, as was seen in Venezuela in the late 1990s (J.P. Rodriguez 2000). The trade of live animals and biological products (e.g. medicinal plants) is estimated at 3% of the total global wildlife trade (www.traffic.org), and includes 350 million live tropical fish, 640 000 live reptiles, 4 million live birds and 40 000 live primates traded globally each year (www.wwf.org.uk).

Worldwide, between 1.6 and 3.2 million birds are taken annually from wild populations for the pet trade (Beissinger 2001). Most (95%) of these birds are finches and parrots, including endangered species (Figure 6.14). These numbers may be grossly underestimated because many birds remain uncounted (e.g. due to illegal capture) or die before reaching the point of official tallying, or are sold in local markets. Indeed, it is estimated that up to 60% of birds trapped may

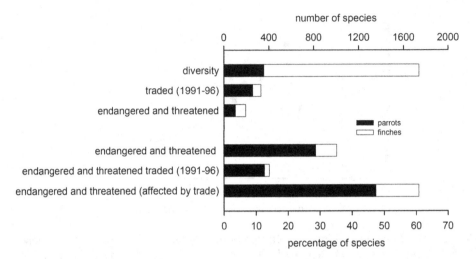

Figure 6.14 Levels of diversity, trade and threat of parrots and finches. Endangered and threatened species refer to those listed by the World Conservation Union as critically endangered, endangered or vulnerable. (Data derived from Beissinger 2001.)

die before reaching the export market (Inigo-Elias and Ramos 1991). Tropical countries alone account for 92% of the volume of birds exported, with African countries topping the charts (Figure 6.15) (Mulliken *et al.* 1992). The major bird importing countries are the European Union countries, the USA and Singapore (Mulliken *et al.* 1992).

Southeast Asia is a major hub of the global trade in wildlife and wildlife products, functioning as both consumer and supplier (Table 6.2). A rampant trade in live birds operates in Southeast Asia, affecting many rare or little-known species, and even those protected by national laws (Corrigan 1992). Within the region, Indonesia represents a 'hotspot' of bird capture. From 1981 to 1992, roughly 100 000 birds were reported to be exported from Indonesia, but this number is, in all likelihood, a gross underestimation of the true numbers of wild birds caught [Director General of Forest Protection and Nature Conservation (PHPA) 1998]. As with the bush meat issue, the biological information required to guide the sustainable extraction of wild birds is generally lacking (Beissinger 2001; see below).

The critically endangered Bali starling (*Leucopsar rothschildi*) epitomizes how excessive capture for the pet trade can lead to a species' endangerment. Currently, there are only six individuals of this species left in the wild, and they are restricted to Bali's (Indonesia) Barat National Park (van Balen *et al.* 2000; www.birdlife.org). Rampant trapping coupled with habitat loss resulted in the precipitous population decline of the Bali starling. The species was listed on Appendix I of CITES in 1970, and subsequently received official protection in Indonesia in 1971. Despite legal protection, 19 individuals were witnessed being sold illegally in shops in Singapore in 1979, and 16 individuals were observed in cages in Denpasar (Bali) in 1982 (van Balen *et al.* 2000). An extremely small population size, a distribution limited to one site, continued destruction of its

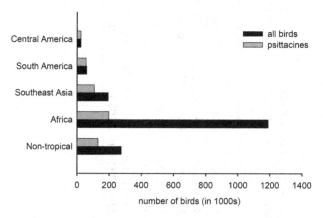

Figure 6.15 Regional estimated gross number of all birds and psittacines (parrots) exported only from the top 20 exporting countries. (After Mulliken *et al.* 1992. Copyright, Traffic International.)

natural habitat and possible ongoing illegal trapping have all conspired to drive this species to the brink of extinction (Brook and Sodhi 2006). If this species is to have any realistic chance of long-term persistence, efforts such as 'real' (enforced) protection, habitat preservation and enhancement through revegetation, and possibly a release of captive-bred individuals to bolster wild numbers, are urgently needed.

As with the Bali starling, captures of the yellow-crested cockatoo (*Cacatua sulphurea*) is the main factor causing its decline in Indonesia (PHPA 1998). The recovery of this species can be brought about by curtailing or halting illegal capture, and through the provision of adequate habitat (e.g. cavity-producing trees for nesting). It is rather ironic that accidental or deliberate releases of yellow-crested cockatoos have led to its establishment in new areas outside of its native range, such as Singapore and Hong Kong. There has also been concern that, in these new areas, this species can outcompete and displace native birds (Sodhi and Sharp 2006). This situation is mirrored by the deliberate introduction of a wild cattle species native to Southeast Asia, the banteng (*Bos javanicus*), to northern Australia in the mid-nineteenth century (Calaby 1975). Although the wild form of the species is now endangered in its native range (Indonesia, Malaysia, Vietnam, Cambodia), it thrives in one small area of northern Australia (Bradshaw *et al.* 2006a). The establishment of yellow-crested cockatoos and banteng in new areas does show that captive release can be a viable strategy to reinforce wild populations, provided that wild capture is prevented and adequate habitat is available (Bradshaw *et al.* 2006a).

Native birds, macaques and gibbons (especially the lar gibbon, *Hylobates lar*) are commonly kept as pets in both rural and urban areas of Thailand, with abandoned pets often ending up in temples or government agencies. Forest loss has made gibbons particularly vulnerable to trapping (Eudey 1994), but it is possible that reintroducing rehabilitated gibbons might boost wild populations (Eudey 1994).

Table 6.2 Net legal export and import of selected wildlife products for 2003 (source of species: wild; purpose of transaction: commercial, personal, transit). Net legal exports and imports in wildlife products obtained from World Conservation Monitoring Centre (WCMC) Convention on International Trade in Endangered Species of Wild Fauna and Flora (CITES) trade database; numbers in parentheses represent percentage of each item exported in the world.

Regions		Live lizards	Live snakes	Live primates	Live parrots	Lizard skins	Snake skins	Crocodilian skins	Cat skins
Neotropics	Exports	38952 (14.9)	2217 (3.1)	347 (5.7)	35026 (16.5)	857545 (45.3)	5955 (0.4)	192743 (63.0)	8 (<0.1)
	Imports	2361 (0.9)	106 (0.1)	110 (1.8)	484 (0.2)	31856 (1.7)	25107 (1.7)	9557 (3.1)	1627 (1.9)
Sub-Saharan Africa	Exports	180404 (68.9)	19784 (27.7)	770 (12.7)	108158 (50.8)	184118 (9.7)	1821 (0.1)	12900 (4.2)	37 (<0.1)
	Imports	6835 (2.6)	0 (0)	24 (0.4)	1789 (0.8)	73315 (3.9)	4994 (0.3)	1573 (0.5)	0 (0)
Southeast Asia	Exports	22793 (8.7)	43110 (60.4)	3004 (49.5)	4905 (2.3)	716333 (37.9)	977431 (67.1)	15409 (5.0)	1 (<0.1)
	Imports	4859 (1.9)	6736 (8.9)	8 (0.1)	16615 (7.7)	509902 (27.0)	275327 (18.9)	24024 (7.9)	0 (0.0)

For centuries, humans have been hunting reptiles for food, skin and medicinal purposes. More recently, however, reptiles have also been extracted for the pet trade. Chameleons (family: Chamaeleonidae) in particular have been hunted for this purpose since the early 1800s (Carpenter *et al.* 2004). Between 1977 and 2001, 845 013 chameleons were traded, with Togo and Madagascar trading the highest number (Figure 6.16) (Carpenter *et al.* 2004), and nearly 70% of all chameleons being imported by the USA. The number of species traded (Figure 6.17a) and exporting countries have been gradually increasing. There has been concern that habitat loss and extensive harvesting is placing many chameleon species at risk of extinction. However, the number of captive chameleons traded is also increasing steadily (Figure 6.17b), which may help to alleviate wild harvests through the provision of individuals via ranching (extraction of wild eggs and rearing in controlled conditions) and captive breeding (specimens produced in controlled environments) (Carpenter *et al.* 2004).

Some experts have also expressed concern that wildlife trade enhances the prospect of disease transmission, to both humans (zoonoses) and domestic livestock (Karesh *et al.* 2005). For example, a highly pathogenic bird flu virus was recently discovered in two crested hawk-eagles (*Spizaetus nipalensis*) illegally imported to Belgium from Thailand (Van Borm *et al.* 2005). Karesh *et al.* (2005) recommended that focusing efforts on regulating, reducing and eliminating wildlife markets could provide a cost-effective way to reduce the spread of zoonotic diseases.

6.3 Medicinal and other uses

Not all capture of wildlife is for the pet trade. Many animal and plant products form components of traditional medicine dating back more than 5000 years. Approximately 900 animal species harvested for Chinese medicine may already be responsible for the depletion of wild populations of species such as tigers,

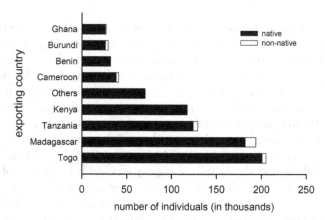

Figure 6.16 Number of native (filled bars) and non-native (unfilled bars) chameleons exported between 1977 and 2001 from African countries. (Data derived from Carpenter *et al.* 2004.)

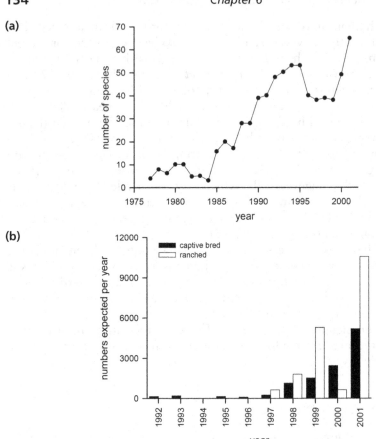

Figure 6.17 (a) Rapid increase in the number of chameleon species traded from 1977 to 2001 and (b) quantity of captive chameleons exported in the past decade. (After Carpenter *et al.* 2004. Copyright, Elsevier.)

bears, rhinos and swiftlets (Lever 2004). This overharvest is exemplified by the Sumatran tiger (*Panthera tigris sumatrae*), whose body parts (e.g. bones and penis) are used in traditional Asian medicine. The continued medicine-related demand for this species is extremely worrying because it is now listed by the IUCN as critically endangered, with fewer than 500 individuals thought to remain in the wild (Seidensticker *et al.* 1999). Although Sumatran tigers enjoy full legal protection, with poaching attracting penalties of imprisonment and heavy fines, a survey in 2002 revealed a continuing substantial commerce in tiger parts and products in Sumatra (C.R. Shepherd and Magnus 2004). Of 24 towns and cities surveyed, 17 (71%) had shops selling tiger parts, with a quarter of 117 shops and dealers located therein having tiger parts for sale. Most tigers were killed by professionals or semiprofessional hunters using inexpensive and simple-to-use wire-cable snares. Although it could be argued that at least some tigers were killed legitimately because they attacked humans or livestock (see section 6.5), most were taken for commercial purposes (78% of 51 estimated tigers killed per year in Sumatra). Because there is no programme in place to compensate local people for loss of livestock or life due to tiger attack, the illegal sale of tigers killed after they have come into conflict with humans remains the only way for people to recuperate their losses (C.R. Shepherd and Magnus 2004).

Between 1975 and 1992, South Korea imported an astounding 6128 kg of tiger bones, an average of 340 kg per year (J.A. Mills 1993). The majority (61%) of these imported bones originated from Indonesia. In general, economics is postulated to govern hunting effort (J.M. Robinson 2001), so, intuitively, the scarcer a species, the more expensive it becomes (Brook and Sodhi 2006; Courchamp *et al.* 2006). Therefore, scarcity tends to enhance the incentives for poaching and creates a vicious feedback loop that may ultimately end in the harvested species' demise. This scenario seems to be the case for tiger bone exports; there is a strong negative correlation between the price of bones imported into South Korea from Indonesia and their total weight (Figure 6.18). C.R. Shepherd and Magnus (2004) recommended that better protection of wildlife reserves for tigers, public education, proper law enforcement and addressing social issues (e.g. alternative employment/livelihoods for traditional people) hold the only hope of curtailing tiger hunting in Sumatra and will aid in the recovery of this magnificent animal.

Like tiger body parts, rhino horn has been used in Chinese medicine since at least 2600 BC, with particularly heavy trading occurring during the T'ang Dynasty (618–907 AD), and has led to the near extinction of the Javan (*Rhinoceros sondaicus*) and Sumatran rhinos (*Dicerorhinus sumatrensis*) (Schafer 1963; Nowell *et al.* 1992). Despite the fact that the medicinal properties of rhino horns are questionable (Lever 2004), the continued supply of rhino horn via poaching, coupled with habitat loss, has driven rhinos to the brink of extinction in Southeast Asia. Although protected by law in both Indonesia and Malaysia, legislation has been largely ineffective in curbing the poaching of Sumatran rhinos (Rabinowitz 1995). Inadequate monitoring and subsequent protection in key areas, and futile efforts at captive breeding, have all failed to halt the decline. The inability of responsible national governments and international funding agencies to take

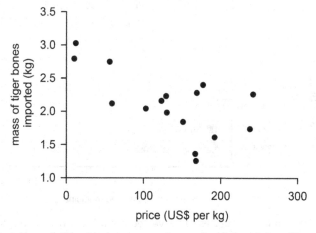

Figure 6.18 Negative relationship between price and weight of tiger bones imported into South Korea from Indonesia. (Data derived from Shepherd and Magnus 2004.)

tangible steps towards enforcement are likely to make extinction a reality for this iconic species (Rabinowitz 1995).

Similar predicaments exist for other animals as well. Nijman (2005) reported up to an 80% decline in Hose's langur (*Presbytis hosei*) in East Kalimantan (Indonesia) from 1996 to 2003. This sharp decline was primarily caused by hunting for bezoar stones (visceral excretions), which are used in traditional medicine. In fact, Nijman (2005) reported that a merchant guaranteeing purchase of bezoar stones was instrumental in causing excessive hunting of the species.

Farming of species used in traditional medicine and finding alternatives to the use of animal parts of threatened species are two possible alternatives to hunting that may reduce the pressure on precarious wild populations (Mainka and Mills 1995; von Hippel and von Hippel 2002; see also section 6.4). However, if farming is undertaken with this aim, efforts must be established to monitor animal welfare. Further, public education and proper wildlife protection schemes are also required for the effective alleviation of poaching pressure for medicinal use on the threatened wildlife species of the region.

In addition to animals, medicinal plants are regularly used by various people throughout the tropics. It is estimated that 80% people worldwide use medicinal plants (Bermudez *et al.* 2005), with rain forest plants responsible for about a quarter of the drugs used in Western medicine (www.mongabay.com). However, traditional medicine remains the only source of healthcare for

(a)

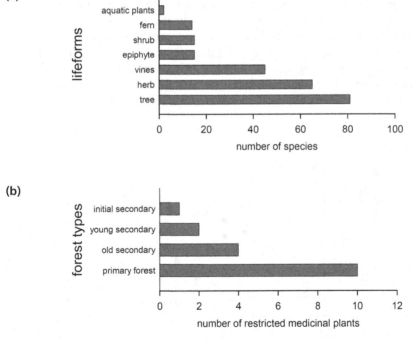

(b)

Figure 6.19 (a) Number of species of harvested plant life forms with medicinal purposes and (b) their corresponding forest types. (Data derived from Caniago and Siebert 1998.)

many local communities. In a village in Kalimantan (Indonesia), 237 medicinal plants were recorded by Caniago and Siebert (1998) as being exploited for such purposes, of which tree species (81) were the most frequently used (Figure 6.19). Compared with other forest types, primary forests had the highest number of exclusive medicinal species (Figure 6.19). There may be few alternative species to medicinal plants found only in primary forest, so the loss of this habitat type through logging and other human activities (see Chapter 1) rather paradoxically poses a grave challenge to their ongoing use for traditional medicine across the tropics and the rest of the world (Caniago and Siebert 1998; Giday *et al.* 2003; Bermudez *et al.* 2005; see also Chapter 2).

6.4   Commercial exploitation

6.4.1   Wildlife products

Overexploitation of animal populations in tropical regions is not always driven by a need for meat or medicines. For instance, over 500 000 wild reticulated pythons (*Python reticulatus*) are harvested each year for the snake leather industry, which places a high value on their attractive skins for luxury goods (Groombridge and Luxmoore 1991) (Figure 6.20). Shine *et al.* (1999) collected information on the sizes, sexes, reproductive status and diet of 784 slaughtered pythons in Sumatra and concluded that because of intrinsic features of their biology (e.g. rapid growth, early maturation, high fecundity and their ability to evade detection), even relatively high levels of hunting are unlikely to cause their extirpation from Sumatra. However, because pythons also act to keep rodent populations in check, harvests that lead to a decline in python densities may have the unwanted side-effect of concomitantly increasing damage to crops caused by their rodent prey (Shine *et al.* 1999).

Figure 6.20   Harvest of wild reticulated pythons. (After Shine *et al.* 1999. Copyright, Elsevier.)

Prior to the introduction of plastics, elephant ivory was widely used for billiard balls, piano keys, buttons and ornamental items, and the trade in ivory is widely recognized as the single most important cause of substantial decline in elephant populations (Stiles 2004). Recognizing this fact, CITES banned the international trade in Asian (*Elephant maximus*) and African (*Loxodonta africana*) elephants by listing them on Appendix I in 1973 and 1989, respectively. Considerable research has been done to identify genetically the origin of ivory (Wasser *et al.* 2004) so that enforcement agencies can target illegal sources. However, as elephant poaching has continued (Figure 6.21), there have been concerns that the ban is counterproductive. Many southern African countries have argued that their effective wildlife management policies have resulted in an increase in elephant populations to a point at which sustainable offtake for the ivory trade is now feasible. They argue that they should be legally allowed to sell ivory and use some of the generated funds for conservation. Populations have indeed increased substantially in some areas, but the established reserve system in Africa, which effectively limits migration to within confined boundaries (Homewood 2004), has also resulted in elephant populations being genetically bottlenecked and achieving densities leading to destruction of the very habitats that support them. The corollary of this phenomenon is that many elephants die slowly and painfully of starvation (Neumann 2001).

Many groups have complained that failing to involve local people in the sustainable use, management and protection of indigenous wildlife such as elephants results in extensive poaching and clashes with reserve authorities (Homewood 2004). CITES succumbed to this pressure by allowing a limited auction of existing ivory stores, but the CITES general ban remains in place (Maguire 1996). With the debate on the ivory ban still raging, Homewood (2004) has argued that unless there is more acceptance of local community custodianship in nature conservation (including the opportunity to benefit financially and legally from the wildlife that occupies their shared lands), there will be little

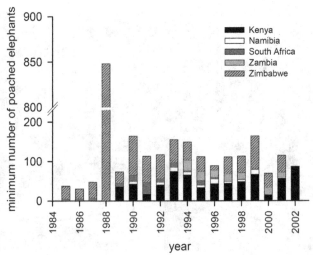

Figure 6.21 Persistence of elephant poaching in African countries. (Data derived from Stiles 2004.)

public desire to preserve what the international conservation community values the most – the wildlife itself. Indeed, the recommendations by Bowen-Jones and Entwistle (2002) that successful flagship species represent those that are valued culturally by local people instead of reviled as a non-useable resource is most applicable to the African elephant situation (indeed, *L. africanus* is designated as a WWF flagship species – www.panda.org) where preservationist philosophy ends up being counterproductive to conservation goals. However, some form of intensive control on future ivory trade is essential and will be effective in maintaining international conservation and local values of wildlife only if accompanied by appropriate public education to prevent a second wave of mass elephant slaughter (Stiles 2004; R.T. Naylor 2005).

A number of organizations have been established with the aim of reducing the illicit wildlife trade and protecting endangered biotas. TRAFFIC [the Worldwide Fund for Nature–World Conservation Union (WWF-IUCN) wildlife trade monitoring network] was established to ensure that wild resources extracted by humans are not threatened by unsustainable trade (www.traffic.org). Similarly, CITES, an international regulating body covering most of the countries in the world, was created in response to declining wildlife populations suffering from overexploitation for international trade (www.cites.org; see also Chapter 10). CITES lists wildlife threatened by international trade in three appendices in decreasing order of conservation urgency. Species listed in Appendix I are threatened with extinction and their trade is allowed only under exceptional circumstances. Species listed in Appendix II are not necessarily threatened but trade is controlled to minimize any detrimental effects on their recovery or long-term persistence. Species listed in Appendix III are protected in at least one member country, with the member country seeking assistance from other countries to control trade.

Species listed in Appendix I are accorded an international trade ban. However, trade bans can have mixed effects on poaching (see above): although they can reduce international demand and official production, black markets can still encourage poachers to continue illegal harvesting (Heltberg 2001). Wild animal numbers, demographics and social effects should be carefully considered before banning trade in specific species (Jepson *et al.* 2001). Sustainable harvesting is certainly possible, with maximum allowable offtake targets calculated using information such as population size, habitat requirements, resilience to human disturbance, demographics (e.g. age-specific mortality) and factors affecting population fluctuations (e.g. weather). Such biological information for most of harvested species has been lacking (Beissinger 2001); however, in many instances simple abundance or demographic data can provide precautionary harvest targets to wildlife managers (Boot and Gullison 1995; Fryxell *et al.* 2001; Brook and Whitehead 2005; Bradshaw *et al.* 2006b).

Perhaps one of the best examples of the success of CITES legislation to alleviate poaching pressure is the case of the estuarine crocodile (*Crocodylus porosus*) in northern Australia. Intensive commercial hunting and poaching of wild animals for the skin trade began in Australia in 1945 and resulted in the rapid depletion of crocodile numbers across northern Australia (Messel and Vorlicek 1986). This predominantly uncontrolled exploitation continued until the early 1970s when an export ban was imposed and full legal protection was established in 1972 (Messel and Vorlicek 1986). Subsequent transfer of the species to CITES Appendix I, then II, saw the establishment of intensive crocodile farming ('ranching'; Figure 6.22) and a reduction in wild harvest (Webb *et al.* 1984). These actions effectively flooded the international market with carefully monitored skins supplied by ranching operations, thereby removing the demand for wild-caught skins (Webb and Manolis 1993). As a result, crocodile populations across most of northern Australia have recovered to the point where they could even sustain some well-regulated wild harvest (Bradshaw *et al.* 2006b; Figure 6.23).

Figure 6.22   Estuarine crocodiles (*Crocodylus porosus*) farmed in northern Australia. (Photo by Grahame Webb, Wildlife Management International.)

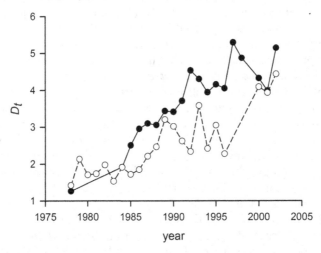

Figure 6.23   Recovery of estuarine crocodiles (*Crocodylus porosus*) in two rivers in the Northern Territory of Australia. D_t = the linear density of crocodiles observed per kilometre of river surveyed at year t. (After Bradshaw *et al.* 2006a. Copyright, Ecological Society of America.)

While the conservation issues surrounding tropical commercial fisheries are discussed in detail in Chapter 8, several examples of failures and successes in the commercial exploitation of some iconic tropical aquatic species are discussed here. We focus on several examples of some well-known tropical marine taxa that have, in some areas, been exploited excessively for commercial purposes – marine turtles, sharks and seahorses.

Most populations of marine turtles around the world have been depleted to the point of local extirpation (Spotila *et al.* 2000; Jackson *et al.* 2001; Hays 2004), resulting in the listing of six [loggerhead (*Caretta caretta*), green (*Chelonia mydas*), leatherback (*Dermochelys coriacea*), hawksbill (*Eretmochelys imbricata*], Kemp's ridley (*Lepidochelys kempii*) and olive ridley (*L. olivacea*)] of the seven extant species (the flatback turtle, *Natator dupressus*, is listed as 'data deficient') as endangered or critically endangered by the World Conservation Union (IUCN 2005). Whereas species such as the leatherback turtle are feared to be declining as a result of incidental capture by pelagic fisheries (Spotila *et al.* 2000), others (mainly green and hawksbill turtles) have been exploited historically for meat, eggs and carapaces (National Research Council 1990). The Caribbean is probably the region that has experienced the most cataclysmic declines in species such as green and hawksbill turtles. Jackson *et al.* (2001) estimated that the region's green turtle population, which formerly numbered in the tens of millions, has been reduced by 93–97% since the arrival of Europeans in the fifteenth century. McClenachan *et al.* (2006) demonstrated that 20% of historic nesting sites have been lost entirely during the past millennium, and 50% of the remaining sites have been extensively depleted.

Despite these worrying declines, some populations appear to be locally abundant, prompting calls for delisting of certain populations and species from the World Conservation Union's *Red List of Threatened Species* (Mrosovsky 2000, 2003), and reinitiation of sustainable harvest for the international trade of products such as hawksbill carapacial scutes (the so-called 'tortoiseshell' or 'bekko', valued as raw material by artisans) (Mrosovsky 2000). However, others have identified flaws in this reasoning (J.G. Robinson and Thorbjarnarson 2000), and the international community generally supports a continued moratorium on the trade of hawksbill scutes based on insufficient evidence of recovery (Heppell and Crowder 1996).

Although marine turtles have conventionally been regarded as particularly sensitive to overexploitation given their low fecundity, late maturation and long generation time (National Research Council 1990; Jackson 2001), many populations are now increasing with the cessation of heavy exploitation and the implementation of active conservation management programmes (Hays 2004). For example, green turtles in Hawaii are now recovering following the introduction of conservation measures in the 1970s (Balazs and Chaloupka 2004), and other populations in the Caribbean, Australia and Ascension Island are showing signs of positive trends (Chaloupka and Limpus 2001; Godley *et al.* 2001; Tröeng and Rankin 2005). Sustainable harvest of marine turtle eggs in Costa Rica has also been touted as one reason for longer-term population stability of some nesting sites through community support based on the economic incentive of keeping the productive nesting population intact (Campbell 1998).

Sharks are also known to be particularly vulnerable to overfishing (T.I. Walker 1998; Musick *et al.* 2000) due to their relatively low reproductive rates and poor recovery potential (S.E. Smith *et al.* 1998; Frisk *et al.* 2001; Cortés 2002; Otway *et al.* 2004). Not only does overexploitation affect shark abundance, but there is evidence that the selective removal of top marine predators such as sharks results in food web collapse driven by the strong feedback mechanisms between predators and prey (Stevens *et al.* 2000; Bascompte *et al.* 2005). Of particular concern is that, with the depletion of many commercially exploitable fish species (Pauly *et al.* 1998), sharks have generally suffered increasing exploitation in recent decades from by-catch in pelagic longline and directed commercial and artisanal fisheries (Baum *et al.* 2003b). A major study by Baum *et al.* (2003b) analysing logbook data for fisheries within the western Atlantic (including the Gulf of Mexico and regions offshore from northern South America) found that most recorded shark species (coastal and pelagic) had declined by more than 50% in the past 8–15 years. Baum and Myers (2004) also estimated that oceanic whitetip (*Carcharhinus longimanus*) and silky (*Carcharhinus falciformis*) sharks in the Gulf of Mexico had declined by over 99% and 90%, respectively, even though these species were formerly the most commonly caught.

Perhaps a greater concern is the increasing demand for shark fin in countries with large ethnic Chinese populations. Shark fin is one of the most valuable food items in the world (a 25-cm-long dried caudal fin retailed for US$415 in 1998 – Fong and Anderson 2000), and an increasingly affluent society in many parts of Asia has resulted in a much greater demand for the product in recent years (Fong and Anderson 2002; Clarke 2004). These factors make it a particularly lucrative enterprise for illegal, unregulated and unreported (IUU) fishing (Shivji *et al.* 2005). Because of chronic under-reporting of total take (Clarke 2004), misidentification of traded species (Clarke *et al.* 2004, 2006a; Abercrombie *et al.* 2005) and the general susceptibility of sharks to overexploitation, this remains an important conservation concern. There is particular concern that the fishing industry represents one of the major threats to sharks in the tropics and elsewhere (Fong and Anderson 2002; Clarke *et al.* 2006a). This concern is even held by the major importers and processors of shark fin in Hong Kong (the centre of trade and consumption of shark fin) that over-harvesting may compromise the industry (Fong and Anderson 2000).

Seahorses (*Hippocampus* spp.) are another marine taxon that is highly valued by Chinese medicine for its curative and tonic properties, and seahorses are also popular as aquarium pets and curios (Foster and Vincent 2005). As such, many seahorse populations are threatened by direct over-fishing, incidental catch in trawl nets and habitat destruction (Vincent 1996; Foster and Vincent 2005). In contrast to sharks, which are at risk largely from broad-scale directed commercial and illegal fishing ventures, many seahorses are exploited by some of the world's poorest fishers, who make their living by selling their catches to brokers for the Chinese market (Vincent 2006). This industry, in combination with a large incidental catch (Vincent 1996; Baum *et al.* 2003a), accounts for approximately 20 million dried seahorses traded annually, representing one of the largest volume trades governed under CITES (Foster and Vincent 2005).

There is good evidence that the international trade has depleted many seahorse populations around the world, with lower catch rates occurring in spite of increasing effort in recent years (Foster and Vincent 2005). The *H. comes* artisanal fishery in the Philippines, which comprises nearly 200 fishers spread along 150 km of coastline, is responsible for depleting local populations and endangering the fishers' own livelihood (Martin-Smith *et al.* 2004). However, concerted efforts to create no-take marine protected areas (MPAs; see Chapter 7) and implement minimum size limits are largely supported by the industry and are considered the most efficient means to continue sustainable harvests of seahorses (Martin-Smith *et al.* 2004; Foster and Vincent 2005). Likewise, substantial declines in seahorse abundance resulting largely from incidental catches in shrimp trawl fisheries have been reported in Latin America (Baum and Vincent 2005). The seahorse fishery in East Africa is currently relatively small, but there is anecdotal evidence of some local declines that will require close attention and mitigating management policy to avoid the overexploitation observed in other regions (McPherson and Vincent 2004).

6.4.2 Non-timber forest products

Forested lands set aside as extractive reserves are generally seen as promising conservation strategies because they allow for the profitable harvest of 'sustainable' non-timber forest products such as latexes, fruits, nuts and other commercially important plants while maintaining forested habitats (Fearnside 1989; Moegenburg and Levey 2002). Indeed, over 50 million people in India alone are believed to be dependent on non-timber forest products for their subsistence (Shaanker *et al.* 2004). Extractive forest reserves are generally thought to provide long-term economic benefits in excess of those achieved through timber extraction and conversion to pasture (Salafsky *et al.* 1993); however, the effects of product enrichment and extraction on biodiversity are largely assumed or based only on theoretical models (Moegenburg and Levey 2002). A good example is the extraction of Brazil nuts (*Bertholletia excelsa*), which are rich in nutrients, including protein, fibre, selenium, magnesium, phosphorus and thiamine. Brazil nuts are also considered to contain important antioxidant compounds, which are believed to be protective against both coronary disease and cancer. Brazil nuts taste sweet, and their texture is similar to that of coconut. These nuts can be eaten raw or roasted, and in various desserts such as ice-cream and baked goods. Over 45 000 tonnes of Brazil nuts are harvested in the Brazilian Amazon alone, contributing US$33 million to local and regional commerce (Clay 1997).

There is mounting empirical evidence that non-timber forest product enrichment and extraction can negatively affect biodiversity, even though these impacts may be far less than those resulting from more destructive conversion of forest landscapes (Trauernicht and Ticktin 2005). Peres *et al.* (2003) found that exploitation of the Brazil nut has been unsustainable in the Amazon because populations subjected to persistent harvest lacked a sufficient number of juvenile trees [< 60 cm in diameter at breast height (dbh)] to sustain the population over the long term. However, Zuidema and Boot (2002) concluded that Brazil nut harvest in the Bolivian Amazon appears to be sustainable over decadal time scales.

Palms are used by local people for food and beverages. Plantations of the palm *Chamaedorea hooperiana* in old-growth tropical rain forests of southern Mexico affected plant community composition and structure (Trauernicht and Ticktin 2005). These enrichment activities resulted in lower stem density, species richness and basal area of woody species in smaller size classes (< 10 cm dbh). Plantation sites also reduced the density of other species, such as the mid-storey palm *Astrocaryum mexicanum*. Moegenburg and Levey (2002) also demonstrated that intensive palm fruit extraction and enrichment rapidly reduced bird diversity and changed the bird community to one that was dominated by frugivorous species.

Despite these examples of the ecological costs associated with non-timber forest product extraction and enhancement, there is great prospect for managing such industries to maintain long-term sustainability and maintain biodiversity. Shaanker *et al.* (2004) recommended that focus be put on reducing the ecological cost of dependence on these products. They suggested that the adoption of ecologically friendly methods of harvesting, monitoring of harvest trends, aligning property rights with incentives for long-term management, the formulation of working plans, semi-domestication and non-exploitative harvest regimes with respect to the collectors themselves are all potentially effective ways of reducing the ecological footprint of harvest.

6.5   Nuisance control

The reliance of carnivores on other animals for food has brought many large predators into direct conflict with humans, especially where native prey have been replaced by domesticated livestock or when some carnivores target humans as prey (Saberwal *et al.* 1994; Patterson *et al.* 2004). For example, the lion (*Panthera leo*) has suffered perhaps some of the greatest declines in range and population size of any major extant terrestrial carnivore (Patterson *et al.* 2004). Prior to human colonization of the western hemisphere, lion distribution included areas from southern Africa to northern Europe, most of Asia and North America, and ranged as far south as Peru in South America (Kurtén and Anderson 1980; Turner and Antón 1997). Today, however, lions persist in only a small proportion of this former range as a result of habitat loss, reductions in prey density and direct persecution by humans (Patterson *et al.* 2004). A recent study in Kenya determined the extent of depredation of domestic stock by lions and the associated economic costs (Patterson *et al.* 2004). Lions were responsible for 86% of all recorded predator attacks on livestock (mainly cattle), which represented 2.4% of range stock and cost ranchers approximately US$290 per lion kill. It is for these tangible economic losses that many ranchers and herders in East Africa are now intensifying the snaring or poisoning of predators such as lions (Patterson 2004).

Tigers represent another well-known example of predators persecuted by humans who fear attack and economic losses of domesticated livestock. The Sumatran tiger (*Panthera tigris sumatrae*) is now classed as critically endangered due to a number of factors – the trade in its bones for their perceived benefits in Asian medicines represents perhaps the greatest threat (C.R. Shepherd and

Magnus 2004; see section 6.3). In addition to the direct poaching of Sumatran tigers for their economic value, there is also a long history of human–tiger conflict in the region, with many people having been killed or wounded by tigers, and tigers having taken livestock. These threats lead many villagers to support the eradication of 'problem' tigers, even though live trapping and removal options are often available (C.R. Shepherd and Magnus 2004). As such, the multitude of threats faced by this enigmatic species paints a bleak picture for its future persistence in Southeast Asia. There are other examples of animals, e.g. elephants, being frequently captured and killed as a result of conflict with humans, especially in agricultural areas (Sukumar 2003).

Finally, we describe another example of a marine predator facing persecution by humans. The use of shark nets ('shark meshing') to reduce the probability of shark attacks at popular swimming beaches throughout the world was pioneered in Australia in the 1930s and has been touted as a successful tool in saving human lives. Despite these obvious benefits, there have been repeated calls to examine the impact of such devices on local populations of sharks and other marine species (Paterson 1979, 1990; D.D. Reid and Krogh 1992; Krogh and Reid 1996; Paxton 2003). While some studies have not observed declines in the relative abundance of targeted species (e.g. Simpendorfer 1992), a large study in South Africa indicated that catch rates of four shark species declined over more than 20 years (Dudley and Simpendorfer 2006), suggesting the potential for depletion of local populations at least for some species.

6.6 Summary

1   Wildlife (bush meat) hunting for sustenance and cash income is probably depleting populations of some tropical animals and is causing a crisis not only for biodiversity, but also for local people who rely on these for food. Deeper biological as well as sociological understanding should be used to find solutions to this dangerous predicament.
2   Capture of wildlife for the pet trade and medicinal uses has also endangered some tropical animals and plants. Sustainable harvesting and farming should be pursued in some cases to promote wildlife recovery.

6.7 Further reading

Grigg, G. C., Hale, P. and Lunney, D. (1995) *Conservation through Sustainable Use of Wildlife*. Centre for Conservation Biology, University of Queensland, Brisbane.
Milner-Gulland, E. J. and Bennett, E. L. (2003) Wild meat: The bigger picture. *Trends in Ecology and Evolution* **18**, 351–357.
Stiles, D. (2004) The ivory trade and elephant conservation. *Environmental Conservation* **31**, 309–321.

7

Threats in Three Dimensions: Tropical Aquatic Conservation

In the not-too-distant past, many people regarded the seas and large freshwater ecosystems throughout the world as limitless suppliers of resources for human exploitation (Roberts 2003). However, the rapidly expanding human population has put immense pressure on the resources these habitats support through the direct exploitation of species and the extraction of freshwater for consumption. These threats to aquatic environments mirror, and sometimes exceed, the conservation crises observed in terrestrial systems, and they are particularly acute in tropical areas given high human populations, intense coastal development and the lack of infrastructure and finances to deal with intense human pressures. So far, we have described many of the processes threatening biodiversity in the tropics. However, many unique characteristics of aquatic biodiversity and the particular conservation challenges they face deserve a dedicated discussion. Here we describe some of the major conservation issues facing marine, freshwater and estuarine systems throughout the tropics.

7.1 Tropical fisheries exploitation

7.1.1 Commercial fisheries

The human population explosion over the last century has resulted in huge pressure on the planet for increased food production, with the quest for cheap sources of protein perhaps having some of the most profound effects on biodiversity (e.g. clearing forests for agriculture – Chapters 1 and 3; and the rise in the bushmeat trade – Chapter 6). Of particular concern over the last few decades is the overexploitation of fish and marine invertebrates driven by open-access fishing policies and overcapitalization supported by government subsidies (Garcia and Newton 1997; Pauly *et al.* 2002; Gewin 2004). Indeed, over the last 50 years there has been a large downward shift in the average size and trophic guild of species caught in fisheries around the world, providing evidence that many of the world's fisheries are unsustainable in the long term (Pauly *et al.* 1998). The idea that removing predation pressure from higher trophic-level species allows larger species to recover after fisheries targets have changed (Daskalov 2002) is a fallacy given the complex and finely meshed food webs in which targeted species live (Pace *et al.* 1999, 2002). Indeed, overexploitation can threaten fish species with extinction in spite of their generally high fecundity and the perceived limitlessness of available habitat (Casey and Myers 1998; Hutchings 2000; Pauly *et al.* 2002; Dulvy *et al.* 2003; Kappel 2005). For example, tropical groupers (subfamily Epinephelinae of the Serranidae), once thought to be unlikely candidates for extinction given their high fecundity, large range size and dispersal capacity, contain a high proportion of threatened species across the tropics (Morris *et al.* 2000). It is believed that the rapid intensification of fisheries in the tropics using habitat-destroying gear is the main culprit threatening this taxon (Morris *et al.* 2000) and most likely many others.

Most of the evidence for these worrying trends, along with the associated scientific and management foci, has come from temperate systems, mainly in the northern hemisphere, but there is comparatively little scientific interest in fisheries biology and management in the tropics (Albaret and Lae 2003). However, based on marine fisheries data from the United Nations Food and Agriculture Organization (FAO) Fisheries Areas landings statistics, Pauly *et al.* (1998) found some ambiguous results for much of the tropical region. The Central Eastern Pacific, Southern and Central Eastern Atlantic and the Indo-Pacific showed no clear trends in the mean trophic level caught from 1950 to the late 1990s (Figure 7.1). However, some of these data may mask declines seen elsewhere given the relatively recent development of new tropical fisheries targeting higher trophic levels, and for some regions the data are inadequate (Pauly *et al.* 1998). Indeed, there is considerable under- and misreporting of fisheries landings data, especially from artisanal fisheries (small-scale fisheries for sustenance and/or commercial

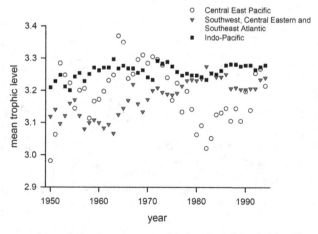

Figure 7.1 Temporal trends in the mean trophic level of fisheries landings in tropical marine areas (and surrounding waters) in Food and Agriculture Organisation Fisheries Areas. [After Pauly *et al.* 1998. Copyright, American Association for the Advancement of Science (AAAS).]

purposes) in poorer tropical countries (Pauly *et al.* 1998; Caddy and Garibaldi 2000), so broad-scale trends may not highlight local depletions. Furthermore, it is only relatively recently that mass commercial exploitation of traditionally artisanal fisheries areas off West Africa, in Southeast Asia and elsewhere have been established (Caddy and Garibaldi 2000; Pauly *et al.* 2002), resulting in the general trend of increasing landings across most tropical fisheries (Figure 7.2) (Caddy and Garibaldi 2000).

There have been some well-publicized commercial fisheries problems in the tropics, some with global economic implications. The massive fishery targeting the Peruvian anchoveta (*Engraulis ringens*) off the eastern coast of South America (estimated catches once exceeding 18 million tonnes annually) collapsed spectacularly in 1971–72. The collapse was originally attributed to a severe El Niño event (Brainard and McLain 1987), but it has been now accepted that much of the collapse was driven by overfishing as well (Pauly *et al.* 2002).

The Lake Victoria freshwater fishery in East Africa has had an interesting and eclectic history of human modifications and overexploitation. Perhaps one of the best-known examples of freshwater fish endemism in the world (Matsuishi *et al.* 2006), the fish community of Lake Victoria was seriously compromised in the 1950s with the introduction of the Nile perch (*Lates niloticus*), resulting in the estimated loss of some 200–400 endemic species, which compromised about 90% of the lake's fish biomass (Witte *et al.* 1992; Matsuishi *et al.* 2006; see also Chapter 5). This dramatic change to the fish community resulted in a shift of the local fisheries from one dominated by tilapia (*Tilapia* spp.) to one comprising mainly Nile perch (Matsuishi *et al.* 2006). However, a sustained and rapid expansion of the number of fishers has resulted in a fishery no longer dominated by Nile perch, and there are serious concerns about the degradation of the entire fishery of the lake (Matsuishi *et al.* 2006). Many analogous issues face fisheries management and sustainability goals in Africa's other great lakes such as Lake Tanganyika (Mölsä *et al.* 1999).

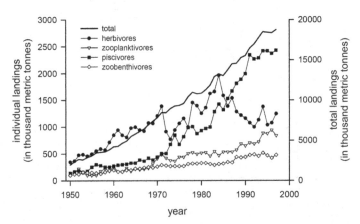

Figure 7.2 Total landings (right axis) and totals of the four trophic fish categories (left axis) – piscivores, zooplanktivores, zoobenthivores, herbivores – in tropical fisheries areas (Food and Agriculture Organisation Areas 31, 51, 57, 71). (After Caddy and Garibaldi 2000. Copyright, Elsevier.)

7.1.2   Artisanal fisheries

Many people living in tropical countries depend on fishing to provide some, if not the majority, of their protein (Figure 7.3). Indeed, Indonesia alone is estimated to have 1.3 million fishers (Gewin 2004), most of them artisanal. In the past it was assumed that increasing pressure from artisanal fishers was unlikely to have any significant impact on local fish biodiversity and abundance. Good data are still comparatively rare (Johannes 1998a), but more studies are focusing on quantifying the impacts of artisanal fisheries on marine and freshwater ecosystems. Despite the global importance of artisanal fisheries, stocks are perhaps most severely depleted in tropical nearshore fisheries, which are particularly intractable to effective management (Johannes 1998a).

In the mid-1980s, the traditional grouper fishery of Palau collapsed

Figure 7.3   Artisanal fishers. (Photos by Arvin Diesmos.)

precipitously as a result of overfishing (Sadovy 1994; Johannes 1998a). Another artisanal fishery in the Gulf of Mexico targeting white shrimp (*Penaeus setiferus*) has been shown to compete with commercial fisheries for the same species and result in overexploitation of the stock (Gracia 1996). The *P. setiferus* fishery is one of the three most important commercial fisheries in the Gulf of Mexico. The offshore fishery coexisted with an inshore artisanal fishery until the latter was banned in 1974. A new drift-net artisanal fishery operating in nearshore waters began in 1983 and quickly became highly profitable as a result of low investments and high revenue, resulting in artisanal landings equivalent to, or even greater than, the commercial catch. The combined fishing pressure of the two industries, coupled with the fact that only incomplete and poor landing statistics necessary for managing the fishery were available for the artisanal fishery, resulted in overexploitation and near-collapse of the entire industry in the early 1990s (Gracia 1996).

S. Jennings *et al.* (1999) examined artisanal fishing impacts in Fiji and found that the abundance of many exploited reef fishes [parrotfishes (family Scaridae), groupers (Epinephelinae) and snappers (Lutjanidae)] decreased as fishing intensity increased. Not only were decreases detected, but the vulnerability of exploited species appeared to be related to fish size. The grouper and snapper species that declined in abundance had greater maximum sizes than close relatives that did not decrease in abundance (S. Jennings *et al.* 1999). The investigators concluded that in the case of some of the more heavily exploited species, fishing, and not environmental variation, was the key determinant of their dynamics and structure. Similar declines in the artisanal catches of fish and shellfish in the lagoons of Samoa have also been reported (King and Faasili 1999). Likewise, artisanal fishing practices in the Galápagos Islands have resulted in local depletions of target species accompanied by a reduction in natural spatial variation in the fish communities (Ruttenberg 2001).

Despite the bleak prospects for traditional fisheries management in the tropics, there are some examples of potentially sustainable practices. Espino-Barr *et al.* (2002) examined the temporal trends in fisheries landings from a small (approximately 1000 boats) marine artisanal fishery off the Pacific coast of Mexico [landing over 140 species, with snapper species (*Lutjanus peru* and *L. guttatus*) being the most abundant] and although they found a decline in the average species diversity of landings over time (1983–98), they attributed the change to environmental factors rather than fisheries-related modifications of the fish community. However, they admitted that if cyclic changes in landings composition did not improve in the near future, fishing pressure might ultimately be involved.

A village-based management programme to promote the sustainable take of trochus shellfish (*Trochus niloticus*) in Vanuatu has demonstrated some progress towards effective self-management in a data-limited environment. Working with local government, the management programme worked on the precautionary principle (if an action or policy might be harmful to the environment, in the absence of a scientific consensus that harm would not occur, the burden of proof falls on those who advocate not taking the action) of harvesting stocks only once every 3 years (Johannes 1998a,b). News of the success in some villages resulted in more villages adopting the practice and applying it to other fished species such as finfish, lobsters and octopus (Johannes 1998a,b). Similar success stories have also been reported for village-based management in the Philippines (Bolido 2004; Rajamani *et al.* 2006).

7.1.3 Bycatch

In fisheries parlance, the term 'bycatch' refers to the incidental take of undesirable (i.e. commercially unviable or non-targeted) size or age classes of the target species or, more commonly, to the incidental landing of other non-target species (Lewison *et al.* 2004a). Non-target species can include invertebrates (Freese *et al.* 1999), sharks (Stevens *et al.* 2000; M.J. Barker and Schluessel 2005), turtles (Lewison *et al.* 2004b), sea snakes (Milton 2001), seabirds (Majluf *et al.* 2002; Furness 2003; Lewison and Crowder 2003) and marine mammals (Majluf *et al.* 2002; de Thoisy *et al.* 2003; Lewison *et al.* 2004a; H. Marsh *et al.* 2005; Read *et al.* 2006) (Figure 7.4). Bycaught species are not necessarily destroyed, and some are released unharmed; however, the majority of caught species are either injured or dead when released. Rough estimates of the size of the discarded bycatch (mostly fish) in the world's commercial fisheries exceed 27 million tonnes per year (Alverson *et al.* 1994; Pauly and Christensen 1995). Although there was some controversy surrounding that estimate (S.J. Hall and Mainprize 2005), subsequent revisions and more recent estimates suggest that the value has declined to approximately 20 million tonnes (FAO 1999; S.J. Hall and Mainprize 2005).

The massive growth of global fisheries in recent decades has led to an explosion of bycatch and although this problem has been an acknowledged component of fisheries management for some time, establishing bycatch mortality rates and their implications for non-target species have traditionally fallen outside the realm of typical management focus (Alverson and Hughes 1996). The issue of bycatch first came to the public eye through mass media coverage of the dolphin bycatch in

Figure 7.4 Bycatch (a manta ray and a leatherback turtle) from a fishery. (Photo by Hélène Petit, petit.helen@wanadoo.fr.)

the tuna purse-seine fishery in the eastern Pacific Ocean during the 1960s (S.J. Hall and Mainprize 2005). Annual incidental dolphin landings and subsequent mortality were estimated at several hundreds of thousands of individuals, and it is believed that the massive public outcry resulted in key legislation (e.g. the US Marine Mammal Act) that effectively curtailed the killings (M.A. Hall 1998; S.J. Hall and Mainprize 2005). More recently, however, the conservation implications for bycaught species have emerged as serious lines of investigation given the large volumes discarded and the general lack of good data for management purposes (Alverson and Hughes 1996). A recent assessment among marine taxa concluded that bycatch is an important contributor of extinction risk in the case of approximately 50% of the species listed as threatened under World Conservation Union (IUCN) criteria (Kappel 2005; Figure 7.5).

Tropical prawn trawling is perhaps one of the world's most destructive fisheries with respect to bycatch. It is the least selective of commercial fisheries, and total landings of non-target species outweigh those of commercially important prawns – the industry is responsible for a third of all fisheries discards worldwide (Saila 1983; Andrew and Pepperell 1992; Alverson et al. 1994; Stobutzki et al. 2001a). Another problem associated with this bycatch issue is that, in the case of most tropical prawn trawl fisheries, the incidental catch is highly diverse. This, combined with a general lack of historical data, makes quantitative stock assessments difficult (Stobutzki et al. 2001a). For example, in Australia's northern prawn fishery, the bycatch is dominated (73% of catch) by over 400 species of bony fish (teleosts) (Stobutzki et al. 2001b).

7.1.4   Illegal, unreported and unregulated fishing

IUU fishing, although of concern throughout the world, is causing particular unease in some biodiverse tropical regions. Indeed, it is estimated that IUU fishing

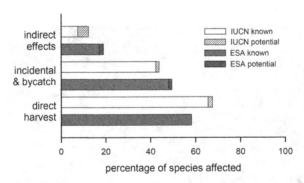

Figure 7.5   Relative importance of the top threat – overexploitation – to marine species at risk of extinction under World Conservation Union (IUCN) criteria (225 species) and the US Endangered Species Act (ESA – 168 species). The percentages of species affected by direct, targeted harvest versus incidental catch and bycatch, or indirect effects such as habitat degradation, competition for prey or trophic cascades are shown. (After Kappel 2005. Copyright, Ecological Society of America.)

exceeds permitted fishing quotas by 300% globally (Gewin 2004), a figure that holds some credence given the size of lucrative markets for products such shark fin. A recent analysis based on trade data and genetic information estimated that the total shark biomass destroyed for the fin trade is three- to fourfold higher than previous estimates (Clarke *et al.* 2006b). The importance of IUU fishing as a major threat to many of the world's tropical shark species was examined in Chapter 6, so here we elaborate on several other examples of IUU fishing threats elsewhere in the tropics.

Recognized as perhaps one of the world's greatest biological treasures, the Galápagos Islands' marine biodiversity has been of particular concern since at least the 1980s (Camhi 1995). In addition to a large IUU shark fishing operation for the Asian shark fin trade discovered in 1988 and 1991 (Camhi and Cook 1994), an industrial-scale IUU sea cucumber (*Holothuria* spp.) fishery began in 1992. After successful lobbying by local and international interests, commercial harvesting of several marine species, including sea cucumbers, was permitted. However, the large and immediate reduction in the local population, which led to an outcry from scientists and tourist interests, resulted in an almost immediate closure of the fishery. This escalated into a near-lethal stand-off between local fisherman, scientists and Ecuadorian authorities (Camhi 1995; Stone 1995), although blatant illegal fishing of this species continues (Camhi 1995).

Another burgeoning area of conflict and ecosystem collapse is the recent massive increase in the number of IUU fishing vessels entering the Exclusive Economic Zone (EEZ) of northern Australia in the Timor and Arafura Seas and the Gulf of Carpentaria. In 1974, Australia and Indonesia signed a Memorandum of Understand (MoU) to allow artisanal Indonesian fishermen to continue traditional fishing within a defined area within the Australian EEZ north of Broome, Western Australia (the so-called 'MoU Box'). In recent years, however, traditional Indonesian fishing has increasingly expanded beyond the MoU Box into Australian territorial waters (K. Saunders 1997), and there has been a noticeable shift away from 'traditional' fishing boats to more technologically advanced and industrial-scale IUU fishing vessels, which the Australian authorities perceive as

Figure 7.6 Illegal Indonesian fishing vessel apprehended by Australian authorities. (Photo by Mark Meekan, Australian Institute of Marine Science.)

an important threat to many commercial shark and fin fish fisheries managed locally (Figure 7.6). Research is currently progressing on this issue and will help to quantify the degree of apparent threat in the near future.

7.1.5 Tropical fisheries management

Tropical fish differ from their temperate counterparts in many ways, with complex repercussions for tropical fisheries management. Owing to their relatively warmer environment, tropical fish tend to have higher metabolic and growth rates, leading to higher food consumption rates. The corollary of this is that natural mortality also tends to be higher, and tropical fish generally have smaller asymptotic sizes (Pauly 1998). Tropical fish species also appear to occupy higher trophic levels than do colder-water fish of equivalent mass – a feature driven by the high species richness and connectivity of tropical fish communities (Pauly 1998). As a result, natural mortality rates also tend to be higher because their (mainly) fish predators (Christensen 1996) must consume more prey per unit time than their colder-water counterparts (Pauly 1998).

Unfortunately, most past fisheries management in tropical countries has relied on the temperate model of maximum sustainable yields within a centralized administrative framework (Pomeroy 1995). The differences in the sociopolitical systems between temperate and tropical countries and, more importantly, the large ecological differences in their fish communities have stimulated many fisheries scientists to adopt different approaches to modelling fisheries harvest to suit the peculiarities of tropical systems. For example, because most coldwater fish can be aged using the straightforward technique of reading annuli (rings) in the otoliths (modified ear bones), scales or other hard parts, age-classified models can easily be parameterized using this information. However, many of the initial attempts to age tropical fish species using scales and otoliths provided ambiguous age data (Munro 1983; Longhurst and Pauly 1987), which has engendered the widespread belief that ageing tropical fish is impossible in most cases (Choat and Robertson 2002). Even though the approach can be more difficult for many tropical species, the detection of consistent annuli can be demonstrated (Thorrold and Hare 2002). However, it is usually necessary to examine large samples of otoliths to achieve reliable data (Choat and Robertson 2002).

The main problems associated with using age–structure data to manage tropical fisheries are that (1) many species demonstrate a 'square' growth curve and (2) multispecies assemblages make age validation difficult. Rapid initial growth to asymptotic size followed by an extended lifespan with little somatic growth characterizes the typical growth curve for many tropical fish families (Choat and Robertson 2002). This phenomenon effectively decouples the relationship between age and size, making inference about age structure from length data impossible (or at least highly suspect) for many taxa. The high biodiversity of tropical reefs and the resultant multispecies takes of fisheries based on these systems mean that validating age data through the tagging and retrieval of marked individuals becomes extremely laborious (Choat and Robertson 2002). This is because the probability of retrieving sufficient numbers of marked individuals for any one species is extremely low given the broad range of species harvested (many of them considered to be rare). This is also complicated by the small scale of (largely artisanal) fisheries in the tropics, which often makes centralized management and policy enforcement intractable.

Additionally, tropical oceans exhibit an extremely varied suite of environments and communities, ranging from relatively species-depauperate expanses to ultraspecies-rich coral reefs (Longhurst and Pauly 1987; section 7.2). The variety of ecological systems is mirrored by the diversity of the socioeconomic and cultural conditions of the (mostly) developing countries surrounding tropical oceans (Pauly 1998). As a result, in tropical systems, 'classic' fisheries management based on modelling techniques designed for arguably more predictable temperate systems requires mathematical approaches adopting a higher level of complexity and parameterization (Pauly 1998).

It must be acknowledged, therefore, that models traditionally used to inform fisheries management policies (e.g. catch limits) are inherently prone to imprecision and error, especially in highly variable tropical ecosystems, given their reliance on explicit knowledge that is rarely available (Caddy 1986; Pauly *et al.* 2002). These uncertainties may require alternative models based on simpler data (e.g. Bradshaw *et al.* 2006b), and a dedicated acceptance and implementation of the precautionary principle (Harwood and Stokes 2003; Beddington and Kirkwood 2005) will lead to more sustainable practices. In addition, Pauly *et al.* (2002) recommended several other approaches that may help reverse the global declines in fisheries landings: (1) a huge reduction in fishing effort worldwide via the decommissioning of a large proportion of the contemporary fishing fleet; (2) closing sections of currently fished regions to prevent overexploitation; and (3) maintaining adequate enforcement of these regulations.

Another impetus for enhancing effective and sustainable management of tropical fisheries is the direct link between the supply of exploitable marine and freshwater fish populations and the conservation of terrestrial fauna. In contrast to more developed, temperate nations, where fluctuations in commercial fish supply do not overtly influence the supply and demand of terrestrial food production, there is strong evidence that reductions in fish supply result in a shift to hunting terrestrial wildlife to replace the required protein (Brasharcs *et al.* 2004; Wilkie *et al.* 2005; see Chapter 6).

7.2 Coral reefs in peril

Coral reefs represent some of the most productive ecosystems on the planet, and are certainly the most biodiverse and ecologically complex marine ecosystems (Sebens 1994; Reaser *et al.* 2000). This productivity results from a concentration of biodiversity in relatively restricted areas. Globally, the benefits of ecosystem and other services provided by coral reefs are numerous (e.g. fisheries production, shoreline protection, tourism, medicinal products, ecosystem indicators – Reaser *et al.* 2000; see also Chapter 2). Coral reefs harbour dynamic biological communities with highly stochastic recruitment patterns (Bellwood *et al.* 2006); however, coral reefs worldwide are declining (e.g. it has been established that a massive decline in coral reef area has occurred over the last 30 years in the

Caribbean – Figure 7.7; Gardner *et al.* 2003). Additionally, there is evidence for longer-term declines. An examination of the change in ecological status (pristineness) of coral reefs from pre-human to modern times has revealed sharp declines (Figure 7.8; Pandolfi *et al.* 2003). Pandolfi *et al.* (2003) reported that large reef species tend to decline faster than smaller species, and most guilds are now no longer in a pristine state in any of the world's coral reefs. Even the best-protected reefs on the Great Barrier Reef in Australia are 25–33% on their way towards ecological extinction, with some more southerly reefs in the Great Barrier Reef nearly as compromised as the badly degraded reefs of Panama and the Virgin Islands (Pandolfi *et al.* 2003). These results suggest that recent highly publicized events of coral bleaching (section 7.2.1) and outbreaks of disease (Harvell *et al.* 1999) are masking some of the more chronic declines due to overfishing (section 7.2.2).

7.2.1 Bleaching

Severe environmental stress, caused mainly by extreme and sustained fluctuations in water temperature, disrupts the delicate symbiotic balance that exists between the resident algae (zooxanthellae) and their coral animal hosts. Symbiotic algae provide much of the corals' nutrition and colour, so the loss of algae can lead to reductions in tissue growth, skeletal accretion and sexual reproduction (Glynn 1996; Reaser *et al.* 2000). This loss of algae and the associated colour change is commonly known as 'bleaching' and is a clear indication of widespread coral mortality (Figure 7.9). The bare skeleton of bleached corals succumbs to rapid

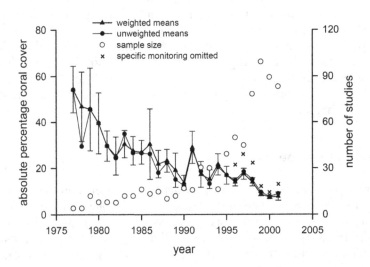

Figure 7.7 Absolute percentage coral cover across the Caribbean basin since the late 1970s. Also shown are unweighted means for each year, the unweighted means a specific monitoring programme omitted, and the sample size (number of studies) for each year. [After Gardner *et al.* 2003. Copyright, American Association for the Advancement of Science (AAAS).]

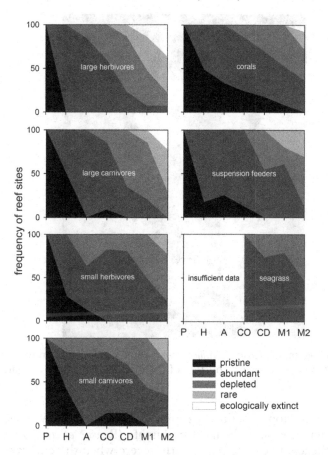

Figure 7.8 Change in the ecological status (pristine to ecologically extinct) of coral reef guilds (carnivores, herbivores, corals, suspension feeders and seagrasses) from prehuman to modern times from 14 regions in the tropical western Atlantic, Red Sea and Australia. Cultural periods: P, prehuman; H, hunter–gatherer; A, agricultural; CO, colonial occupation; CD, colonial development; M1, early modern; M2, late modern to present. [After Pandolfi *et al.* 2003. Copyright, American Association for the Advancement of Science (AAAS).]

colonization by seaweeds and other opportunistic colonizing organisms; however, without sustained production of coral larvae from nearby to renew the reef, slow degradation makes many reefs vulnerable to storm surges (Eakin 1996). Once reduced to bottom rubble by wave action, its complex biological community of fish and invertebrates cannot be supported, so fisheries stocks, protection of shoreline erosion and tourism industries are severely compromised (Reaser *et al.* 2000).

Large (≥1°C) and sustained (2–3 days) rises in sea temperature are usually enough to bring about some bleaching, but reefs generally survive and recover provided the higher temperatures do not last more than 1 month (Goreau and Hayes 1994). Other causes of bleaching include large fluctuations in salinity, bouts

Figure 7.9 Bleached coral (*Acropora millepora*) from Keppel Islands on the southern Great Barrier Reef, Australia. (Photo courtesy of the Australian Institute of Marine Science.)

of intense solar radiation, sustained exposure to air during low tides, lowered sea level, sedimentation and chemical pollutants (Reaser *et al.* 2000; see also section 7.2.3). Bleaching can be particularly severe when these factors interact; for example, elevated sea temperatures coupled with high solar irradiance are particularly damaging (Glynn 1996; Reaser *et al.* 2000).

When first described last century, coral bleaching was considered a natural event because of its limited extent, rarity and the bleached reef's tendency to recover eventually (Goreau and Hayes 1994). However, by the 1980s bleaching had become widespread and affected large sections of many of the world's coral reefs (Goreau 1990; Glynn 1991), leading to the suggestion that global warming was at least partly responsible (Glynn 1991). Since the late 1980s, coral bleaching has occurred almost annually and has affected every coral reef system in the world (Goreau *et al.* 2000; Reaser *et al.* 2000), with some large reef systems experiencing complete coral mortality over a single season (B.E. Brown and Suharsano 1990). During the severe El Niño event of 1998, the most geographically extensive and severe coral bleaching in recorded history took place in over 60 countries in the Pacific Ocean, Indian Ocean, Arabian Sea, Red Sea and Caribbean (Goreau *et al.* 2000; Reaser *et al.* 2000). The Indian Ocean was the most severely affected region (Sheppard 1999; Reaser *et al.* 2000). As such, it has generally been accepted that global climate change poses a serious threat to coral reef systems worldwide (Knowlton 2001; T.P. Hughes *et al.* 2003), especially in light of the minimum temperature rises predicted over the coming century (IPCC 2002). Indeed, mathematical models constructed by Hoegh-Guldberg (1999) predicted that in the face of continued sea temperature rises at the current estimated rates, many coral reefs will not be able survive beyond 2050; however, modest acclimation or adaptation by corals to such rises would substantially ameliorate extinction predictions (T.P. Hughes *et al.* 2003; Sheppard 2003). (See also Chapter 8 for discussion of climate change issues in the tropics.)

In sharp contrast to the sensitivity of corals to climate change scenarios and poor recovery after severe bleaching events, some reef fish communities have been reported to be relatively resilient to disturbances associated with bleaching alone (Booth and Beretta 2002) (although they are susceptible to the combined effects of bleaching, invasive species and sedimentation – G.P. Jones *et al.* 2004; Munday 2004). A study of coral reef fish communities of the Great Barrier Reef in Australia revealed that coral bleaching did not result in a large change in fish abundance, species richness or species diversity (Bellwood *et al.* 2006). However, that study did observe a profound change in the fish community structure from a prebleaching assemblage comprising predominantly habitat and diet specialists to a post-bleaching assemblage consisting mainly of generalists. While the implications of these sorts of shifts are unknown, the loss of specialists may reduce the potential for positive associations between corals and commensal fish species (Pratchett 2001; Bellwood *et al.* 2006). Likewise, the drastic reduction in coral reefs in the Seychelles resulting from the 1998 bleaching led to profound changes in the associated reef fish assemblages (Graham *et al.* 2006). Post-bleaching assemblages were taxonomically depauperate, with local extinctions of four species recorded.

7.2.2 Susceptibility to overfishing

Geography is also contributing to the basket of threats faced by coral reefs – over 75% of coral reefs occur in developing countries with rapidly expanding human populations. The concentrated productivity of coral reefs (Sebens 1994; Reaser *et al.* 2000) and their coastal distribution means that tens of millions of people obtain a substantial part of their protein and artisanal livelihood from these productive ecosystems (Spalding *et al.* 2001; Pauly *et al.* 2002). Although originally believed to be relatively immune to overfishing because of the high levels of primary productivity and the previously held notion that reef fishes have fast turnover rates in general (Lewis 1977), there is increasing evidence that coral reef fish communities are susceptible to overfishing (Polunin and Roberts 1996; Pauly *et al.* 2002; Roberts *et al.* 2002). Indeed, many tropical reef systems have experienced profound habitat destruction (Figure 7.10), reduced fish production, changes in community structure and local extinctions (S. Jennings *et al.* 1999).

For example, high fishing pressure has probably been responsible for a large reduction in reef fish biomass in the Hawaiian Islands in recent history. DeMartini *et al.* (1996) compared fish biomass at remote shallow reefs where no fishing took place with that observed at fished reefs in the main Hawaiian Islands and determined that biomass was twofold higher when fishing was absent. Elsewhere in the islands, however, lower fishing rates appear to indicate that overfishing

Figure 7.10 Derelict fishing gear caught on coral. [Photo courtesy of the National Oceanic and Atmospheric Administration (NOAA) Pacific Islands Fisheries Science Center.]

did not occur, although the small size of landed fish may suggest a more chronic reduction in total biomass and shifts in the fish community (Friedlander and Parrish 1997; S. Jennings *et al.* 1999). Analogous declines in reef fish abundance have been reported in Samoa (King and Faasili 1999), Fiji (S. Jennings and Polunin 1997; S. Jennings *et al.* 1999), the Seychelles (S. Jennings *et al.* 1995) and Kenya (McClanahan 1994).

Another problem associated with overfishing on coral reefs is the growing international trade in live reef fish (R.J. Jones and Steven 1997; Sadovy and Vincent 2002). There is a long history of trade in live fish as luxury food items (mainly groupers) in many parts of Asia, but the trade entered the international market only in the late 1980s (Sadovy *et al.* 2003). Although banned in many parts of Asia, the illegal use of cyanide as a fishing technique is widespread throughout Southeast Asia and beyond, and is still one of the major means of collecting specimens for the live fish and aquarium trade (R.J. Jones and Steven 1997). Cyanide (dispersed in water or placed in baits) temporarily asphyxiates fish, so that they become immobilized and easy to collect among the corals. Other disadvantages of poisoned baits are that they are regurgitated by the stunned fish and then reingested by other non-target fish, which soon die (Sadovy *et al.* 2003), and that they sink to the bottom, where they slowly release cyanide for some time (R.J. Jones and Steven 1997). Additionally, cyanide causes the dissociation of the symbiosis between coral algae and their animal hosts (bleaching) by reducing the capacity of zooxanthellae to photosynthesize (R.J. Jones and Steven 1997; R.J. Jones and Hoegh-Guldberg 1999). Fishing in this and other manners for the live fish trade has caused serial depletion of many large coral reef species such as the humphead wrasse (*Cheilinus undulates*), groupers and snappers (Pauly *et al.* 2002). Most of the reefs near Hong Kong were depleted in the 1990s, so this

(a)

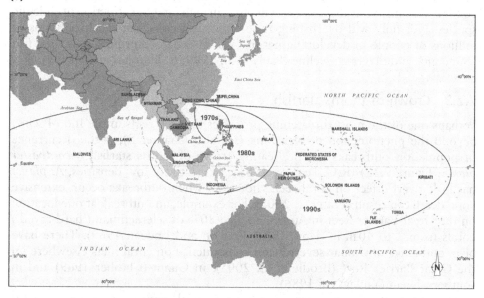

(b)

Figure 7.11 Expansion of the live reef fish trade fishery into the Pacific and Indian Oceans. (a) The western extent of the fishery in the Indian Ocean is the Seychelles (not shown). (b) The sale of big groupers in an aquarium in a Hong Kong market. (After Sadovy *et al.* 2003. Photo by Yvonne Sadovy. Copyright, Asian Development Bank.)

poorly regulated industry now continues in a 'boom-and-bust' manner, serially depleting reefs throughout Southeast Asia and the western Pacific of large species (Figure 7.11; Sadovy *et al.* 2003).

A global analysis of the geographic ranges of reef fish found that 26.5% of species examined were highly endemic (Roberts *et al.* 2002). This result suggests that the continued loss of coral reefs and overfishing from artisanal subsistence, commercial and live-trade fisheries will result in high rates of extinction of endemic species. Not only will overfishing raise extinction risks, but the livelihoods of millions of people in developing nations will also be compromised as total fish abundance and diversity decline (Pauly *et al.* 1989; S. Jennings *et al.* 1999).

7.2.3 Crown-of-thorns starfish

Perhaps one of the most threatening processes to coral reefs in the Indo-Pacific, beyond the phenomenon of bleaching (see also Chapter 8), is the occurrence of population outbreaks of coral-eating crown-of-thorns starfish (*Acanthaster planci*; Figure 7.12) (R.G. Pearson 1981). At relatively low densities, *A. planci* has little overt effect on coral reefs; however, when outbreaks occur, extensive coral death can result (Pratchett 2005). For example, an outbreak at one location on the Great Barrier Reef in Australia killed 80% of scleractinian ('hard-rayed') corals from 2 to 40 m in depth (R.G. Pearson and Endean 1969). There have been outbreaks of similar severity and consequence on coral reefs elsewhere on the Great Barrier Reef (Brodie *et al.* 2005), in Guam (Chesher 1969) and in southern Japan (Yamaguchi 1986).

The causes of these outbreaks are controversial, with two main hypotheses commanding support: (1) outbreaks are a natural phenomenon driven by the species' high fecundity (Vine 1973); and (2) anthropogenic changes to the marine environment periodically enhance the conditions favourable to rapid population growth. Other supporting hypotheses include the removal of starfish predators (fish and gastropods), terrestrial run-off of pesticides, altering predator abundance and community structure, construction on reefs leading to elevated

Figure 7.12 Crown-of-thorns starfish (*Acanthaster planci*) on a coral of the Great Barrier Reef, Australia. [Photo courtesy of the Australian Institute of Marine Science (AIMS).]

mortality of starfish larvae predators, and enhancement of larval food supply via eutrophication from terrestrial nutrient enrichment (Brodie *et al.* 2005). Indeed, support for the last hypothesis as the cause of the Great Barrier Reef outbreaks is substantial, provided by the fourfold increase in nutrient discharges to the reef over the last century, the resulting increase in phytoplankton densities and the measured increases in *A. planci* fecundity with higher phytoplankton density (Brodie *et al.* 2005).

7.2.4 Sedimentation and pollutants

Corals are affected not only by activities in and under the water – landscape modification of terrestrial environments on the coasts adjacent to coral reefs can contribute to their decline. Terrestrial agriculture, deforestation and development introduce large quantities of sediment, nutrients and other pollutants into coastal waters, causing widespread eutrophication and degradation of biologically productive habitats (Pimentel *et al.* 1995; see section 7.6). Coastal coral reefs are being exposed to increasing levels of nutrients, pollutants and sediments originating from land (Figure 7.13; Fabricius 2005).

A review compiled by Fabricius (2005) identified some of the major local-scale effects of sedimentation and pollution on coral reefs:

1   High nutrient run-off can cause algal blooms, leading to coral mortality and reduced reef calcification (e.g. Red Sea – D.I. Walker and Ormond 1982; Loya 2004).
2   Sediment dredging, pollutants and nutrients can reduce coral recruitment (e.g. Hong Kong – B. Morton 1994).
3   High sedimentation resulting from logging, agricultural development and vegetation loss during drought reduces coral cover, coral diversity and inhibition of coral settlement (e.g. Papua New Guinea – Munday 2004; Philippines – Hodgson 1990).

Figure 7.13   Coral reef affected by sedimentation and algal growth. (Photo by Karenne Tun.)

4   Excess nutrients can reduce coral skeletal density and coral cover (e.g. Indonesia – Edinger *et al.* 1998, 2000; Hawaii – Hunter and Evans 1995; Costa Rica – J.N. Cortés and Risk 1985).
5   Increased turbidity and nutrients from river discharges can reduce octocoral species richness and decrease crustose coralline algae (Great Barrier Reef, Australia – Fabricius and De'Ath 2001, 2004).
6   Increased eutrophication reduces species diversity, gamete formation, larval development and settlement, juvenile density, adult density and survival (Barbados – Tomascik and Sander 1987a,b; Hunte and Wittenberg 1992).
7   Untreated faecal sewage exposure can increase internal bioerosion by the boring sponge, *Cliona delitrix* (Grand Cayman Island – Rose and Risk 1985).

It has recently been estimated that 22% of the world's coral reefs are greatly or moderately threatened by anthropogenically derived sedimentation and pollution (D.G. Bryant *et al.* 1998). Countries with the highest level of threat are those with the highest rates of land clearing and terrestrial run-off, such as Taiwan, Vietnam and the Philippines (Bourke *et al.* 2002). As such, sedimentation and pollution rank closely with the other major threats to coral reefs worldwide (Spalding *et al.* 2001), and these processes can even dominate at local scales (Fabricius 2005).

7.2.5   Recreational impacts

Finally, a seemingly subtle threat that is gaining in importance as the number of pristine reefs declines (and, ironically, as interest in their plight increases) is the negative impact of recreational activities on coral reefs. The extremely lucrative tourist industry, worth several billions of US dollars annually (e.g. Driml 1994), that promotes both above- and below-water viewing of coral reefs has recently been identified as a key damaging process on some well-visited reefs (Hawkins *et al.* 1999; N.H.L. Barker and Roberts 2004). The type of damage done by divers depends on the types of corals present at a particular site. Branching corals tend to sustain the highest damage (Figure 7.14; Rouphael and Inglis 1997), although the growth rates of this group mean that recovery is usually swift (Hawkins *et al.* 1999).

A study in the Caribbean found that high-intensity diving caused a 19% loss of old colonies of massive coral species, but even low levels of diving had pronounced effects on coral community structure (Hawkins *et al.* 1999). N.H.L. Barker and Roberts (2004) reported that divers using a camera contacted the reef more than those not taking photographs, and shore-based divers also had higher contact rates than boat-based divers. Briefing divers about the potential impacts of their activities had no effect on the diver's probability of breaking corals, but intervention by the dive leader was effective in reducing contact rates.

7.3   Marine reserves

The setting aside of regions within a species' range that are protected from

Figure 7.14   A diver contacting and potentially damaging a coral reef. (Photo by Karenne Tun.)

harvesting activity [often termed 'no-take reserves', 'reserves' or 'marine protected areas' (MPAs)] has been used to manage fisheries since prehistoric times (Johannes 1978, 1998a; see also Chapter 10). The concept has generated a large body of theoretical and empirical work regarding the optimal size, connectivity and distribution of reserves for effective conservation (Neubert 2003). The main advances in this field have been made for the marine environment as a result of its attractiveness as a practical management tool for overexploited fisheries stocks (Roberts 1997a; Sala *et al.* 2002; Botsford *et al.* 2003; Gerber *et al.* 2003; Lubchenco *et al.* 2003; Neubert 2003). Despite theoretical support for the benefit of marine and estuarine reserves, limited empirical support for their ability to improve fisheries yields (Halpern 2003) and the debatable evidence provided thus far has made the topic controversial in the past (Hilborn 2002; Tupper 2002; Wickstrom 2002).

Empirical evidence in support of marine reserves to improve fisheries yields and promote stock recovery has increased in recent years (Mosquera *et al.* 2000; Murawski *et al.* 2000; Roberts *et al.* 2001; Ley *et al.* 2002; Pauly *et al.* 2002; Bolido 2004; Rajamani *et al.* 2006; see Spotlight 7: Daniel Pauly), including strong evidence that no-take harvest reserves reduce extinction risks of patchily distributed organisms (Fryxell *et al.* 2006). Even artisanal fisheries, such as those in small islands of the Philippines (Bolido 2004; Rajamani *et al.* 2006), which are at near-collapse, have benefited remarkably from the simple application of no-take marine reserves to prevent overexploitation. Despite these obvious successes and the woeful state in which global fisheries are found (section 7.1), less than 0.01% of the world's oceans is protected effectively from harvest (Pauly *et al.* 2002), and few new marine reserves have been established in recent years (Gewin 2004).

Spotlight 7: Daniel Pauly

Biography

After completing my doctorate studies in Germany in 1979, I spent many years at the International Centre for Living Aquatic Resource Management (ICLARM), then in Manila, The Philippines, where I developed methods for tropical fish stock assessment, which I applied and taught in many tropical developing countries. I became a professor at the University of British Columbia's Fisheries Centre in 1994, and its Director in 2003. My scientific focus has mainly been on the management of fisheries and ecosystem modelling, comprising over 500 contributions to peer-reviewed journals, authored and edited books, reports and popular articles. The concepts, methods and software I have (co-)developed are in use throughout the world. This applies notably to the ecosystem modelling approach incorporated in the Ecopath software (see www.ecopath.org), to FishBase, the online encyclopaedia of fishes (see www.fishbase.

org), and the global mapping of fisheries trends (see www.seaaroundus.org). My work has received numerous awards, notably the Cosmos Prize (2005, Japan) and the Volvo Environment Prize (2006, Sweden). Profiles on me and my work were published in *Science* on 19 April 2002, *Nature* on 2 January 2003, the *New York Times* on 21 January 2003, and in other publications.

Major publications

Pauly, D. and Christensen, V. (1995) Primary production required to sustain global fisheries. *Nature* **374**, 255–257.

Pauly, D., Christensen, V., Dalsgaard, J., Froese, R. and Torres Jr, F. C. (1998) Fishing down marine food webs. *Science* **279**, 860–863.

Pauly, D. Christensen, V., Guénette, S. Pitcher, T. J., Sumaila, U. R., Walters, C. J., Watson, R. and Zeller, D. (2002) Towards sustainability in world fisheries. *Nature* **418**, 689–695.

Pauly, D., Alder, J., Bennett, E., Christensen, V. Tyedmers, P. and Watson, R. (2003) The future for fisheries. *Science* **302**, 1359–1361.

Watson, R. and Pauly, D. (2001) Systematic distortions in world fisheries catch trends. *Nature* **414**, 534–536.

Questions and answers

Which type of fisheries – commercial, recreational or artisanal – represents the greatest exploitative threat to tropical marine ecosystems?

All fisheries have the potential to deplete the resources they exploit. Industrial fisheries, however, are extremely effective at what they do, and even over a short period they have a devastating effect on their resource base.

What fisheries management practices can be used to counter the phenomenon you have described as 'fishing down marine food webs'?

Establishing large marine protected areas, and strict controls over the remaining fished areas.

Why are freshwater and lacustrine systems so sensitive to human-induced environmental change?

Because they are small systems compared with the reach of our industries (fishing, pollution, habitat modification, etc.). The oceans are larger, and hence the human impacts appeared later.

How effective are marine protected areas (MPAs) in conserving tropical biodiversity, and should alternative solutions also be pursued?

MPAs should never be seen as sufficient by themselves. Conventional management is needed too.

How can scientists work to overcome misconceptions among policy makers and the public that arise from the 'shifting-baseline' syndrome?

We should use old records and data routinely, and always refer to the earliest time for which data are available. We should use a wide range of data, not only those compatible with the model currently fashionable.

Not only do marine reserves act to reduce mortality from fishing, but the individuals they harbour are larger on average because most fisheries tend to be size selective (Roberts 1997a). Individuals within reserves tend to grow to larger sizes and produce more offspring (egg production is at times up to an order of magnitude higher than among exploited populations – Roberts *et al.* 2001; Halpern 2003). However, in marine reserves lower trophic-level species may sometimes suffer reductions where large predator abundance is allowed to recover (Tupper and Juanes 1999); despite this, the benefits to fish communities of a healthy guild of large predators are indisputable (Stevens *et al.* 2000; Bascompte *et al.* 2005).

The increased egg production of large females from reserves results in higher recruitment of juvenile fish in adjacent exploited populations via their export in ocean currents (Roberts 1997b). Areas adjacent to marine reserves may also benefit from the surplus production of individuals that spill over as reserve populations approach carrying capacity as a result of density-dependent mechanisms (Bohnsack 1998; Lizaso *et al.* 2000). However, the evidence for this phenomenon and the degree to which it can supply adjacent areas with sufficient individuals for exploitation is still controversial (Rudd *et al.* 2003) and its contribution appears to be highly subject to reserve design (Sladek Nowlis and Roberts 1999; Hyrenbach *et al.* 2000; Hastings and Botsford 2003; Lubchenco *et al.* 2003) and size (Halpern 2003). Marine reserves may also preserve ecosystem resilience to environmental perturbation by protecting key species and habitats (A.L. Green *et al.* 1999; Guénette and Pitcher 1999).

Tropical reef systems are typically characterized by variable recruitment dynamics driven by high environmental stochasticity and spatial structure (Roughgarden *et al.* 1988; Bailey and Houde 1989), which are thought to exert strong selective pressures for particular growth and fecundity patterns (Conover and Munch 2002; Heino and Godo 2002; Vigliola *et al.* 2007). In contrast, modern size-selective fisheries impose directional selection in favour of genotypes for short lived, early maturing individuals (Heino and Godo 2002). As such, reserves that reduce or prevent such strong selection driven by fishing are hypothesized to maintain the genotypes that provide most compensation to harvest in adjacent areas subject to intense fishing pressure. Another advantage of marine reserves is the potential maintenance of functional breeding sex ratios. Many tropical species demonstrate size-dependent sex shifts (Gust 2004), so intensive and selective fishing of the largest individuals may eliminate one adult sex, thereby severely limiting egg production and recruitment. Establishment of reserves as havens for these larger individuals should, in practice, prevent the catastrophic reduction of recruitment due to disruptions to operational sex ratios and other components of mating systems (Rowe and Hutchings 2003).

The economic and conservation benefits of marine reserves go beyond the straightforward increase in production of harvestable fish within adjacent exploited areas. Healthy marine reserves with abundant, charismatic species can maintain much higher species diversity than more intensively fished areas (Worm *et al.* 2003), so their role as conservation tools is unquestionable. Highly species-diverse and healthy reefs are attractive to non-extractive industries such as snorkelling and diving tourism (Williams and Polunin 2000; Rudd *et al.* 2003). In most tropical fisheries, artisanal interests will probably continue to outweigh commercial ones (section 7.1.3) for some time. In addition to the problems associated with obtaining sufficient information with which tropical fisheries managers can plan harvest strategies (Johannes 1998a) and the challenges of appropriate governance (Rudd *et al.* 2003), the establishment of marine reserves can also incur substantial costs (Milon 2000). The majority of these costs tend to be associated with the planning of the reserve itself and the associated initial loss of income of people who have traditionally fished in the designated areas (Rudd *et al.* 2003). Substantial costs can also be incurred with enforcement, especially in poorer countries, where poaching can be rife (Maliao *et al.* 2004).

7.4 Megafauna

The rapid and systematic extermination and depletion of large terrestrial vertebrates due to human hunting is well documented in recent centuries (H.F. James 1995; Ceballos and Ehrlich 2002), with good evidence for major extinctions due to human hunting since the Pleistocene (Jackson *et al.* 2001; Barnosky *et al.* 2004; Brook and Bowman 2004; Martin 2005). In this section we discuss the plight of several large tropical marine and freshwater vertebrates that have not been covered elsewhere. In Chapter 6 we discussed the overharvesting of marine turtles and tropical sharks, and in section 7.1.4 we briefly mentioned how directed catch and bycatch in fisheries is threatening many invertebrate, fish, bird, mammal and reptile species. In this section, we provide examples of large sea-dwelling species that are of conservation concern within the tropical environment – tropical marine mammals of the orders Sirenia (dugongs and manatees) and Cetacea (dolphins and whales), the world's largest fish, the whale shark (*Rhincodon typus*), and one of the world's largest freshwater fish, the Mekong giant catfish (*Pangasianodon gigas*).

7.4.1 Sirenians

There are four extant sirenian species, and all are listed as vulnerable by the World Conservation Union (IUCN 2005) and in Appendix I of the Convention on International Trade in Endangered Species of Wild Fauna and Flora (CITES): the dugong (*Dugong dugon*), distributed from south-eastern Africa through to the South Pacific; the (West) African manatee (*Trichechus senegalensis*), distributed from Senegal to Angola; the Amazonian manatee (*T. inunguis*), restricted to the Amazon region; and the West Indian (American) manatee (*T. manatus*) of the Gulf of Mexico to central South America.

Of these, the dugong is perhaps the least vulnerable to extinction given its broad distribution (Figure 7.15) and its relatively large population size (approximately 85 000 in Australia alone – H. Marsh *et al.* 1999, 2003). However, the dugong is thought to be at risk from unsustainable harvest practices (largely artisanal), fisheries bycatch and loss of its coastal seagrass habitats (H. Marsh *et al.* 2003). Although the species has relatively good prospects for persistence in its distributional stronghold of Australia, most of the remaining populations are relictual and separated by vast distances (H. Marsh *et al.* 2003).

Dugongs are seagrass-specialist herbivores (phanerogamous seagrasses of the families Potamogetonaceae and Hydrocharitaceae), and so threats to their food source are perhaps of greatest concern. Seagrass beds around the world are under threat from nearshore mining activities, destructive trawl fishing, damage from boat propellers, seabed dredging, land reclamation and rising sedimentation and turbidity from coastal and inland vegetation clearing (e.g. logging and agricultural development) (S. Shepherd *et al.* 1989). Of these, the loss of photosynthetic activity due to increased turbidity from sedimentation and eutrophication from nutrient loading is the most acute threat to seagrass beds around the world (H. Marsh *et al.* 2003). Extreme weather events such as cyclones also episodically destroy large areas of seagrass in a single season (Poiner and Peterken 1996),

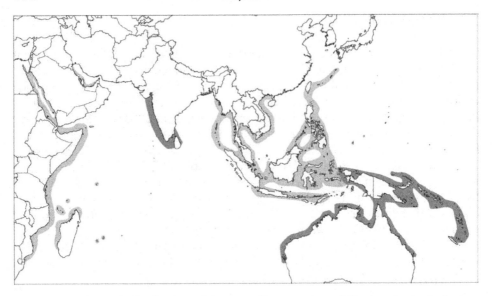

Figure 7.15 Current distribution of dugongs (*Dugong dugon*). Darker shades indicate higher relative densities. [After H. Marsh *et al.* 2003. Copyright, Commonwealth Scientific and Industrial Research Organisation (CSIRO) Publishing.]

necessitating local dugong populations to relocate en masse or starve (H. Marsh *et al.* 2003).

Dugongs are an important food, medicinal and cultural source for many human societies throughout their range (H. Marsh *et al.* 2003). Although in many countries dugong hunting is prohibited, many indigenous groups maintain dugong hunting rights as an important part of their culture (Heinsohn *et al.* 2004; Kwan *et al.* 2006; Rajamani *et al.* 2006). For example, in the western islands of the Torres Strait between Papua New Guinea and Cape York (Australia), dugong harvest can be in excess of 1000 individuals (H. Marsh *et al.* 1997), and subsequent modelling has indicated that these rates may be unsustainable (Heinsohn *et al.* 2004). Other likely, but largely unquantified, threats include incidental entanglement and drowning in mainly artisanal fishing nets, vessel strike and the accumulation of chemical pollutants (H. Marsh *et al.* 2003, 2005).

Although much less is known regarding the population status of other sirenians, there is evidence that most species face similar threats to dugongs. The West Indian manatee was overharvested in many parts of its range, and despite legal protection its recovery has been slow (IUCN 2005). In French Guiana and Brazil, the species is still actively hunted for meat, and sedimentation, pollution and clearing of important mangrove habitats are growing (de Thoisy *et al.* 2003). Similar processes threaten the species in Costa Rica (Smethurst and Nietschmann 1999). Hydroelectric development is perhaps the greatest threat to Amazonian river manatees and other Amazonian aquatic mammals, although bycatch and habitat degradation are also of concern (Vidal 1993).

7.4.2   Cetaceans

The broad-scale exploitation of whales between the eighteenth and twentieth centuries by temperate and high-latitude populations is well documented (H. Whitehead 2002; Roman and Palumbi 2003), with many populations recovering (Gerber and DeMaster 1999; Stevick *et al.* 2003; Hucke-Gaete *et al.* 2004) despite some continued exploitation (e.g. C.S. Baker *et al.* 2000). The history of whaling and the artisanal exploitation of other cetacean species is not as well known (Reeves 2002), although large reductions in tropical cetacean populations are known for some species (e.g. sperm whales, *Macrocephalus physeter*, in the tropical Pacific Ocean – H. Whitehead *et al.* 1997). Furthermore, more evidence is becoming available that incidental and direct catch is threatening many cetacean species in the tropics and elsewhere (Lewison *et al.* 2004a; Read *et al.* 2006).

There many examples of aboriginal subsistence hunting of cetaceans in the tropics. Species such as humpback (*Megaptera novaeangliae*), sperm and short-finned pilot (*Globicephala macrorhynchus*) whales are hunted in Lesser Antilles (Caribbean – Figure 7.16), humpbacks are taken in Equatorial Guinea, sperm and other toothed whales and various baleen whales are hunted in Indonesia, and Bryde's whales (*Balaenoptera edeni*) are targeted in the Philippines (Reeves 2002). However, the harvest rates in many of these ventures appear small, although insufficient reporting and the inadequacy of population data, which are needed to assess the acceptable level of sustainable takes, suggest that more research is required.

Perhaps a more pressing conservation problem for the world's tropical cetaceans is the incidental catch of smaller coastal and river species in artisanal and commercial fisheries. The Irrawaddy dolphin (*Orcaella brevirostris*) is found

Figure 7.16   Short-finned pilot whales taken in aboriginal subsistence whaling in Lesser Antilles. (After Reeves 2002. Copyright, Blackwell Publishing Limited.)

from Bengal to the Philippines in tropical and subtropical coastal marine, estuarine and freshwater habitats. In Australia and Papua New Guinea, the endemic *O. brevirostris* has recently been renamed *O. heinsohni* to distinguish it from its more widely distributed congeneric relative (Beasley *et al.* 2005). In the Mekong River region of Cambodia, Vietnam and Laos, Irrawaddy dolphins suffered heavy hunting pressure in the 1970s, and now even the main population within the Cambodian region (100–150 individuals) is extremely small and showing evidence of decline, largely induced by incidental capture in fishing nets (Baird and Beasley 2005). The species was formerly abundant in Vietnam, but has now been largely extirpated from the region (Baird and Beasley 2005). There are proposals for continued development of large dam projects along the Mekong River, which will, if implemented, pose even greater threats to the entire region's population (Baird and Beasley 2005).

Incidental catch of many dolphin species also occurs in some tropical commercial fisheries. For example, Van Waerebeek *et al.* (1997) reported central Peruvian gillnet fisheries taking high numbers (877) of dusky (*Lagenorhynchus obscurus*), long-beaked common (*Delphinus capensis*) and bottlenose (*Tursiops truncatus*) dolphins and Burmeister's porpoise (*Phocoena spinipinnis*) during 87 days of fishing in 1994. Although the effects of these takes are unknown, the catch of 227 bottlenose dolphins in the inner estuary of the Gulf of Guayaquil, Peru, represented a catch that was nearly double the estimated production of young dolphins in that population (Van Waerebeek *et al.* 1997). Bycatch of dolphins and porpoises in the Peruvian drift gillnet (Majluf *et al.* 2002) and other Latin American fisheries (Vidal 1993; D'Agrosa *et al.* 2000) continues to be an issue of concern.

Other anthropogenic dangers in addition to bycatch threaten other tropical cetaceans. The endangered Ganges river dolphin (*Platanista gangetica*) is distributed throughout the Ganges/Brahmaputra/Megna and Karnaphuli river systems in India, Bangladesh and Nepal and is threatened by a host of processes including dams, dredging, reduced water flow, incidental catch, extensive pollution, reduced food availability from the overfishing of prey species and even directed catch (B.D. Smith *et al.* 1998). Small populations of Indo-Pacific (*Tursiops aduncus*) and humpback dolphins (*Sousa chinensis*) are also highly threatened by direct and indirect takes in Zanzibar, eastern Africa (Stensland *et al.* 2006).

7.4.3 Whale sharks

The world's biggest fish (up to 18 m long; Figure 7.17), the whale shark (*Rhincodon typus*), is widely distributed throughout tropical and temperate seas (Colman 1997; Stewart and Wilson 2005). Yet surprisingly, detailed life history and population data are rare for the majority of known populations (Colman 1997). A popular food item in parts of Asia (especially Taiwan), increasing demand for the meat during the 1990s saw the intensification of existing and the development of new targeted fisheries, leading eventually to the listing of the species in CITES Appendix II (CITES 2002). Today, the fishing of whale sharks is

Figure 7.17   Whale shark. (Photo by David Ross, Ningaloo Reef Dreaming.)

illegal in many countries, although anecdotal suggests that the larger Indo-Pacific population may be declining (CITES 2002; Meekan *et al.* 2006).

It is considered unlikely that whale sharks can sustain heavy exploitation as they are thought to share the typical life history characteristics of other large sharks that make many species slow to recover from overharvesting (Meekan *et al.* 2006). Whale sharks may not achieve sexual maturity for over 30 years (Colman 1997), they are thought to grow slowly (Wintner 2000), and the frequency of breeding is unknown (Bradshaw *et al.* 2007b). These traits, combined with the species' capacity for long-distance migrations within the Indo-Pacific region (Eckert and Stewart 2001; S.G. Wilson *et al.* 2006), may make it vulnerable to local extirpation if only certain parts of its range are protected. For example, the large whale shark aggregation that occurs at Ningaloo Reef, Western Australia, may be undergoing a decline as a result of overfishing elsewhere. Since the mid-1990s the population of 300–500 animals has shifted towards one composed of a higher proportion of young males (Meekan *et al.* 2006), and recent population models for this aggregation suggest that the population is declining (Bradshaw *et al.* 2007b). Concerted efforts among neighbouring countries within the Asia-Pacific region and elsewhere to reduce or ban fishing of this species (as has occurred recently in Taiwan) are need if there is to be any chance of recovery (Meekan *et al.* 2006).

7.4.4   Mekong giant catfish

One of the world's largest freshwater fish, the Mekong giant catfish (*Pangasianodon gigas*), can reach lengths of 3 m and weigh over 300 kg (Figure 7.18); in Cambodia it is known as the 'king of fish' (Hogan *et al.* 2004). A little over a century ago this migratory fish's range spanned the entire Mekong River and its tributaries from southern China to Vietnam. A popular and highly sought food item, the

Figure 7.18　One of the world's largest freshwater fish, the Mekong giant catfish. The specimen shown weighs 153 kg, but specimens weighing double this and measuring up to 3 m in length have been reported. The species is currently listed as critically endangered by the World Conservation Union. (Photo by Zeb Hogan.)

species has experienced a massive 90% decline in the last three decades. Indeed, post-1999 the landings of the species plummeted to the point where it could no longer be called a fishery, although many are still caught as bycatch in nets targeting other fish species (Hogan *et al.* 2004).

Although a large captive-rearing and release programme was begun in 1985 in Thailand to re-stock that country's population, there were concerns about the genetic implications for the wild stock through accentuated in-breeding (over 95% of the hatched fingerlings shared the same parents) (Hogan and May 2002; Hogan *et al.* 2004). Although numbers have increased in some parts of its range, genetic erosion coupled with continued unsustainable harvesting resulted in the species being accorded critically endangered status by the World Conservation Union in 2004 (IUCN 2005). The construction of dams on some tributaries of the Mekong has already resulted in plummeting fisheries catches, so maintaining the connectivity between spawning and nursery areas for this and other similarly threatened catfish species (So *et al.* 2006) will require a reduction in the number of dams proposed for the main river (Hogan *et al.* 2004). Additional management actions that will need to be considered include fishing moratoria, no-take reserves (see section 7.3) and the maintenance of captive rearing to meet market demand (Hogan *et al.* 2004).

7.5 Tropical freshwater ecosystems – water for life

In addition to the threats to human health and well-being that degrading freshwater sources generate (see Chapter 2), surface freshwater ecosystems are estimated to support over 10 000 fish species (Lundberg *et al.* 2000) and a large proportion of amphibians and aquatic reptiles, indicating that approximately one-third of global vertebrate biodiversity is confined to freshwater (Dudgeon *et al.* 2006). However, much of the vertebrate biodiversity in freshwater systems is still undescribed, and this is even more pronounced in the tropics. New species of freshwater organisms are being discovered each year; for example, the Mekong River system is thought to support up to 1 700 fish species (Rainboth 1999; Sverdrup-Jensen 2002), which is more than double previous estimates (Dudgeon *et al.* 2006). This places the Mekong among the top rivers in the world for fish species richness and makes it comparable to the Amazon and Congo as an important global hotspot for river fish biodiversity (Dudgeon *et al.* 2006).

Our knowledge of tropical invertebrate, microbial and plant biodiversity supported by freshwater ecosystems is intolerably incomplete; adequate data on tropical invertebrate diversity are missing for most taxa (Dudgeon *et al.* 2006). This lack of information is a major impediment to biodiversity conservation in the tropics because of the high levels of local endemism and species richness among decapod crustaceans, molluscs and aquatic insects (Dudgeon 2000a; Benstead *et al.* 2003). Again, the Mekong River represents an important centre for global freshwater invertebrate diversity, with over 100 species of endemic species already described (Dudgeon *et al.* 2006). Likewise, plant biodiversity in adjacent wetlands can be incredibly diverse. For example, about 20% of the Amazonian tree species occur in floodplain forests although these constitute a mere 5% of the total rain forest area (Junk 2002).

In gross contrast to the huge biodiversity supported by these systems, freshwater habitats contain only about 0.01% of the world's water and they cover only 0.8% of the planet's surface (Gleick 1996). These figures demonstrate that these ecosystems are some of the most endangered in the world (Dudgeon *et al.* 2006). Indeed, the biodiversity supported by surface freshwater is declining at far greater rates than in even the most threatened terrestrial systems (Sala *et al.* 2000), and these declines are typically more pronounced in the tropics than at higher latitudes (Groomsbridge and Jenkins 1998). For example, the global decline of amphibians that started over 50 years ago (Houlahan *et al.* 2000; Stuart *et al.* 2004) is much more pronounced in tropical streams than elsewhere (Stuart *et al.* 2004).

The relatively predictable cycle of wet and dry seasons in much of the tropics, coupled with seasonal inundations from rivers, has benefited humans for millennia (Junk 2002). For example, the floodplains of the Nile, Euphrates, Tigris and Indus Rivers provided early human civilizations with the economic prosperity they needed to advance (Boulé 1994; Junk 2002). Today, tropical wetlands continue to support the cultivation of rice paddies and other aquatic crops, fish culture, vegetables, fruits and livestock that sustain dense human populations. Wetlands are estimated to cover large areas of most continents – for example, around 20% of tropical South America is considered to be wetlands (Junk 1993).

As outlined in Chapter 2, wetlands provide essential functions in the landscape, mostly for water retention, but also as filters and habitats for a huge diversity of plants, animals and humans (Junk 2002).

Human influences on tropical freshwater systems often alter the balance of natural forcing factors such as altering the normal variation in flow and flooding dynamics, changing the proportional abundance of substratum types, modifying the annual temperature regime, adding toxic components and removing microhabitat types altogether (Junk 2002; Malmqvist and Rundle 2002). Some of the more obvious impacts derive from point-source discharge of sewage or factory effluent, but more subtle and gradual impacts result from sustained nutrient input from intensive agriculture and terrestrial vegetation removal (Figure 7.19; Malmqvist and Rundle 2002). Furthermore, links between the proximate causes of freshwater ecosystem degradation and their ultimate impacts can be complex (Figure 7.19).

Dudgeon *et al.* (2006) broadly classified the threats to freshwater ecosystems into five major interacting categories: (1) overexploitation, (2) pollution, (3) flow modification, (4) habitat degradation and (5) invasion by exotic species. The first section of this chapter dealt with overexploitation in aquatic habitats in detail, and we described invasive species issues in Chapter 5. In the following sections we examine the remaining issues of pollution, flow modification and habitat degradation with respect to their effects on biodiversity in tropical freshwater systems.

7.5.1 Pollution

Although many industrialized (largely temperate) countries have made considerable progress in reducing water pollution or even restoring freshwater habitats to some semblance of a 'pristine' state, freshwater pollution is pandemic

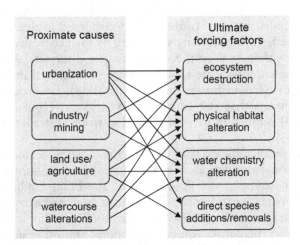

Figure 7.19 The four main causes of ecosystem change in freshwater systems and their links with factors that ultimately lead to environmental change. (After Malmqvist and Rundle 2002. Copyright, Cambridge University Press.)

in most of the tropical world (Dudgeon *et al.* 2006). Indeed, around 90% of all wastewater is discharged directly into rivers and streams in developing countries (N. Johnson *et al.* 2001). Excessive nutrient enrichment and other chemicals, such as endocrine-disrupting compounds, are also of increasing concern (Colburn *et al.* 1996; Dudgeon *et al.* 2006). Many tropical river and wetland systems have extensively modified nitrogen and phosphorus nutrient budgets, due mainly to intensive agriculture and deforestation. Humid tropical wetlands are particularly vulnerable to excess nutrient loading because, although they have rich plant growth and wildlife, this characteristic is the result of efficient nutrient use and recycling rather than high fertility (Figure 7.20). As such, these essentially oligotrophic systems are easily upset by the addition of nutrients from agriculture and wastewater disposal (Junk 2002).

Borbor-Cordova *et al.* (2006) examined the nutrient budget in a major agricultural catchment of the Pacific coast of South America, the Guayas River basin of Ecuador. They found that synthetic fertilizers accounted for the greatest input to the basin of both nitrogen and phosphorus, and that near-zero nutrient balances resulted from crop export and heavy riverine losses. Not only do these modifications upset the fine nutrient balances in productive agricultural land, but the large deposition to major river systems also has important implications for the biota that they support (Mokaya *et al.* 2004; Dudgeon *et al.* 2006). For example, the Njoro River in Kenya has been extensively modified by poor farming practices and wastewater from organic matter-loading factories such as canneries. Mokaya *et al.* (2004) found that the river was stressed by high nutrient loads and predicted that pollution would increase as the human population continued to expand.

Lake Victoria, infamous for its catastrophic loss of endemic cichlid fish species after the introduction of the exotic predator, the Nile perch (Witte *et al.* 1992; see also Chapter 5 and section 7.1.1), has also been extensively modified by

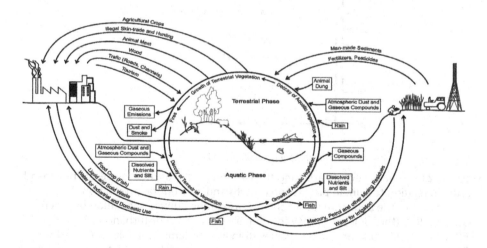

Figure 7.20 Nutrient cycles and major human impacts on tropical wetlands. (After Junk 2002. Copyright, Cambridge University Press.)

high nutrient loading. The lake is surrounded by five countries where intensive agriculture and habitat loss through logging is rife (Scheren *et al.* 2000), leading to excess nutrient loading and a replacement of diatoms by cyanobacteria as the dominant planktonic algae (Kling *et al.* 2001). Verschuren *et al.* (2002) examined sediment cores in Lake Victoria to reconstruct the recent history of plankton community changes with respect to human settlement and development. They found evidence that changes in phytoplankton productivity and community composition were caused mainly by excess nutrient loading rather than food web changes resulting from the introduction and proliferation of Nile perch. Trade between the resident Buganda people and Europeans and Arabs, which began in the late 1880s, followed by completion of the Uganda railroad in 1930, resulted in rapid human settlement and establishment of plantation agriculture that led to an exponential increase in primary productivity (Figure 7.21).

Continued and rapid population growth through the twentieth century stimulated broad-scale deforestation and the conversion of tropical forests to agriculture (Verschuren *et al.* 2002). This rapidly induced change in the

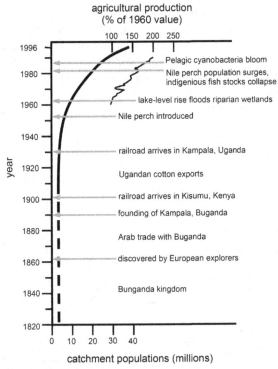

Figure 7.21 Principal events affecting the Lake Victoria ecosystem since the nineteenth century in relation to human population growth and agricultural production in its drainage basin. The approximate sixfold increase in the human population between 1932 and 1995 (thick line) is a proxy for human-induced soil disturbance prior to 1960, when agricultural production values (thin line) became available. The trends are indicative of the inferred rise in primary productivity in the lake owing to higher nutrient loading over this period. (After Verschuren *et al.* 2002. Copyright, The Royal Society London.)

phytoplankton community, resulting in persistent deep-water anoxia in the 1970s, lends credence to the hypothesis that deep-water oxygen loss facilitated the rapid decline of certain fish species by Nile perch (Kaufman and Ochumba 1993). With the deep water refugia from intensive predation by Nile perch effectively rendered inaccessible to many prey species, they were soon eliminated by the efficient predator (Verschuren *et al.* 2002). This hypothesis may also explain the rapid proliferation of Nile perch in the 1980s (Verschuren *et al.* 2002).

Fertilizer use is the greatest driver of nutrient loadings in freshwater systems worldwide and is predicted to increase by 145% over the next half-century (Figure 7.22; Kroeze and Seitzinger 1998), which is equivalent to an increase in average per capita fertilizer use from 15 to 21 kg (Malmqvist and Rundle 2002). Although highly variable worldwide, these predicted rates of increase are highest in countries bordering the western Pacific (e.g. China) and the Indian Ocean, with large increases also predicted for Asia, Africa and Latin America (Kroeze and Seitzinger 1998). Increasing development and industrialization within the tropics and elsewhere is also predicted to result in higher nitrogen inputs into freshwater catchments from the atmosphere and other point sources, although these will be substantially lower than inputs from fertilizers (Figure 7.22; Kroeze and Seitzinger 1998).

Although chemical contaminants may be degraded faster in tropical systems (Viswanathan and Krishna Murti 1989; Bhushan *et al.* 1997), they are usually also assimilated more easily by tropical biota (G.E. Howe *et al.* 1994). Wetlands in particular act as periodic or permanent sinks for many compounds, so the addition of toxic substances is particularly difficult to remedy (Junk 2002). For example, the accumulation of lead shot in tropical wetlands in northern Australia after successive years of hunting magpie geese (*Anseranas semipalmata*) was thought to be the cause of higher mortality in this species. P.J. Whitehead and Tschirner (1991) found that accumulated shot was highly dense (330 000 shot/ha) at depths of < 20 cm, and that it overlapped in size and distribution with a favoured plant food, the sedge tuber, *Eleocharis dulcis*. Although the use of steel shot has curtailed much of the ongoing accumulation, residual lead concentration in some wetlands is likely to remain high for many years. Likewise, Buss *et al.* (2002) found that certain tropical macroinvertebrate taxa (e.g. Plectoperan

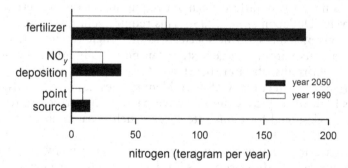

Figure 7.22 Predicted increase in nitrogen inputs from synthetic fertilizer, NO_y atmospheric deposition and point sources to world catchments from 1990 to 2050. (After Malmqvist and Rundle 2002 and Kroeze and Seitzinger 1998. Copyright, Springer Science and Business Media.)

insects) were particularly sensitive to chemical degradation of freshwater streams in Brazil.

A review of pesticide effects in aquatic ecosystems in Central America indicated that there is still a high dependency on chemicals to control insects and other pests, with high residual concentrations in many water bodies (Castillo *et al.* 1997). However, there was only circumstantial evidence for negative biodiversity effects, owing mainly to a lack of dedicated research. Likewise, Abdullah *et al.* (1997) examined the findings regarding pesticide effects in rice paddies and found strong evidence for important reductions in fish biodiversity, although the evidence for changes to microalgae and cyanobacteria was found wanting, again largely because of a lack of research. Pesticide and heavy metal contamination in Brazil's Pantanal wetlands is also thought to affect many fish and other wildlife species (Alho and Vieira 1997), and there is also good evidence for a reduction in diatom diversity with increasing levels of industrial and urban pollution in rivers in Malaysia (Khan 1991).

7.5.2 Flow modification

Although water course exploitation, such as damming, channelization or abstraction (removal) of water that directly alters the passage of water in river channels and adjacent wetlands (Junk 2002; Malmqvist and Rundle 2002), is ubiquitous, it tends to be most problematic in systems with highly variable flow regimes (Dudgeon *et al.* 2006), a characteristic of many tropical rivers. Extensive damming worldwide (existing dams retain approximately $10\,000\,km^3$ of freshwater, which is five times greater than all the water contained within all of the world's rivers combined; Nilsson and Berggren 2000) mean that even some of the world's largest rivers (many of them tropical) run dry for part of the year (Postel and Richter 2003; Dudgeon *et al.* 2006). These massive fluctuations in water wreak havoc with freshwater biodiversity, with many species unable to migrate out of an area during periods of low flow (Dudgeon 2000b; Xenopoulos *et al.* 2005).

Because rivers offer a broad array of potential habitats given the vast altitudes, substratum types and vegetation communities through which they flow, many riverine taxa have vastly different habitat requirements during certain phases of their life cycle (Dudgeon *et al.* 2006). This requirement gives rise to sometimes vast migrations by many riverine species within a single catchment, so stresses to particular habitats within a large system can potentially affect the biota over a population's entire distribution (Benstead *et al.* 1999). As such, a single dam in a tropical river can affect many of its resident species, especially considering that most dams built in tropical countries have passage infrastructure (e.g. fishways or fish passes) designed for temperate species such as salmonids (Dudgeon *et al.* 2006).

The increase in the number of dams projected for many tropical rivers is particularly worrying. Most proposals for dams in tropical Asia target large rivers such as the Mekong and its tributaries, with severe implications for the freshwater biota that these systems support (Dudgeon 1999). Indeed, 12 hydropower dams are planned for the Mekong River alone (Malmqvist and Rundle 2002), and these will probably further affect many species already threatened by other human activities such as overharvesting and bycatch (section 7.4). The resulting socioeconomic effects of reduced fish biodiversity are also likely to be high (Dudgeon 1999; Malmqvist and Rundle 2002).

Structural modifications such as damming, flow regulation, channelization and bank stabilization can also reduce the incidence and severity of flooding to adjacent wetlands, or prevent it entirely (Malmqvist and Rundle 2002). This reduction in seasonal flooding has important implications for many freshwater plant and animal species that depend on annual fluctuations in water availability (Junk 2002). Lateral migrations from the main channel to adjacent floodplains and forests inundated during tropical monsoon seasons are essential breeding habitat for many tropical freshwater species (J.V. Ward *et al.* 2002). Therefore, human activities such as riparian and floodplain farming (e.g. rice paddies), deforestation and vegetation modification (e.g. logging), as well as flood control infrastructure (e.g. levees, dykes), seriously threaten important habitats for many freshwater species (Dudgeon 2000a; Dudgeon *et al.* 2006). The corollary is that flow modification can seriously reduce fish landings as a result of the direct impediment to migrations and the degradation of important breeding habitats (Welcomme 1979).

Like humans, who depend on freshwater sources such as rivers and lakes for the provision of drinking water and protein, many terrestrial species depend on rivers and lakes for their own needs. In many tropical areas with pronounced monsoons punctuated by dry seasons, water shortage at certain times of the year can severely stress many herbivores and other species. This dependency makes the provision of year-round access to freshwater habitats and the maintenance of their flow patterns essential for the survival of many terrestrial, riparian and amphibious species such as frogs, water dragons, snakes and water birds (Dudgeon *et al.* 2006). Likewise, many bird species either live permanently in or visit seasonally tropical wetlands during long-distance migrations from higher latitudes (Shukla and Dubey 1996; Junk 2002).

Water abstraction, mainly for the irrigation of agricultural lands, is a problem for many of the world's tropical rivers (Postel 1998; Malmqvist and Rundle 2002; Dudgeon *et al.* 2006). For example, dam-regulated abstraction of water from a tropical river in Puerto Rico in the Caribbean resulted in obstruction of amphidromous freshwater shrimp migration and elevated mortality (Benstead *et al.* 1999). However, the study suggested that periodic interruptions of water abstraction during peak larval drift times would improve overall survival prospects. In the Ethiopian Rift lake system, water abstraction for agricultural use has resulted in large reductions in water levels and increased salinity, with serious consequences for the fragile rift ecosystem (Legesse and Ayenew 2006). Water abstraction can be particularly acute during the dry season in tropical rivers – in the Njoro River, Kenya, up to 20% of the flow volume can be extracted in a single day, with strong effects on the distribution and composition of the macrozoobenthos and aquatic insects (Mathooko 2001). In Zanzibar, excessive water abstraction from tourists was also predicted to cause deterioration of rare freshwater habitats (Gossling 2001).

7.5.3 Habitat degradation

In addition to direct modification of flow dynamics, habitat degradation of freshwater systems can result from modification of habitats used by a variety of taxa (e.g. sand removal, dredging), or such systems can be eroded as a result of changes to the system elsewhere in the catchment, such as downstream sedimentation resulting from upstream deforestation (Dudgeon *et al.* 2006). Habitat degradation in tropical freshwater systems is thought to have potentially severe implications for biodiversity (W.V. Reid 1992), although there is currently a lack of information regarding the magnitude of predicted effects (Malmqvist and Rundle 2002). High endemism in many tropical systems also increases the likelihood of species extinctions in this region as a result (Malmqvist and Rundle 2002).

While deforestation can lead to extensive habitat degradation, even in situations where stream habitats are left largely intact after vegetation removal, there can still be profound effects on fish community structure. For example, Bojsen and Barriga (2002) found that a reduction in canopy cover and leaf litter input from deforestation seriously altered the fish community structure in the Ecuadorian Amazon.

7.5.4 Conservation actions and catchment management policy

The undeniably high value of maintaining healthy freshwater ecosystems for biodiversity conservation and human welfare, combined with the crisis facing many of the world's freshwater systems, makes sensible conservation actions paramount activities. In areas where the opportunities exist to establish 'pristine' lake and river reserve systems, urgent funding to support these ventures is required. However, because few, if any, areas of the planet's freshwater systems will not have important value to humans outside of biodiversity conservation, trade-offs between biodiversity conservation and human use of the ecosystem services and goods these systems provide will be unavoidable (Dudgeon *et al.* 2006). Hence, scientists and managers will have to embrace an approach that represents a compromise between biodiversity conservation and human livelihood maintenance (Moss 2000; Dudgeon *et al.* 2006).

Because freshwater ecosystems include a vast network of drainages, water channels, floodplains and underground aquifers, a catchment-scale approach is the most effective way forward for conserving and managing freshwater habitats (Moss 2000). Such broad-scale approaches to defining the problems these systems face and mitigating the cascading effects of environmental degradation help to resolve the fine-scale conflicts of interest among competing human and biodiversity requirements (Dudgeon *et al.* 2006). The disadvantage of such an approach is that it usually requires management capacity over sometimes vast landscapes that are normally under the jurisdiction of many different political and socioeconomic systems. Therefore, effective collaboration among disparate organizations that integrates both terrestrial and freshwater management will have the added advantage of broadening the traditional focus from river channels to entire freshwater systems, including floodplains and seasonally inundated forests (Dudgeon *et al.* 2006).

However, for management to work effectively, even if coordinated over the appropriate catchment scale, sound biological knowledge is needed to dictate action. There is unfortunately a great disparity in the ecological understanding of freshwater ecology between developed and developing nations (Wishart *et al.* 2000; Malmqvist and Rundle 2002). Greater focus on tropical freshwater systems will be needed to offset this trend and provide sound data for effective management. In the meantime, many international conventions and organizations have recognized the importance of conserving tropical freshwater habitats for the immensely diverse biota they support. The Convention on Wetlands of 1971 (Ramsar) is an international treaty whose mandate is to conserve and use wetlands sustainably. The Ramsar Convention is also supported by other conventions such as CITES, the Convention on Migratory Species (CMS) and the Convention on Biological Diversity (CBD) (Junk 2002).

7.6   Where marine and freshwater habitats merge: coasts and estuaries

Thus far we have discussed the pervasive threats to marine and freshwater systems with respect to overharvesting, habitat degradation, pollution and flow modification, but we have not examined the particular threats to aquatic systems that define the juxtaposition of marine and freshwater habitats. Brackish estuaries and coastal zones are subject to most of the same threats described in previous sections (Lotze *et al.* 2006), but there are also a number of specific conservation issues that are peculiar to these transitional habitats. An essential precursor to understanding these threats is linking human activities on land to processes occurring in brackish and coastal systems, such as eutrophication, siltation, pollution and habitat modification (Lotze *et al.* 2006).

As with freshwater habitats, coastal water quality is changing rapidly in response to excessive fertilizer use and deforestation from agricultural development and logging (Vitousek *et al.* 1997; Tilman *et al.* 2001). Fertilizer use in the tropics is increasing at an alarming rate (Matson *et al.* 1997; Junk 2002; Borbor-Cordova *et al.* 2006), deforestation continues at greater rates in the tropics than elsewhere (Chapter 1), and coastal urbanization is expanding disproportionately to human population growth (Brook *et al.* 2006a; Lotze *et al.* 2006). For example, coastal seafloor zones are becoming increasingly anoxic as a result of river-borne agricultural nutrient loading and the ensuing eutrophication of coastal waters (Fabricius 2005). In Chapter 1 we described how estuarine and coastal mangroves also provide essential breeding habitat for commercially exploited fish (Baran and Hambrey 1998; Mumby *et al.* 2004) and protect coastlines from extreme climatic or geologic events (Danielsen *et al.* 2005). In the next sections, we describe some particularly sensitive transitional aquatic zones and the threats they face in addition to the ones outlined in previous sections. First, we describe the conservation status of coastal seagrass beds, and then we follow on with a discussion of estuaries and their threats.

7.6.1 Seagrass beds

Seagrass beds cover large areas of shallow subtropical and tropical seas around the world (Ogden 1980), where they once supported large populations of sirenians, marine turtles, sharks, fishes and invertebrates prior to overexploitation in many systems (Ogden 1980; Jackson *et al.* 2001; see also Chapter 6). In recent decades, large expanses of seagrass beds have disappeared (Kirkman 1978; Abal and Dennison 1996) as a result of sedimentation, turbidity and possibly disease, as well as the removal of large herbivores, such as sirenians and marine turtles, that increased seagrass susceptibility to other environmental stressors (Jackson *et al.* 2001).

The mass reduction in marine turtle abundance throughout much of the tropics (Chapter 6) has also resulted in a reduction in organic matter and nutrient flux to sediments in seagrass beds (Ogden 1980; Thayer *et al.* 1982). Seagrasses in many areas have responded by growing longer blades that impede light penetration to bottom substrata, with a concomitant increase in the slime moulds that cause seagrass wasting disease (Zieman *et al.* 1999). Similarly, dugongs can remove up to 96% of above-ground and 71% of below-ground seagrass biomass (Preen 1995) resulting in the removal of dominant species and the proliferation of less competitive grasses. Dugong grazing also releases nutrients to the water column via debris and faeces that flow to adjacent ecosystems (Jackson *et al.* 2001).

7.6.2 Estuaries

Much of the evidence for the massive degradation of the world's estuaries comes from the temperate zone (Jackson *et al.* 2001). However, many of the world's great estuaries are found in the tropics, such as the Amazon and Orinoco in South America, the Congo, Zambezi and Niger in Africa, the Ganges in India and the Mekong in Southeast Asia (Blaber 2002). Many of these estuaries experience similar, if not greater, threats than their temperate counterparts as they generally support much greater human populations. As with marine and freshwater systems, overfishing is perhaps one of the greatest threats to estuaries worldwide, but other activities such as aquaculture are proving to be particularly damaging with respect to nutrient overloading.

The live rearing of commercially important fish and invertebrates in estuarine or coastal enclosures (aquaculture) has been touted as a possible replacement activity for typical fisheries. For example, in tropical Asia, massive fisheries overexploitation has led to the proliferation of aquaculture ventures that produce lower trophic-level species such as milkfish (*Chanos chanos*) and tilapia (*Oreochromis* spp.), although the farmed production of larger species such as groupers and snappers for the live fish trade is also proliferating (Marte 2003; see also section 7.2.2).

The proliferation of large-scale aquaculture in tropical estuaries has usually been done at the expense of mangrove protection (Chapter 1 and section 7.6.3) and regeneration and the maintenance of sustainable wild fisheries (Blaber 2002). Most aquaculture operations in tropical regions rear carnivorous fish, which require massive quantities of fish meal to achieve marketable size and quality.

Although efficiency has increased in recent years, the conversion factor of fish and now some vegetable meal to marketable aquaculture fish is approximately 1.4 (Gewin 2004). This demand for fish meal, coupled with the extreme nutrient loading resulting from the faecal rain emanating from concentrated densities of farmed fish, is resulting in pollution of many tropical estuaries (Gewin 2004). As with any other intense livestock farming practice, there are also many concerns regarding the accumulation of toxic compounds in captively reared fish (Hites *et al.* 2004) and the overuse of antibiotics.

Finally, it should be noted that most large tropical estuaries are adjacent to highly populated cities, which generally dump large quantities of industrial effluent and raw sewage (see also section 7.5.1) into these sensitive ecosystems. These inputs, combined with discharges from shipping ballast, have resulted in many tropical estuaries becoming badly polluted where they enter the sea (Blaber 2002).

7.7   Summary

1   Massive overexploitation of marine and freshwater fisheries around the tropics has resulted in major reductions in fish biodiversity.
2   Fisheries exploitation occurs as a result of commercial, artisanal and illegal, unreported and unregulated (IUU) activities.
3   Coral reef reefs throughout the tropics are being lost at alarming rates due to 'bleaching' from the combined effects of elevated temperatures, increased exposure to UV, changing water levels, overfishing, sedimentation and recreational activities.
4   In addition to fisheries, many large aquatic organisms, such as dugongs and manatees, river dolphins, whale sharks and giant catfish, are vulnerable to extinction in many tropical aquatic ecosystems.
5   Other threats to freshwater systems throughout the tropics include nutrient overloading from agricultural development and deforestation, flow modification from dams and channelization and habitat loss from direct removal of substrata and catchment-level sedimentation.
6   Coastal and estuarine environments at the juxtaposition zone between marine and freshwater are also under threat through intense food production (aquaculture), nutrient overload and pollution from densely populated tropical cities.

7.8   Further reading

Dudgeon, D., Arthington, A. H, Gessner, M. O., Kawabata, Z. I, Knowler, D. J., Leveque, C., Naiman, R. J., Prieur-Richard, A. H., Soto, D., Stiassny, M. L. J. and Sullivan, C. A. (2006) Freshwater biodiversity: Importance, threats, status and conservation challenges. *Biological Reviews* 81, 163–182.
Jackson, J. B. C., Kirby, M. X., Berger, W. H., Bjorndal, K. A., Botsford, L.W., Bourque, B. J., Bradbury, R. H., Cooke, R., Erlandson, J., Estes, J. A., Hughes, T. P., Kidwell, S., Lange, C. B., Lenihan, H. S.,

Pandolfi, J. M., Peterson, C. H., Steneck, R. S., Tegner, M. J. and Warner, R. R. (2001) Historical overfishing and the recent collapse of coastal ecosystems. *Science* **293**, 629–637.

Pandolfi, J. M., Bradbury, R. H., Sala, E., Hughes, T. P., Bjorndal, K. A., Cooke, R. G., McArdle, D., McClenachan, L., Newman, M. J. H., Paredes, G., Warner, R .R. and Jackson, J. B. C. (2003) Global trajectories of the long-term decline of coral reef ecosystems. *Science* **301**, 955–958.

Pauly, D., Christensen, V., Guenette, S., Pitcher, T. J., Sumaila, U. R., Walters, C. J., Watson, R. and Zeller, D. (2002) Towards sustainability in world fisheries. *Nature* **418**, 689–695.

Roberts, C. M., Bohnsack, J. A., Gell, F., Hawkins, J. P. and Goodridge, R. (2001) Effects of marine reserves on adjacent fisheries. *Science* **294**, 1920.

8

Climate Change: Feeling the Tropical Heat

Earlier chapters explored the multitude of factors now threatening tropical biotas, but so far we have given only cursory mention to potentially one of the greatest, yet least understood processes threatening biodiversity – global climate change caused by human agency. The subtlety with which climate change gradually alters biological systems, coupled with its complex interactions with other, more overt impacts of humans such as habitat loss, overharvesting and the introduction of invasive species, has made the phenomenon something of an intangible concept for governments and conservation practitioners. Yet climate change is a very real, pervasive and accelerating threat to the world's biodiversity and ecosystem functioning, from the tropics to the poles. As such, conservation biologists are compelled to consider its effects in concert with other drivers of species decline and ecosystem degradation. Much of the current evidence for the impact of anthropogenic climate change on biodiversity has come from temperate and high-latitude ecosystems (Flannery 2005). Nonetheless, in this chapter we summarize the main lines of evidence for the biotic response to climate change in the tropical realm, and discuss some of the options that we, as planetary custodians, have at our disposal for mitigating the worst of its effects.

8.1 Overwhelming evidence for human-mediated climate change

8.1.1 Global warming

Warnings of human-mediated climate change have been around for over 50 years, although many scientists were initially sceptical of the reported and predicted trends. However, by 1988, the general consensus among scientists was that climate change was a very real and worrying phenomenon with the potential to affect global ecosystems. The establishment of the Intergovernmental Panel on Climate Change (IPCC) was a response to this concern, and resulted in the publication of a series of landmark reports (www.ipcc.ch). These sounded the climate change alarm in political and public arenas. Yet even today many governments, including major pollutors such as the USA and Australia, have failed to commit to taking serious action to avert the drastic warnings described in these reports (Flannery 2005).

Physical evidence for a general trend of warming since the industrial revolution of the early nineteenth century is indisputable. In particular, data on recent temperature anomalies relative to climatology from 1951 to 1980 indicate that the current warming trend (Figure 8.1a) is nearly ubiquitous and unprecedented in magnitude over human time scales, although greater warming has occurred over land than over the ocean. The largest changes are occurring in the high latitudes of the northern hemisphere (Figure 8.1b; J. Hansen *et al.* 2006).

(a)

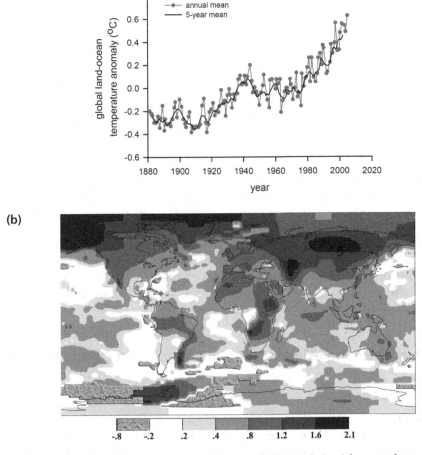

(b)

Figure 8.1 Temperature anomalies relative to 1951–80, derived from surface air measurements at meteorological stations and ship and satellite sea surface temperature measurements. (a) Global annual mean anomalies. (b) Temperature anomaly for the first half decade of the twenty-first century. (After J. Hansen *et al.* 2006. Copyright, National Academy of Sciences, USA.)

The global distribution of the climate change is congruent with climate models linked to changes in the composition of atmospheric gases, particularly the large increase in CO_2 caused by the burning of fossil fuels for industry and transportation (J. Hansen *et al.* 2005). Over the past century, the planet has warmed by 0.8 °C, of which 0.6 °C has occurred in the last three decades (J. Hansen *et al.* 2006). This trend is also clearly visible in the tropics; for example, the Indian Ocean is now warmer than at any time in the past 10 000 years (J. Hansen *et al.* 2006). These observed changes are real, not an artefact of measurement in urban areas. Confirmation comes from borehole temperature profiles, the retreat of alpine glaciers and the progressively earlier seasonal break-up of ice on rivers and lakes (IPCC 2001a; J. Hansen *et al.* 2006). Additionally, the tropical Pacific is a primary driver of the global atmosphere and ocean dynamics, so a warming of this region has worldwide implications – for not just tropical but all biodiversity (J. Hansen *et al.* 2006; see sections 8.2 and 8.3).

8.1.2 Super El Niño

In previous chapters we described several instances in which the deterministic drivers of species decline and habitat degradation have been influenced by severe climate phenomena such as extreme El Niño–Southern Oscillation (ENSO) events (see Chapters 2–5 and 7 and section 8.3.4). Indeed, the occurrence of so-called 'super' El Niños, such as those that occurred in 1982–83 and again in 1997–98, has been particularly influential on the patterns of tropical biodiversity loss and change (Barber and Chavez 1983; Siegert *et al.* 2001; Cochrane 2003; Charrette *et al.* 2006). Thus, the influence of climate change on the incidence and severity of extreme ENSO events is an essential consideration for tropical biodiversity conservation, beyond the more obvious deterministic trends.

Climate change is predicted to alter the pattern of such oscillatory climatic phenomena. For example, J. Hansen *et al.* (2006) predict that the warming of the Western Equatorial Pacific will increase the likelihood of super El Niños. Although there is no clear answer to the question of the effects of global warming on El Niños (Fedorov and Philander 2000; Cane 2005), many climate models predict a higher frequency of extreme events (Collins 2005). This is also analogous to the inferred effect of global warming on the incidence and severity of tropical storms (Emanuel 1987). It has even been hypothesized that during the Pliocene (5.3–1.8 million years ago) when the earth was 3 °C warmer than today, a permanent El Niño condition persisted (Ravelo *et al.* 2004).

8.2 Past evidence for climate change effects on tropical biodiversity

Tropical ecosystems, and the evolution and extinction of tropical biodiversity, have been moulded by climate change throughout geological time. The major drivers of climate change in deep time (Zachos *et al.* 2001) include: (i) plate tectonics causing continental drift (which redirects oceanic currents, distorting terrestrial topography and altering global rainfall patterns and winds; Ramstein *et al.* 1997); (ii) major periods of volcanic activity and asteroid/comet impacts (both of which can release vast quantities of greenhouse gases such as methane, water vapour and CO_2, but can also induce climatic cooling through global dimming from particulates and sulphur; Raup 1991a); (iii) variation in solar output (e.g. stellar evolution towards a hotter sun over the long term and fluctuations in sun spot activity over the short term; Mann *et al.* 1998); (iv) orbital variations and orbital precession, known as Milankovitch cycles, which lead to cyclic variation in the amount of sunlight reaching the earth's surface – this is thought to be the primary cause of the ice ages and the resultant large changes in sea level observed in the Pleistocene (K.D. Bennett 1990); and (v) mediation of atmospheric carbon dioxide (CO_2) by photosynthesizing organisms (algae and green plants), which produce sugars by combining water and CO_2 and release oxygen (O_2).

Unlike the situation we now face, past climatic shifts have largely resulted from long-term physical changes to earth's environment, which have proceeded at a sufficiently slow pace to allow organisms to adapt to altered conditions. That said, the long-term biological turnover associated with the progression to new climatic regimes is often profound. For instance, changing climate has probably been a primary driver of the dramatic waxing and waning of the extent of tropical reefs over periods of many millions of years. At certain times in earth's past, reefs were almost entirely absent (e.g. late Cambrian, late Devonian, early Jurassic, early Tertiary), being confined to just a few localized occurrences. At other times (e.g. mid-Cambrian, Permian, late Cretaceous, present day), reefs were fully developed (Copper 1988). The identity of the reef-building organisms has also shifted over time, from calcareous algae, to sponges, rudist clams and, finally, scleractinian corals. Recent work has shown that tropical seas are an evolutionary cradle of global biodiversity (Jablonski *et al.* 2006), and as such, past changes in tropical communities in response to climate and other physical disturbances have shaped both the tropical and temperate biotas we see today.

Tropical rain forests tell a similarly dynamic tale as coral reefs, with palaeogeographic work showing that over the past 350 million years rain forest has been the dominant tropical terrestrial biome for only about a quarter of that span (Raup 1991b). There is also patchy evidence that suggests that the Amazon and West African rain forests cooled, dried and retracted, then expanded again multiple times during the late Pleistocene (within the last 50 000 years) in response to global climatic shifts (Simberloff 1986). Temperature-related changes in the distribution of animal habitats (e.g. sea level changes altering the depth of water along continental margins; latitudinal shifts in the extent and degree of fragmentation of forests), have also been effective evolutionary forces, acting to isolate once-continuous populations and foster allopatric speciation (Cronin and Schneider 1990).

8.3   Effects of recent and projected anthropogenic climate change on tropical biotas

The physical evidence for recent climate change is overwhelming, with a high probability that pollution, rather than natural climatic variability, is the main cause (section 8.1). However, the implication of these trends for the future of biodiversity is less well understood. There is evidence that climate warming can affect species in five principal ways: (1) alterations of species densities; (2) range shifts, either poleward or upward in elevation; (3) behavioural changes, such as the phenology (seasonal timing of life cycle events) of migration, breeding and flowering; (4) changes in morphology, such as body size; and (5) shifts in genetic frequencies and reduction in genetic diversity (Root *et al.* 2003; see Spotlight 8: Stephen Schneider). It is worth emphasizing that, although large fluctuations in climate have occurred regularly throughout the earth's history, the implications of anthropogenically driven climate change for current biodiversity are particularly bleak. This is due to both the rapidity with which change is now occurring (on a scale of decades rather than millennia) and the fact that landscapes through which organisms must disperse to remain within favourable bioclimatic conditions have already been heavily modified by humans.

From an evolutionary perspective, gradual climate change over geological time has favoured some species and promoted speciation, with the inevitable loss of others. However, median projections of climate change models suggest a rise in average global temperature of ~3 °C within a century (IPCC 2001a). The end of the last ice age saw a 5 °C change, but this rise took at least 7000 years. So the coming change over the twenty-first century is almost as large as major prehistoric climatic shifts, but it will unfold many times faster. Such rapid climate change is predicted to outpace the rates at which many species can adapt via natural selection and range shifts. It is the unique combination of the magnitude and rate of modern climate change that will drive widespread extinctions. Although evidence for some of these effects in tropical areas is currently sparse, there are many examples and predictions that support the view that rapid climate change, in concert with other drivers of species loss and habitat degradation, is perhaps one of the most pressing conservation issues facing tropical species over the coming centuries.

8.3.1 Range expansion and changing phenology

Species living at higher altitudes in the tropics are particularly vulnerable to climate change due to the disruption or loss of specific microclimates and the higher likelihood of invasive species influx from lower elevations (see Chapter 5). Increases in the levels of CO_2 in the atmosphere are predicted to reduce cloud contact with high-elevation habitats and increase the rate of evapotranspiration in tropical montane forests, thus threatening the integrity of these unique and highly endemic ecosystems (Still *et al.* 1999; Soh *et al.* 2006). The increased risk of invasion from lower-altitude species (shifting or expanding their altitudinal range as temperatures rise) is also thought to result in increased competition for montane species. However, Pounds *et al.* (2006) found that the percentage of species lost in the toad genus *Atelopus* in South America was the highest between elevations of 1001 and 2339 m (Figure 8.2), a trend for which no clear explanation was found. Most of the local amphibian extinctions they observed occurred immediately following a relatively warm year, suggesting also that their vulnerability was a direct consequence of sensitivity to a warmer climate (Pounds

Spotlight 8: Stephen H. Schneider

Biography

I am the Melvin and Joan Lane Professor for Interdisciplinary Environmental Studies, Professor of Biological Sciences and Professor by Courtesy of Civil and Environmental Sciences Engineering at Stanford University. I am Co-Director of the Center for Environmental Science and Policy in the Freeman-Spogli Institute and a Senior Fellow in the Woods Institute for the Environment. I received my PhD in Mechanical Engineering and Plasma Physics from Columbia University, USA, in 1971. When considering research areas then, I became aware that anthropogenic dust can cool the climate, and that greenhouse gases can warm it, and thus decided to switch to studying climate science. Today, my global change interests include the ecological and economic implications of climatic change; integrated assessment of global change; climatic modelling of palaeo-climates and human impacts on climate (e.g. carbon dioxide 'greenhouse effect'); dangerous anthropogenic interference with the climate system; food/climate and other environmental science/public policy issues; and environmental consequences of nuclear war. I am also dedicated to advancing environmental literacy in all levels of education.

I co-founded the Climate Project at the National Center for Atmospheric Research (NCAR) in 1972 and founded the interdisciplinary journal *Climatic Change* in 1975, which I continue to edit today. I was honoured in 1992 with a MacArthur Fellowship for my ability to integrate and interpret the results of global climate research through public lectures, seminars, classroom teaching, environmental assessment committees, media appearances, Congressional testimonies and research collaboration with colleagues. I was elected to membership in the US National Academy of Sciences in 2002, and received both the National Conservation Achievement Award from the National Wildlife Federation and the Edward T. Law Roe Award from the Society of Conservation Biology in 2003, and the Banksia Foundation's International Environmental Award in Australia in 2006. I have served as a Coordinating Lead Author in Working Group II of the Intergovernmental Panel on Climate Change (IPCC) from 1997 to the present. My recent work has centred on the identification and classification of 'key vulnerabilities' in the climate system and the role of risk management in climate policy decision-making. I continue to serve as an advisor to decision-makers and stakeholders in industry, government and the non-profit sectors. I am also engaged in improving public understanding of science and the environment through extensive media communication and public outreach.

Major publications

Mastrandrea, M. D. and Schneider, S. H. (2004) Probabilistic integrated assessment of 'dangerous' climate change. *Science*, 304, 571–575.

Root, T. L., MacMynowski, D., Mastrandrea, M. D. and Schneider, S. H. (2005) Human-modified temperatures induce species changes: Joint attribution. *Proceedings of the National Academy of Sciences of the USA* 102, 7465–7469.

Schneider, S. H. (1990) *Global Warming, Are We Entering the Greenhouse Century?* Lutterworth, Cambridge.

Schneider, S. H. and Mastrandrea, M. D. (2005) Probabilistic assessment of 'dangerous' climate change and emissions pathways. *Proceedings of the National Academy of Sciences of the USA* 102, 15728–15735

Schneider, S. H., and Kuntz-Duriseti, K. (2002) Uncertainty and climate change policy, Chapter 2, In: *Climate Change Policy: A Survey* (eds. S. H. Schneider, A. Rosencranz and J.-O. Niles), pp 53–88, Island Press, Washington DC.

Questions and answers

Climate has varied throughout earth's history. Why is contemporary climate change particularly dangerous to biodiversity?

The current, much-faster-than-natural rate of temperature change, coupled with multiple stressors, makes contemporary climate change particularly threatening to biodiversity. The forecasted global average rate of temperature increase over this century (approximately 1–5 °C/century) greatly exceeds by a rough order of magnitude the rates typically sustained during the last 20000 years. The balance of evidence from meta-analyses of species from many different taxa examined at disparate locations around the globe suggests that a significant impact from recent climatic warming is discernible in the form of long-term, large-scale alteration of animal and plant populations. This evidence takes the form of poleward or upward range shifts and changes in phenology such as dates of migration, breeding and flowering (making spring events for some species 10–15 days earlier over the past few decades). The IPCC has extended climate impact analyses to include such 'environmental systems' as sea- and lake-ice cover and mountain glaciers. Clearly, if such climatic and ecological signals are now being detected above the background of climatic and ecological noise for a twentieth-century warming of 'only' 0.6 °C, it is likely that the combination of highly disturbed landscapes and temperature increases up to an order of magnitude larger by the year 2100 will have a dramatic impact on biodiversity and ecosystem functioning.

Will climate change have less impact in the tropics than at higher latitudes?

There are already clear signs of severe stress in high-latitude and alpine habitats and in coral reefs, showing that these ecosystems are experiencing significant impacts at present levels of climate change. Human-mediated climate change is or is projected to be affecting tropical biotas via range shifts (latitudinal and elevational), changes in phenology, increasing prevalence, distribution and severity of diseases and parasites, coral bleaching, drying of freshwater systems and sea level rise. The magnitude of temperature changes will be less in the tropics, but changes in the hydrological cycle may still be large. Some models suggest that above a few degrees more warming, tropical forests will switch from a sink to a source of CO_2 emissions – a dramatic change if it were to occur as projected. The potential for forest fires under such conditions could become a major threat to forests both in Amazonia and in Southeast Asia because the forests in these regions are not adapted to fire. Species living at higher altitudes in the tropics are particularly vulnerable to climate change due to the disruption or loss of specific microclimates and the higher likelihood of invasive species influx from lower elevations.

How might climate change interact with other threats to tropical biodiversity, such as invasive species, fire and land clearance for agriculture?

Adverse impacts on biodiversity caused by a synergistic suite of threats are already occurring and will continue to intensify climate impacts. It is expected that further warming could substantially rearrange the ranges and interactions of many species. However, because of human land uses such as agriculture, urban settlement and roads, most species no longer have a free range in responding (e.g. by freely migrating) to climatic shifts. The synergism or combined complex interactions of effects among climate changes, land use disturbances, the introduction of exotic species and artificial chemicals will most likely collectively impact on wildlife and terrestrial systems much more significantly than if each of these disturbances were simply considered separately.

Are there any benefits of a warmer world rich in atmospheric carbon for tropical ecosystems?

Undoubtedly some species – particularly those that are adaptable, such as crows or weeds – can flourish in disturbed conditions better than specialists such as warblers or orchids. Thus, although the populations of some well-adapted generalists may expand, the slow rate of speciation and the major threat of endangerment to more vulnerable species have resulted in estimates of 10–50% of species becoming extinct in the next two centuries if warming of more than a few degrees occurs.

Based on current trends, how long will it be before the earth's climate crosses an irreversible and potentially catastrophic tipping point?

It is very difficult to define precise tipping points given remaining uncertainties. Nevertheless, there are potential thresholds for events such as ice sheet disintegration or coral reef bleaching, although most such estimates appear as ranges – for example, 1–3 °C warming for major reef damages and 1.5–4 °C warming for major ice sheet disintegrations. The bottom line is that the harder and faster the system is disturbed, the more likely such catastrophic changes become.

et al. 2006). In Mexico, it was also projected that climate change will reduce the distribution range of two salamanders (*Pseudoeurycea cephalica* and *P. leprosa*) by 75% by the year 2050 (Parra-Olea *et al.* 2005). These salamander species are particularly vulnerable to climatic warming due to their limited dispersal capacity and the ongoing deforestation in the region, which will exacerbate the threat.

The extent and distribution of rain forest in the Southeast Asian region has probably fluctuated considerably over the last 65 million years as a result of periodically changing climate (Heaney 1991). However, the extent to which the

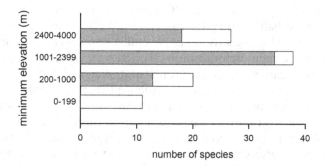

Figure 8.2 Altitudinal patterns in the *Atelopus* extinctions. Bars indicate the number of species known per altitudinal zone (total *n* = 96), and the grey-shaded portions represent the estimated percentage of species lost from each. (After Pounds *et al.* 2006. Copyright, Macmillan Publishers Limited.)

latest and most rapid phase of global warming is affecting the biodiversity of Southeast Asia is poorly known. By comparing data from two field guides of Southeast Asian avifauna published in 1975 and 2000, Peh (2007) found that elevational changes in distribution may have occurred in 94 resident bird species. Changes in the altitudinal distribution were caused by modifications to either (or both) the upper and lower altitudinal boundaries. Three species exhibited a marked response by shifting upwards their upper and lower altitudinal boundaries (e.g. the little forktail, *Enicurus scouleri*). These species were probably forced to shift their ranges towards a higher elevation in response to climate warming, though other explanations, such as improved sampling coverage of their range, are admittedly plausible (Peh 2007). The fact that these elevation-related changes occurred even in habitat generalists reinforces the notion that habitat loss may not be the sole reason for the elevation shifts because these species are known to be relatively insensitive to changes in land use (Peh 2007). Similarly, distributional shifts in the altitudinal range of amphibian species have also been recorded in the rain forests of tropical north Queensland, Australia (Thomas *et al.* 2004). Southeast Asia has warmed by at least 0.3 °C in the past two decades, and temperatures are projected to increase by another 1.1–4.5 °C by 2070 (IPCC 2001b). Although potentially biased due to increased knowledge on species' ranges, Peh's analysis does imply that climatic warming could be an important factor affecting the already imperilled biotas of Southeast Asia, especially in montane areas (Sodhi and Brook 2006). Lower-elevation birds have also been reported to shift towards higher elevations in Costa Rica, probably as a result of climate warming (Pounds *et al.* 1999).

Forests themselves are predicted to respond to climate change in many different ways. Rising temperatures and an increase in CO_2 levels mean that many forest species may spread due to more favourable growing conditions. For example, some have predicted that climate change will invoke the spread of *Podocarpus* species in the Amazon lowland rain forests (Colinvaux *et al.* 2000). Similar patterns of forest expansion have already been observed in northern Australia

(Bowman *et al.* 2001; Brook *et al.* 2005) and are probably due to rising CO_2 levels over the past half-century (Brook *et al.* 2005), which favour C_3 woody plants over C_4 tropical grasses (Bond *et al.* 2003). However, forest expansion is unlikely to offset the rapid deforestation characteristic of many tropical regions.

Climate change can also affect animal and plant phenologies (timing of recurring natural phenomena). There is evidence that in Oxfordshire in the UK, a temperate system, the arrival and departure dates of 20 long-distance migrants have advanced by an average of 8 days over the past three decades (Cotton 2003; e.g. Figure 8.3) due to increased summer temperatures. Likewise, the timing of arrival of migratory birds has advanced due to elevated winter temperatures in sub-Saharan Africa (Cotton 2003). These effects are expected to become more and more prevalent in the scientific literature, especially regarding migrant species that depend on specific climatic conditions for breeding and feeding within their migratory pathways.

8.3.2 Diseases and parasites

The tropical realm is notorious for its diversity of human pathogens and the diseases they cause, of which perhaps malaria, dengue, leishmaniasis, cholera, schistosomiasis and various encephalitic diseases are most common. However, diseases also regulate the dynamics of countless other species (Harvell *et al.* 2002; Lafferty and Gerber 2002), and global climate change has and will continue to alter, and in many cases widen, disease vector distribution and enhance pathogen virulence (Harvell *et al.* 2002; Pascual *et al.* 2006; Patz and Olson 2006). Additionally, changing distributions will result in the generation of new emerging pathogens, especially in plants (Anderson *et al.* 2004). Climate change is hypothesized to affect host–pathogen interactions in three main ways: (1) by increasing pathogen development, transmission and reproductive rates, (2) by relaxing the restrictions imposed by traditionally non-optimal seasons (e.g.

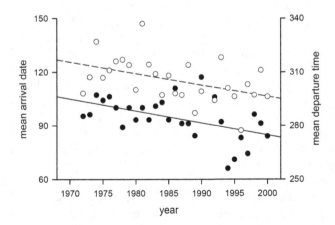

Figure 8.3 The arrival date (filled circles, solid line) and departure date (open circles, dashed line) of migrating northern house martins (*Delichon urbica*), in Oxfordshire, from 1971 to 2000. (After Cotton 2003. Copyright, National Academy of Sciences, USA.)

winter, dry season; Figure 8.4) and (3) by modifying the host's susceptibility to infection (Harvell *et al.* 2002). Indeed, the dynamics of many diseases are shaped by the seasonality of their particular environments; hence, alterations to seasonality predicted from climate change are likely to influence disease and parasite spread (Altizer *et al.* 2006).

The most severe disease outbreaks are predicted to occur if climate change alters the geographic ranges of hosts or pathogens, causing species and populations that were formerly separated to overlap (M.B. Davis and Shaw 2001). For example, the habitat suitability of 73 African tick species improved under each of eight different modelled climate change scenarios (Figure 8.5; Cumming and Van Vuuren 2006). The predicted increase in range and size of tick populations that would result would increase the likelihood of proliferation of tick pathogens that cause diseases such as Lyme disease, Japanese encephalitis and West Nile virus (Cumming and Van Vuuren 2006). Expansion of the geographical ranges of mosquito vectors for malaria and dengue has also been proposed as the main cause of the re-emergence of these human diseases in South America, central Africa and Asia that began in the 1980s (McCurry 1997; Epstein 1998; Spratt 1998). Similarly, the recent invasion of North Africa by a biting midge causing

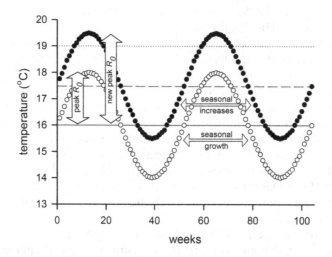

Figure 8.4 Influence of a 1.5 °C temperature rise on the reproductive phenology (per generational reproductive rate, R_0) of a hypothetical pathogen. When R_0 is above 1 (grey solid line), a pathogen will increase. The lower dotted curve (unfilled circles) illustrates the average weekly temperature before climate change; the upper dotted curve (filled circles) illustrates average weekly temperature after climate warming. The pathogen increases in abundance at temperatures above 16 °C (in this example), with the disease problems becoming severe when temperature exceeds the dashed line, and epidemic above the dotted line. This figure also illustrates that increases in temperature also lead to an increased annual duration of the period during which the pathogen is a problem. [After Harvell *et al.* 2002. Reprinted with permission. Copyright, American Association for the Advancement of Science (AAAS).]

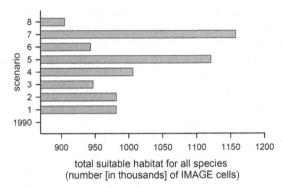

total suitable habitat for all species
(number [in thousands] of IMAGE cells)

Figure 8.5 Summary of the net change in total globally suitable habitat for all 73 tick species, relative to a 1990 baseline for all scenarios. The total amount of suitable habitat shows an increase in all scenarios. (After Cumming and Van Vuuren 2006. Copyright, Blackwell Publishing Limited.)

African horse sickness and bluetongue in livestock is thought to have been a direct result of climate warming in this region (Meltzer 1993).

In the highland forests of Monterverde (Costa Rica), 40% (20 of 50) of frog and toad species have disappeared following synchronous population crashes in 1987 (Pounds *et al.* 1999), with most crashes linked to warming and drying of the local climate. The locally endemic golden toad (*Bufo periglenes*) was one of the high-profile casualities in this area. Similarly, two endemic highland lizard species (*Norops tropidolepis* and *N. altae*) also disappeared from this area by 1996 (Figure 8.6; Pounds *et al.* 1999). Pounds *et al.* (1999) suggested that climate warming resulted in a retreat of the clouds and a drying of the mountain habitats, making amphibians more susceptible to fungus and parasite outbreaks. Indeed, the pathogenic chytrid fungus (*Batrachochytrium dendrobatidis*), which grows on amphibian skin and increases mortality, has been implicated in the loss of harlequin frogs (*Atelopus* spp.) in Central and South America (La Marca *et al.* 2005) and reductions in other amphibian populations elsewhere (Blaustein and Kiesecker 2002; Daszak *et al.* 2003). It is hypothesized that warm and dry conditions may stress amphibians and make them more vulnerable to the fungal infection (Pounds *et al.* 2006).

In addition to the well-publicized case of worsening fungal infections in amphibians, there are many predictions of climate-induced increases in the threats from pathogens to tropical biodiversity. For animal diseases, climate change is predicted to affect a pathogen's free-living, intermediate or vector stages, with vector-borne diseases thought to have the largest response to change (Kovats *et al.* 2001; Harvell *et al.* 2002). Although some claims are controversial, predictions are supported by laboratory and field studies showing that arthropod vectors and parasites die or fail to develop below certain temperatures, and higher temperatures increase reproduction and biting rates and periods of infectivity (Linthicum *et al.* 1999; Harvell *et al.* 2002). For example, there is evidence to suggest that the increased frequency and severity of droughts in Africa resulting from climate change will facilitate the spread of ungulate diseases

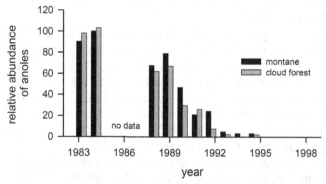

Figure 8.6 Declines in populations of anoline lizards. (From Pounds *et al.* 1999. Copyright, Macmillan Publishers Limited.)

Figure 8.7 The Hawaiian apapne (*Himatione sanguinea*) being attacked by a *Culex* mosquito, vector of avian malaria (*Plasmodium relictum*) and avian pox (*Poxvirus avium*). (Photo by Jack Jeffrey, jackjefferyphoto.com.)

such as tuberculosis as a result of increased population mixing (e.g. Cape buffalo, *Syncerus caffer*) (Cross *et al.* 2004).

Diseases such as avian malaria (*Plasmodium relictum*) and pox (*Poxvirus avium*) have been introduced into Hawaii and caused declines in many endemic birds (Figure 8.7; Atkinson *et al.* 1995, 2000). Disease risk is strongly associated with altitude: disease is most prevalent at mid-altitudes, where the mosquito and bird overlap is highest (Atkinson *et al.* 2000). Thus, the upward shift in mosquito distribution that global warming is predicted to cause is expected to reduce the refuge habitat available to endemic birds, with serious extinction implications for some of the least adaptable species (Harvell *et al.* 2002). Indeed, it has been predicted that a 2 °C rise in global temperatures, a likely proposition by 2050, will effectively eliminate all disease-free forested areas in Hawaii within the next century (Benning *et al.* 2002). However, recent modelling work suggests that reducing the impact of introduced rodents on the survival rates of many endemic

Hawaiian bird species at middle elevations will enable the evolution of resistance to malaria (Kilpatrick 2006).

The distribution and pathogenicity of many plant diseases are also expected to change under climate change predictions. Most research has focused on diseases of commercially important crops such as rice (e.g. Yamamura and Yokozawa 2002) because crop diseases threaten billions of dollars of yield each year (Agrios 1988). For example, the disease known as Panama (a.k.a. Fusarium wilt), caused by the fungus *Fusarium oxysporum*, has resulted in the abandonment of entire banana plantations in Central America (Chakraborty 2000) and seriously affects most banana plantations in the rest of the tropics (Ploetz 2000). The degree to which plant pathosystems in non-crop plants will respond to climate change is largely unknown, but as with crop diseases, it is predicted that the changing physical climate will result in changing distributions of pathogens and altered host plant resistance (Runion 2003), and a strong likelihood of new disease complexes arising (Chakraborty 2000).

The prevalence and severity of disease outbreaks in the marine environment are also predicted to increase with global warming. The reporting of diseases in many marine taxa has increased considerably over the last 30 years (M.L. Hayes *et al.* 2001; Ward and Lafferty 2004). In addition to the catastrophic effects of coral bleaching caused by extreme environmental conditions (Chapter 7 and section 8.3.4), opportunistic infections by *Vibrio* spp. bacteria and other pathogens in corals are thought to contribute to coral death (Harvell *et al.* 1999; E.P. Green and Bruckner 2000; R.J. Jones. *et al.* 2004). Studies have shown that many coral pathogens favour temperatures at or beyond those optimal for coral growth, suggesting that rising sea temperatures will lead to a higher frequency and severity of disease-induced coral death (Harvell *et al.* 2002). Indeed, there is already evidence that viruses and bacteria that thrive at higher temperatures could be responsible for exacerbating coral bleaching itself (Kushmaro *et al.* 1998; W.H. Wilson *et al.* 2001). Many coral diseases are thought to be aggravated by human activities; for example, 97% of Caribbean coral reef locations where disease events have been recorded have also been moderately to severely adversely affected by human activities (E.P. Green and Bruckner 2000).

Another marine disease that has direct implications for both human health and the conservation of marine communities is ciguatera poisoning. Endemic in many tropical marine systems, these highly potent neurotoxins (ciguatoxins) are produced mainly by dinoflagellate microalgae such as *Gambierdiscus toxicus*, and represent the most common and widespread form of seafood poisoning, afflicting over 50 000 people each year (Fleming *et al.* 2000). Climate warming and the subsequent rise in sea temperatures throughout the tropics is predicted to increase the frequency of toxic dinoflagellate blooms (Hales *et al.* 1999; Chateau-Degat *et al.* 2005), with the concomitant negative effects of more human poisonings and perhaps a reorganization of reef fish communities. There is also evidence that fish poisoning increases relative to the amount of disturbance to reefs (Hales *et al.*1999) because warmer waters increase the susceptibility of tropical fish to pathogens because of the high energetic costs associated with coping with thermal stress (Harvell *et al.* 2002).

Changing disease dynamics are likely to be some of the first and most rapidly identifiable impacts of climate change in the tropics. Harvell *et al.* (2002) predicted that climate warming will disproportionately affect pathogens with complex life cycles and those that infect non-homeothermic hosts. However, in some cases global climate warming may aid control of invasive species. Scanlan *et al.* (2006) predicted that a temperature rise of 2.5 °C would increase the effectiveness of pathogens used to control European rabbits (*Oryctolagus cuniculus*) in tropical north Queensland, Australia. Clearly, the enhancement of conditions that support the spread of pathogens and their vectors represents a major concern for human health, prosperity and biodiversity conservation.

8.3.3 Coral bleaching

In Chapter 7 (section 7.2.1) we discussed how the worldwide phenomenon of coral bleaching – a process in which the coral animal host expels its symbiotic algae following changes to the physical attributes of seawater – appears to have been increasing over the last several decades. The factors thought to be responsible for bleaching, such as changing seawater temperatures, rising salinity, increased exposure to air due to tidal changes, sedimentation and reduced light penetration and higher solar radiation, are all predicted to be enhanced under various climate change scenarios (Fitt *et al.* 2001; Knowlton 2001). Mechanistically, bleaching results from three different but sometimes interacting processes:

1   Physiological bleaching relates to the host coral's capacity to maintain high enough densities of symbiotic algae cells to achieve effective photosynthesis, and is affected by variation in water temperature and solar radiation.
2   Algal stress bleaching refers to the symbiont's physiological ability to withstand changing temperature and radiation regimes so as to avoid photoinhibition.
3   Animal stress bleaching is thought to occur when the host cell is more susceptible to environmental stress than its algal symbionts, although symbionts are generally less tolerant than their hosts (Fitt *et al.* 2001).

The extreme bleaching event of 1987 in the Caribbean was originally thought to be temperature related; however, water temperatures did not exceed normal maxima (Attwood *et al.* 1992). However, the duration of maximal temperatures was much longer than previous years (Fitt *et al.* 2001). Thus, the event was probably due to physiological bleaching. Other bleaching events, such as those that occurred worldwide during the intense 1997–98 El Niño, resulted in below-average sea levels and algal stress bleaching in the exposed reefs. As mentioned below (section 8.3.4), the incidence of coral diseases rises with increasing water temperatures, leading to animal stress bleaching. In earlier geological periods, warm global conditions led to massive retractions in the distribution of coral reefs (section 8.2). The likelihood that stress will continue as a result of rapid climate change and a higher frequency of extreme events such as super El Niños (section 8.1.2) is a cause of great concern for all the world's tropical coral reefs in a warmer future world.

8.3.4 Additional physiological, chemical and behavioural implications for aquatic life

The realization that global warming of only a few degrees will have major impacts on ocean temperatures, hydrology and chemical composition has led to predictions of major changes in marine fish, invertebrate and plant species in the coming decades (Roessig *et al.* 2004). Fishes have evolved to withstand a generally narrow range of environmental conditions (Barton *et al.* 2002), so large fluctuations in temperature, dissolved oxygen, salinity and pH will adversely affect many species (Moyle and Cech 2004). These deleterious effects of climate will include reductions in foraging range, growth and fertility rates, maladaptive metamorphosis, disrupted migration and endocrine imbalance (Roessig *et al.* 2004). While the potential effects of climate change on tropical systems are relatively less well documented than for temperate or polar systems, tropical fish are expected to demonstrate similar responses.

Another major change to ocean chemistry predicted to arise from climate change is the acidification of oceans. Although uptake of CO_2 by oceans is an essential moderator of this greenhouse gas concentration by acting as a gas sink (Sabine *et al.* 2004), the associated hydrolysis of CO_2 in seawater increases the concentration of hydrogen ions, $[H^+]$, thus lowering the pH of seawater (Orr *et al.* 2005). This will cause a decrease in carbonate ion concentrations, $[CO_3^{2-}]$, making it more difficult for calcifying marine organisms (e.g. animals that produce shells) to form biogenic calcite and aragonite exoskeletons from calcium carbonate ($CaCO_3$). This shift in the dissolved inorganic carbon (DIC) equilibrium has been hypothesized to have major effects on coral-forming species in the tropics (Kleypas *et al.* 1999) and on pteropods and plankton at higher latitudes (Feely *et al.* 2004; Orr *et al.* 2005). Indeed, Kleypas *et al.* (1999) predicted that by the middle of this century the increased concentration of CO_2 will result in a 30% decrease in the seawater saturation of calcium carbonate needed by reef-building corals in the tropics. Even faster rates of carbonate reduction are predicted for higher latitudes (Orr *et al.* 2005). Another direct impact of seawater chemistry changes is likely to be a reduction in the photosynthetic capability of shallow-water seagrasses that depend on the limited supply of dissolved inorganic carbon.

In addition to changing water chemistry, seagrasses are vulnerable to rising temperatures resulting from climate change. Beyond their obvious role in supporting a suitable foraging environment for many herbivorous species (e.g. dugongs; see Chapter 7), seagrass beds serve as a critical nursery habitat for many fish species, regulate dissolved oxygen content, modify seabed substrata and reduce suspended sediments in the water column. Seagrasses also stabilize bottom sediments, baffle excessive currents and filter suspended matter, which improves water quality (Short and Neckles 1999). Increasing water temperature will compromise many of these functions by affecting seagrass metabolism directly, which in turn will result in many (unpredictable) changes in the seasonal and geographical patterns of species distribution and abundance. In addition to these effects, increased mutagenic and physiological damage from rising ultraviolet radiation are expected for seagrasses in shallow water habitats (Short and Neckles 1999).

8.3.5 Sea-level rise threatens coastal and shallow habitats

Although there is some controversy over the extent to which the sea level will rise as result of large volumes of ice melting under different climate change scenarios (L. Miller and Douglas 2006), there is little doubt that sea level rises have already occurred, and will continue, at least over the next few centuries (Meehl *et al.* 2005; Mitchell *et al.* 2006). We need look no further back than the last 15 000 years to appreciate the dramatic effect that shifts in the polar ice can have on sea levels (section 8.2). In the tropics in particular, many low-lying countries, such as Bangladesh, Guyana, the Maldives, Tuvalu, Tonga and Kiribati, are threatened not only by the expected coastal flooding under normal climatic conditions, but also by their heightened susceptibility to storm surges brought about by extreme climatic events such as cyclones (Ali 1996; Mimura 1999). These trends are exacerbated by a general movement of people towards major coastal urban centres (Mimura 1999) and an economic and subsistence reliance on coastal fisheries in many of these developing nations (Roessig *et al.* 2004).

In addition to the saltwater intrusion of coastal habitats such as freshwater wetlands (e.g. some of the high-value coastal freshwater wetlands of Australia could be replaced by saline marshes – Junk 2002) and beach erosion that would accompany rapidly rising sea levels (Short and Neckles 1999), sea level rise will also increase the depth of water above coral reefs and seagrass beds. This would have important implications for the photosynthetic capacity of symbiotic algae in corals (see Chapter 7 and section 8.3.4) and seagrass species themselves. If water depth increases faster than corals or seagrass beds can grow (and the major concern of many climate change scenarios is the rapid predicted rate of sea level change compared with anything that occurred in the geological past), then entire reefs or seagrass beds could disappear, essentially through starvation (Wells and Edwards 1989). Another indirect effect of rising sea levels is the associated change in the magnitude of tidal ranges in some areas depending on the configuration of coastal geomorphology. Changing the range and extent of tides will probably exacerbate the effects of increasing water depths above seagrass beds and increase the flow rate of currents, which could cause greater submarine erosion and increased suspension of sediments (Short and Neckles 1999).

8.3.6 Drying catchments

Human-mediated climate change is also expected to affect the world's freshwater systems. In concert with the rising demand for freshwater brought about by a rapidly expanding human population (especially in the tropics – Vörösmarty *et al.* 2000), climate change is likely to increase the freshwater demand–supply ratio by up to 5%. Although this change is much lower than that expected to result from the increasing number of people seeking water resources (Vörösmarty *et al.* 2000), the additional pressures imposed by higher evaporation rates will be particularly dramatic in drier areas in tropical Africa and South America (Vörösmarty *et al.* 2000). The harsh reality is that the engineering problems that will result from increased pressure on water infrastructure and service, and the socioeconomic burden they impose in these regions, will in all likelihood subjugate concerns regarding any disruptions to the biological integrity of tropical freshwater systems brought about by reductions in freshwater availability (Malmqvist and Rundle 2002). Paradoxically, even those regions that are predicted to receive more rainfall as the earth warms may nevertheless suffer from worse droughts and soil erosion. This is because new rain has tended to come in the form of more frequent and increasingly intense downpours, with fewer moderate falls, thereby increasing flooding and subsequent runoff (Goswami *et al.* 2006).

Given the importance of pristine areas for the maintenance of the incredible biodiversity supported by tropical freshwater systems (see Chapter 7), these concerns are well supported by evidence from existing studies and modelled predictions. For example, the multiyear drought in the Sahel region of Africa brought about by low rainfall saw a serious reduction in the productivity of the Niger River delta fishery, from 90 000 to 45 000 tonnes/year. Palaeoclimatic studies also show that tropical wetlands are sensitive to changes in hydrology. Rising temperatures will increase evapotranspiration rates and reduce water availability to plants, with a concomitant increase in fire risk during low-water periods (Malmqvist and Rundle 2002). The elevated frequency of extreme weather events such as super El Niños will also strongly affect tropical wetlands by acting synergistically with human landscape degradation (Vörösmarty *et al.* 2000).

8.3.7   Threat synergies

Climate change by itself has major implications for tropical biodiversity; however, many of the negative predictions result from, or are exacerbated by, interactions with other drivers of biodiversity change (see Spotlight 8: Stephen Schneider). For example, the world's smallest butterfly (Sinai baton blue butterfly, *Pseudophilotes sinaicus*), found in Egypt, provides a pertinent case study on how climate change and human actions can together threaten a species. The Sinai baton blue is currently confined to small habitat patches, and is threatened by global warming and direct human activities such as livestock grazing and medicinal plant collection (Hoyle and James 2005). It is dependent on Sinai thyme (*Thymus decussates*), which occurs in discrete patches on the mountains (M. James *et al.* 2003). It has been predicted that only in the absence of global warming could this species persist for at least 200 years despite grazing and plant collections (Hoyle and James 2005).

With the doubling of CO_2 levels in the atmosphere, biodiversity hotspots may also be more vulnerable than predicted by direct human activities alone (see Chapter 1). Malcolm *et al.* (2006) predicted that up to 43% of endemic species in biodiversity hotspots could vanish in 100 years as a result of elevated CO_2 levels, and over 2000 plant species could be extinguished in tropical biodiversity hotspots of the Caribbean, Indo-Burma and the tropical Andes. Other atmospheric changes, such as elevated CO_2 and increases in nitrogen deposition, will also erode biodiversity and disrupt ecological processes. In other tropical and temperate biomes, elevated CO_2 levels have been demonstrated, or are predicted, to reduce species diversity, alter biotic interactions and facilitate the spread of invasive species such as woody weeds (Coley 1998; S.D. Smith *et al.* 2000; Hartley and Jones 2003; Zavaleta *et al.* 2003). It has also been suggested that the negative effects of increased CO_2 levels may be greater in more speciose communities because these may contain more responsive species than species-poor communities (Niklaus *et al.* 2001). The capacity of species to shift their distributional ranges in response to climate change will be curtailed by human-mediated impositions such as habitat fragmentation and the inhospitability of the agricultural and urban matrix surrounding forest remnants (see Chapter 3).

Many negative synergies are also predicted in tropical aquatic realms. Higher rainfall predicted for some areas as a result of climate change will wash more fertilizers and sewage into estuaries and coastal waters, thus triggering algal blooms and an increased mortality of inshore fishes, invertebrates and those that eat them (including humans – Epstein 2000). It is also predicted that the higher future availability of dead corals resulting from human activities and climate change (see Chapter 7 and section 8.3.4) will provide more substrata for colonization by ciguatoxin-producing microalgae (Chateau-Degat *et al.* 2005), thus leading to more frequent and severe blooms. Global climate change and warming freshwater systems are also predicted to interact with increased levels of pollution to reduce tadpole viability. The sensitivity of frog larvae to pesticide contaminants such as endosulfans increases with rising water temperatures, demonstrating how thermal regimes interact with common pesticides to reduce amphibian fitness and increase extinction risk (Broomhall 2004).

8.4   Fighting climate change

The sad truth of climate change is that, regardless of our present and future efforts to reduce greenhouse gas emissions and halt environmentally degrading activities such as mass deforestation and the pollution of freshwater and marine ecosystems, most climate change models predict rapid global change over the coming century, even if a total reversal of current practices were possible (Flannery 2005). This is because of the huge momentum already built up by centuries of industrial pollution. However, global threats to humanity and biodiversity have been overcome by sound policy and better education; for example, the agricultural use of the toxic insecticide DDT (dichloro-diphenyl-trichloroethane) was restricted in much of the world in the 1970s and legislation to ban chlorofluorocarbon (CFC) aerosol propellants was introduced in the 1980s because of the damage they caused to stratospheric ozone concentrations. These successes suggest that the political and moral will to reduce impacts are within our grasp. The short section that follows outlines some of the principal ways in which reductions may be achieved. A more complete list of actions can be found in books such as Tim Flannery's *The Weather Makers* (2005), and films such as Al Gore's *An Inconvenient Truth* (2006).

8.4.1   Carbon credits and carbon-offset funds

Carbon-offset funds are proposed as an effective economic incentive for slowing tropical deforestation and the resultant climate change that this clearance will induce (Santilli *et al.* 2005). Signatories of the 2005 Kyoto Protocol of the

United Nations Framework Convention on Climate Change (UNFCCC) support the concept of carbon trading, whereby a country that produces lower than recommended carbon emissions (i.e. below 1990 levels) can sell its remaining 'carbon credits' to other countries to offset their emissions. Carbon credits can also be purchased, for example, from developing countries by replanting their denuded landscapes. The idea is that carbon would then be stored in trees rather than being released to the atmosphere. It is predicted that such an endeavour should reduce the overall emission and atmospheric concentration of greenhouse gases (W.F. Laurance 2006a). However, some developing countries, such as Brazil, are opposed to carbon credits because they believe that the process will limit their future development of currently 'intact' landscapes such as the Amazon rain forest. The use of carbon credits to halt deforestation nevertheless makes intuitive macroeconomic sense. For example, by clearing 1 ha of tropical forest for slash-and-burn agriculture, a farmer nets a profit of only US$200. However, by leaving the same patch of forest intact, its prevailing carbon value could be worth several thousands of dollars (W.F. Laurance 2006a). There are a number of examples in the tropics that show that halting deforestation and paying locals through carbon credit funds could be feasible; for example, a good working model is being used in Costa Rica to monitor and reward farmers for good land use practices (W.F. Laurance 2006a).

8.4.2 Climate-friendly technology

Climate change is predominantly driven by increased emissions of gases that change the way in which the planet preserves and distributes the heat originating from the sun (the so-called 'greenhouse' effect). As such, any activity that reduces our reliance on technologies that produce greenhouse gases will reduce our climate change footprint as societies and individuals. Simple changes to lifestyles include:

1   reducing the reliance on fossil fuels for transport (e.g. automobiles powered by petroleum products) by switching to renewable technologies (e.g. bicycles);
2   demanding that government representatives provide households with alternative energy sources (e.g. solar energy) that do not rely on the burning of fossil fuels (e.g. coal-powered electricity);
3   embracing non-carbon-polluting energy sources for household power needs (e.g. solar, wind, wave, and even nuclear power);
4   buying more local food and technology products to reduce the need for global transport powered by fossil fuels;
5   recycling materials used in the household;
6   supporting elected leaders who embrace global incentives and protocols to reduce emissions (e.g. the Kyoto Protocol).

8.5   Summary

1   Climate change is a very real and active process that threatens tropical

biodiversity and human welfare, mostly because of the speed with which the current human-mediated changes are taking place in already heavily modified ecosystems.

2   There is ample evidence from the palaeontological record that climate change is a driving force behind tropical biodiversity change.

3   Human-mediated climate change is affecting tropical biotas via range shifts (latitudinal and elevational), changes in phenology, increasing prevalence, distribution and severity of diseases and parasites, coral bleaching, drying of freshwater systems and sea level rise.

4   Carbon-offset funds should be used as an incentive for rural people to preserve forest by halting tropical deforestation.

5   Individuals can offset the negative impacts of climate change by adopting minor lifestyle changes that reduce total emissions and providing momentum for large-scale industrial change through changes in government policy.

8.6   Further reading

Flannery, T. (2005) *The Weather Makers. The History and Future Impact of Climate Change*. Text Publishing, Melbourne, Australia.

Harvell, C. D., Mitchell, C. E., Ward, J. R., Altizer, S., Dobson, A., Ostfeld, R. S. and Samuel, M. D. (2002) Climate warming and disease risks for terrestrial and marine biota. *Science* **296**, 2158–2162.

Intergovernmental Panel on Climate Change (2001) *Climate Change 2001: The Scientific Basis: Contribution of Working Group 1 to the Third Assessment Report of the Intergovernmental Panel on Climate Change*. Cambridge University Press, Cambridge.

Roessig, J. M., Woodley, C. M., Cech, J. J. and Hansen, L. J. (2004) Effects of global climate change on marine and estuarine fishes and fisheries. *Reviews in Fish Biology and Fisheries* **14**, 251–275.

Santilli, M., Moutinho, P., Schwartzman, S., Nepsad, D., Curran, L. and Nobre, C. (2005) Tropical deforestation and the Kyoto Protocol. *Climate Change* **71**, 267–276.

9

Lost Without a Trace: the Tropical Extinction Crisis

In this chapter, we first define what is meant by the term 'extinction'. We then discuss historical extinctions and the various controversies surrounding current and future estimates of biodiversity loss in the tropics. We provide case studies of tropical extinctions and discuss the general traits that increase extinction risk. For example, we explore questions such as: Are naturally rare species more vulnerable to extinction than common ones? Are large-bodied species more vulnerable to extinction than smaller ones? Do certain behaviours (e.g. mixed-species flocking in birds) make a species more prone to extinction? Do extinctions affect ecosystem functioning (e.g. loss of frugivores)? This chapter highlights the extinction proneness of tropical species, and reports on the overall biotic losses and their drivers within tropical realms.

9.1   Defining 'extinction'

Extinction of a population is termed a 'local' or 'population extinction', often referred to by the more specialized term 'extirpation' to distinguish it from true (global) extinction. Local extinctions cannot, of course, be observed directly, but are commonly inferred by comparing results from recent surveys with those made in the past (the historical approach). These population extinctions may result in the erosion of the genetic diversity of the species (e.g. loss of a subspecies) and may eventually translate into species extinction. According to the World Conservation Union (IUCN), if extensive surveys of known or possible habitats done throughout a species' range, and carried out over a reasonable period of time, fail to record an individual, then the species is declared extinct (www.iucnredlist.org). Such surveys should be undertaken over a time frame appropriate to the taxon's life cycle (e.g. covering at least a full generation) and life form (e.g. attempt to compensate for potentially low detectability of evasive or nocturnal species). IUCN maintains a record of the species that are presumed to have become extinct since 1500 AD.

9.2   Historic extinctions

According to the fossil record, only 2–4% of the species that have ever lived persist today (www.iucnredlist.org), and perhaps far fewer (some put the estimate at 0.1%; Raup 1991b). In the past, extinctions were caused by climate change, volcanism, bolide strikes and evolutionary competition (see Chapter 8). However, for over 50 millennia, and most intensively within the last few centuries, emergent threats to the earth's flora and fauna have centred on human agency (Pimm *et al.* 1995). Habitat loss and fragmentation, overexploitation, introduced species and diseases, pollution and global climate change are the major factors threatening global biodiversity, and have probably contributed to an alarming rise in modern extinction rates (see reviews in Lawton and May 1995). Halting these human-caused extinctions at all scales is a primary goal of conservation biology.

Humans may have been in Australia since approximately 45 000 years ago, and their arrival on the continent was coincident with the extinction of 22 genera of megafauna (mammals weighing more than 44 kg) (Brook and Bowman 2002). In the Americas, extraordinary animals such as sabre-toothed cats (*Smilodon* spp.), mammoths (*Mammuthus* sp.) and giant ground sloths (*Megalonyx jeffersonii*) all disappeared between 11 000 and 13 000 years ago, at the same time as Clovis hunters arrived (Burney and Flannery 2005). A large number of South American genera were also lost at this time. Many megafaunal species also disappeared from oceanic islands within a few hundred years after the arrival of humans. Classic examples include the disappearance of the dodo (*Raphus cucullatus*) from Mauritius, moas (e.g. *Dinornio maximus*) from New Zealand and elephantbirds (*Aepyornis maximus*) from Madagascar. Megafaunal collapse during the late Pleistocene following human arrival in previously uninhabited regions of the world can largely be traced to a variety of human impacts such as overharvesting, biological invasions and habitat transformation, although some maintain a role for climate change at the end of the last ice age (Burney and Flannery 2005). Since 1500 AD, at least 844 species are known to have become extinct (McPhee 1999).

9.3   Extinction rates

The rate and extent of human-mediated extinctions is intensely debated (e.g. Heywood *et al.* 1994; Dirzo and Raven 2003), but despite the debate there is general agreement that extinction rates have soared over the past hundred years owing to accelerating habitat destruction and burgeoning human populations (Millennium Ecosystem Assessment 2005; Figure 9.1). Humans are implicated, directly or indirectly, in a 100- to 10 000-fold increase in the 'natural' or 'background' extinction rate that normally occurs due to phenomena such as competitive interactions and local environmental change (E.O. Wilson 1989; Pimm *et al.* 1995; Dirzo and Raven 2003). Further, extinctions of many cryptic or poorly studied taxa, including most invertebrates, may have been gone unnoticed, thus resulting in a gross underestimation of the current human-mediated extinction rate (Dunn 2005). For the well-studied vertebrates,

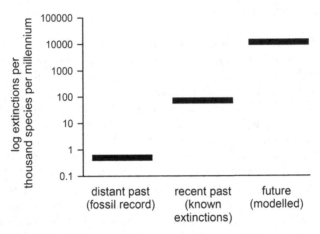

Figure 9.1 Estimated species extinction rates from the distant past, the recent past and the future. The current extinction rate is about a thousand times higher than that indicated by the fossil record, and projected future extinction rates are estimated to be more than 10 times higher than the current rate. (After Millennium Ecosystem Assessment 2005. With permission.)

some 31%, 12% and 20% of known amphibian, bird and mammal species, respectively, are currently threatened with extinction (www.iucnredlist.org). Moreover, the highest percentage of rapidly declining amphibian species occurs in the Neotropics (Stuart *et al.* 2004) (Figure 9.2), with Indonesia, India, Brazil and China among the countries with most threatened bird and mammal species (www.iucnredlist.org). Plant species are rapidly declining in South and Central America, central and West Africa and Southeast Asia (www.iucnredlist.org), and nearly 600 species of threatened birds and mammals are predicted to become extinct within the next 50 years (Dirzo and Raven 2003).

How many species are being lost is also hotly debated (e.g. Heywood and Stuart 1992). Various estimates range between a few thousand to more than 100 000 species being extinguished every year (Ehrlich and Ehrlich 1996; E.O. Wilson 2000). Pimm and Raven (2000) estimated that currently 1000 to 1 000 000 species are being lost per decade, and they predicted that the current rate of deforestation alone may lead to the extinction of 40% of the species in 25 biodiversity hotspots identified by Myers *et al.* (2000) (Chapter 1). Other studies predict likely extinctions in various taxonomic groups. For example, one in eight bird species globally may become extinct over the next 100 years, in 99% of cases as a result of human activities such as deforestation and hunting (BirdLife International 2000). Although all of the above estimates are prone to prediction error, even the most liberal parameters and assumptions point to a looming global biodiversity crisis. This, the 'sixth great extinction', follows the five largest extinction waves to have occurred at the end of the Ordovician [444 million years ago (mya)], Devonian (360 mya), Permian (250 mya), Triassic (200 mya) and Cretaceous (65 mya). The modern crisis may eventually rival the previous five events, and perhaps be more catastrophic in the tropics because this region contains two-thirds of global biodiversity and is heavily affected by humans.

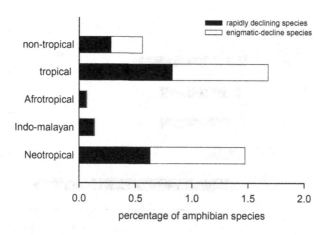

Figure 9.2 Percentage of rapidly declining and enigmatic-decline amphibian species between non-tropical and tropical regions, with the majority of within-tropic amphibian declines coming from the Neotropics. (Data derived from Stuart *et al.* 2004.)

Predictions of future extinction rates have largely depended on the relatively simple species–area relationship. Despite debate over its derivation, forms and uses (and abuses), the species–area relationship can essentially be obtained as an axiomatic result of plotting the number of species against sampling area for a series of samples of increasing size. The curve generally increases with an initially steep slope followed by a flattening out at large sampling areas. This relationship has been used to determine the area required for adequate sampling of species in a community, to quantify community structure and species richness, to measure disturbance effects and to define reserve size (He and Legendre 1996). A more recent application is its wide use in estimating species extinctions in relation to habitat loss (Brook *et al.* 2003a). At the estimated annual loss of 0.8% of forests globally, J.B. Hughes *et al.* (1997) used a species–area equation to predict that between 0.1% and 0.3% of tropical forest species could become extinct every year. In other words, about 14 000–40 000 species may be vanishing from tropical forests annually (J.B. Hughes *et al.* 1997). J.B. Hughes *et al.* (1997) also estimated that 16 million local populations could be lost every year from tropical forests as a result of deforestation. This is particularly disconcerting because many species have distinct subpopulations that are crucial for the maintenance of overall genetic diversity. Another problem is that around 19 of 20 tropical forest species may still be unknown to science (Dirzo and Raven 2003), meaning that it is likely that estimates of extinction risk are vastly underestimated.

Deforestation, however, is not the only driver of biodiversity change (see Chapters 5 and 6). O.E. Sala *et al.* (2000) modelled the sensitivity of biodiversity in various biomes to the major drivers of loss, including changes in land use (habitat loss and degradation), climate change (global warming), increasing concentration of atmospheric CO_2, increase in nitrogen deposition and biotic exchange (deliberate or accidental introduction of non-native biotas). Their broad-brush estimates revealed that major land use changes in the tropics will probably lead to greatly reduced tropical biodiversity by 2100 (Figure 9.3).

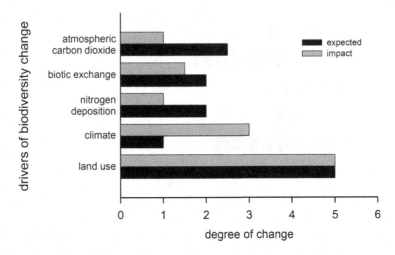

Figure 9.3 Expected changes and likely impacts for the year 2100 in five major drivers of biodiversity change and their influence on biodiversity in the tropics. Ranks are given from 1 (low) to 5 (high) for each driver. (Data derived from O.E. Sala *et al.* 2000.)

Other drivers, such as climate warming and nitrogen deposition, may have lesser effects on tropical biodiversity (Figure 9.3); however, it is highly probable that land use changes will act synergistically with other agents for an accumulative impact on tropical biodiversity (Chapter 8). Therefore, the global estimate that over 85% of amphibian, bird and mammal species are threatened by habitat loss and degradation (www.iucnredlist.org) may well underestimate the eventual extinctions in these taxa.

9.4   Case studies of tropical extinctions

Anthropogenically driven extinctions have occurred on tropical and subtropical islands for thousands of years (Diamond 1989; Steadman 1995). Mechanisms for prehistoric extinctions may have been similar to those operating in the current extinction crisis: overhunting, introduced predators and diseases and habitat destruction (Milberg and Tyrberg 1993), but the scale and intensity of human impacts on tropical island ecosystems seem to be accelerating.

Brook *et al.* (2003a) reported local extinctions of a wide range of taxa (vertebrates, invertebrates and plants) from Singapore, and used these to project the future biotic losses from all of Southeast Asia. Singapore has experienced exponential human population growth, from ~150 subsistence economy villagers in the early 1800s to four million people in 2002 [Corlett 1992; WRI (World Resources Institute) 2003]. In particular, Singapore has transformed itself from a developing country to a developed metropolis of economic prosperity within the past few decades (see Chapter 1). However, the success of Singapore came at a severe cost to its environment and biodiversity. Since the British first established a presence in Singapore in 1819, original vegetation cover has been almost entirely cleared (I.M. Turner *et al.* 1994; Corlett 2000). Deforestation resulted initially from the cultivation of short-term cash crops and subsequently from urbanization and industrialization (Corlett 1992). Brook *et al.* (2003a) showed substantial rates of documented (observed) and inferred (based on what could have occurred in Singapore before habitat loss) extinctions, with most extinct taxa (34–87%) being species of butterflies, fish, birds and mammals (Figure 9.4).

Because they are a relatively conspicuous taxon, birds have been the focus of extinction studies for many years. Sodhi *et al.* (2004a), in their review of tropical bird extinctions, reported that between 1% and 67% of tropical forest avifauna is exterminated following deforestation of an area. Expanding to other taxa, a review by I.M. Turner (1996) showed that there have been reported local losses of species due to fragmentation of tropical forests (bees, termites, butterflies, forest birds and mammals). I.M. Turner (1996) hypothesized that factors such as a reduction in population size and immigration, negative edge effects (Chapter 3) and introduced species (Chapter 5) may be responsible for these losses.

9.5   Extinction lags

Many long-lived species (e.g. tropical trees) may take decades to perish following habitat degradation. This phenomenon evokes the concept of 'living dead' species or those 'committed to extinction' (Simberloff 1986). The eventual loss of such

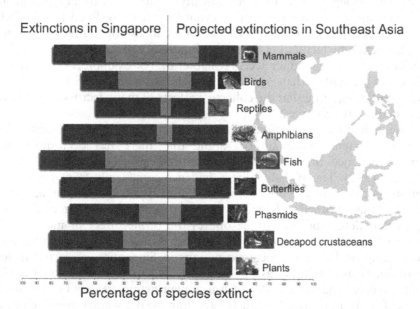

Figure 9.4   Population extinctions in Singapore and Southeast Asia. (After Sodhi *et al.* 2004b. Copyright, Elsevier.). Light and dark bars on the left represent, respectively, recorded and inferred extinctions in Singapore. Light and dark bars on the right represent, respectively, minimum and maximum projected extinctions in Southeast Asia (see Brook *et al.* 2003a.)

species is referred to as the 'extinction debt' due to past habitat loss (Tilman *et al.* 1994). Brooks *et al.* (1999b) found that a half-life (i.e. the time taken to lose half of the species in a particular area) for avifaunal extinctions in fragmented forests in Kenya was on average 50 years. This half-life increased with increases in fragment size (Brooks *et al.* 1999b; Ferraz *et al.* 2003; Figure 9.5). Further, smaller fragments generally contain smaller populations with less genetic diversity, and isolation can reduce immigration and any chance of a 'rescue effect' (immigration by unrelated individuals into isolated populations – J.H. Brown and Kodric-Brown 1977; see also Chapter 3). Half-lives can be shortened further when fragments contain large numbers of predators or parasites, or when they are susceptible to the damage caused by severe catastrophic events such as fires and storms (Brooks *et al.* 1999a). In fact, small fragments of approximately 100 ha can lose half of their bird species in less than 15 years (Ferraz *et al.* 2003; Figure 9.5). The phenomenon of extinction lags can, however, provide conservation practitioners with the opportunity to restore degraded or lost habitats in time to prevent mass extinction. This would depend, of course, on the degree of degradation and the relative sensitivity of the biota to the degrading process (Lens *et al.* 2002).

9.6   Extinction drivers

Ultimately, phenomena that decrease survival and reproductive rates beyond the point at which species are able to replace themselves can cause extinction, and they may act independently or synergistically (Figure 9.6). It may be difficult to identify a single cause of an extinction event (Pimm 1996; Reed 1999) – habitat loss may cause some extinctions directly, but it can also be indirectly responsible for an extinction by facilitating the establishment of an invasive species or disease agent, or by improving hunter access (Chapter 6). For example, avian extinctions in Hawaii were higher than predicted by habitat loss alone (Scott *et al.* 1988; Pimm 1996), suggesting that the collective effects of many factors contributed to the extinction rate observed.

Although habitat destruction is considered a major cause of species losses and endangerment, some authors argue that there has been little direct, empirical evidence that tropical deforestation causes extinctions (Heywood and Stuart 1992). The available direct evidence comes only from well-designed experiments that compare data before and after deforestation (e.g. as has been done in Manaus, Brazil – Bierregaard *et al.* 1992). Mounting evidence also suggests that deforestation may be one of the prime direct or indirect causes of reported extirpations. Brook *et al.* (2003a), using a species–area relationship fitted to empirical data on local extinctions in Singapore, projected losses of 40% of Southeast Asian biotas by 2100 due to deforestation. Likewise, Grelle (2005) projected that up to 18% of Amazonian mammals may be lost by 2020 due to deforestation. Brooks *et al.* (1997) employed a similar approach using a theoretical value for the species–area slope to predict the number of endemic species threatened by deforestation in Southeast Asia.

A more detailed analysis by Brooks *et al.* (1999a) determined the correlation between the degree of deforestation and the existence of threatened endemic

(a)

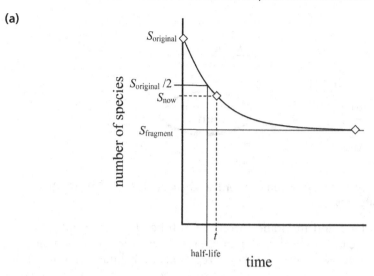

(b)

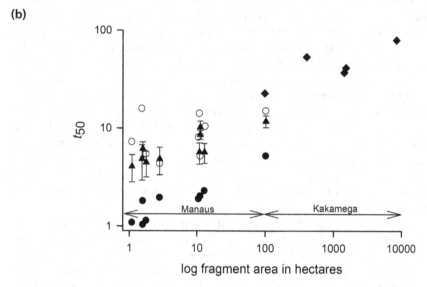

log fragment area in hectares

Figure 9.5 (a) Exponential loss of species from a fragmented forest. The number of species in an area of once-continuous forest ($S_{original}$) declines over time t (S_{now}), eventually declining to a stable number for that fragment ($S_{fragment}$). Estimates of $S_{original}$ and $S_{fragment}$ can be obtained from the formula $S = cA^z$, where z is assumed to be ~0.25. The time taken to lose 50% of the species can be characterized by a half-life because the decay follows an exponential curve. (After Brooks *et al.* 1999b. Copyright, Blackwell Publishing Limited.) (b) Time to lose 50% of the species from forest fragments in Manaus, Brazil (circles and triangles), and Kakamega, Kenya (diamonds). Three different results were shown from Manaus: (1) Bayesian decay with 95% confidence bounds (triangles), (2) jack-knife estimates with exponential fit to all years (empty circles) and (3) jack-knife estimates with exponential fit to initial decay (filled circles). (After Ferraz *et al.* 2003. Copyright, National Academy of Sciences, USA.)

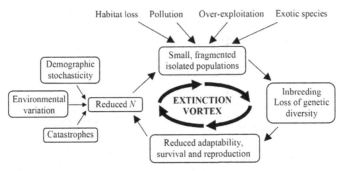

Figure 9.6 Causes of biotic extinctions. (After Frankham *et al.* 2002. Copyright, Cambridge University Press.)

species in the lowland and montane forests of Southeast Asia. Broadly, and as expected, more threatened bird and mammal species were found in heavily deforested areas. However, there were some regional and habitat differences; the number of threatened birds and mammals in montane areas was underestimated by the species–area prediction. This is possibly due to their restricted niches and the constraints on dispersal that make them disproportionately more vulnerable to habitat loss (Brooks *et al.* 1999a). A similar trend was found for mammals of the lowlands of Lesser Sundas (Indonesia). Conversely, lowland avifauna from Lesser Sundas and Java were less threatened than predicted by the same equations and thus apparently more tolerant to habitat loss. However, this result may simply be a manifestation of the more sensitive avian endemics having already been lost prior to the first scientific surveys. The Brooks *et al.* (1999a) analyses did not demonstrate a direct link between deforestation and species endangerment, but they did show that deforested areas generally harbour more threatened species due to habitat loss, degradation or associated causes (e.g. hunting). Therefore, deforestation is likely to be a major direct or indirect force in biotic losses and endangerment in the tropics.

Montane habitats appear to be particularly important reservoirs of tropical biodiversity and unique habitats, yet they are highly threatened. For amphibians, Stuart *et al.* (2004) reported a high percentage of rapidly declining amphibian species in flowing water situated in lowland and montane tropical forests compared with elsewhere, thus highlighting the importance of these habitats for the persistence of this taxon (Figure 9.7). In montane forests of Peninsular Malaysia, 19 bird species responded positively to increases in canopy cover (Soh *et al.* 2006). Their simulation model suggested that at least 80% of the forest canopy must remain intact to guarantee persistence of all montane avian species (Figure 9.8), and a canopy cover of 40% is required to maintain at least two-thirds of these. The factors responsible for this higher sensitivity of Malaysian montane birds to deforestation appears to be restricted ranges, lack of dispersal capacity and/or specialized physiology (Soh *et al.* 2006).

Roads and trails tend to make deforested areas more accessible to human hunters (Chapter 6), which can in turn increase the exploitation of forest species and cause them to decline or become extinct. In Ecuador, hunting may have been partially responsible for the decline of bird species such as the great tinamou

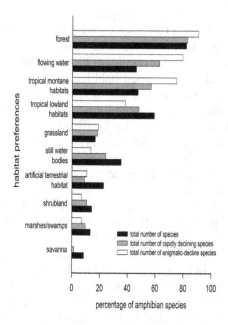

Figure 9.7   Habitat preferences of rapidly declining and enigmatic-decline amphibians in relation to all amphibian species. Rapidly declining species are those that now qualify for listing in a more threatened position on the World Conservation Union Red List scale of endangerment relative to 1980. Enigmatic-decline species are rapidly declining species that have shown dramatic declines, even where suitable habitat remains, for reasons that are not fully understood. (Data derived from Stuart *et al.* 2004.)

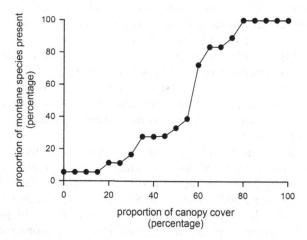

Figure 9.8   Proportion of 19 montane bird species present in simulated proportions of canopy cover remaining. (After Soh *et al.* 2006. Copyright, Elsevier.)

(*Tinamus major*) and crested guan (*Penelope purpurascens*; Leck 1979). In Brazil, there is evidence that hunting pressure reduced the survival rates of forest primates and carnivores living in forest fragments (Michalski and Peres 2005).

Increased accessibility of forests to people may have other negative effects on forest biotas. Diamond *et al.* (1987) speculated that increased human-mediated disturbance (e.g. nest predation by domestic dogs) might have been one factor contributing to the extinction of ground-dwelling bird species such as the banded pitta (*Pitta guajana*) and black-capped babbler (*Pellorneum capistratum*) in the Bogor Botanical Garden (Java). The use of pesticides on tropical agricultural crops may also be detrimental to native biotas. The extirpation of two Mauritian carnivorous forest raptors, the Mauritius kestrel (*Falco punctatus*) and Mauritius cuckoo-shrike (*Coracina typica*), from some areas prior to the 1950s was probably due to the use of organochlorine-based chemicals (Safford and Jones 1997). Fortunately, both species seem to have recovered after the use of organochlorines became less popular after the 1970s.

Deforestation may also facilitate the spread of invasive plants and animals into previously pristine forested areas (Chapter 5), although the magnitude of their impact on native species is not always clear. The introduced brown tree snake (*Boiga irregularis*) was, however, probably responsible for the loss of at least 12 of 18 native bird species in Guam (Fritts and Rodda 1998; Wiles *et al.* 2003). The probability of extinction of bird species on oceanic islands experiencing habitat loss was also elevated by the human introduction of mammalian predators to those islands (Blackburn *et al.* 2004). For example, many ground-nesting birds are negatively affected by feral cats and domestic dogs (Leck 1979; Diamond *et al.* 1987; Kattan *et al.* 1994), although some ground-nesting bird species may be able to avoid excessive predation by cats because of preadaptations to native ground predators such as rodents and land crabs (Mayr and Diamond 2001). On French Polynesian islands, passerine monarchs (*Pomarea* spp.) were rarely found on islands where large populations of the introduced brown rat (*Rattus rattus*) had become abundant (Seitre and Seitre 1992). Additionally, larger invasive birds such as the red-vented bulbul (*Pycnonotus cafer*) can outcompete or actively exclude smaller native species (Thibault *et al.* 2002).

Hawaii has lost a substantial part of its endemic bird fauna since the introduction of avian malaria in the 1920s (Freed *et al.* 2005; see also Chapters 5 and 8). The reason for this devastation was that Hawaiian birds, having evolved in absence of the disease, were immunologically susceptible to this novel parasite. Other pathogens introduced to new areas over the past few decades may have been responsible, in part, for the declines and extinction of many other species. Across locations worldwide, amphibian populations have undergone remarkable declines, with some species being driven to extinction and many populations of other species facing the prospect of a similar fate (Beebee 1992; Blaustein and Kiesecker 2002; Collins and Storfer 2003). Originally, factors such as deforestation, the draining of wetlands, overharvesting and pollution were often attributed to these population declines. However, more recent evidence points to the role of disease and environmental change, perhaps associated with climatic shifts, in causing amphibian population declines in protected and apparently pristine areas (Houlahan *et al.* 2000). For example, the Monteverde harlequin frog (*Atelopus* sp.) and golden toad (*Bufo periglenes*) vanished from the mountains of Costa Rica possibly due to the pathogenic chytrid fungus (*Batrachochytrium dendrobatidis*) (Pounds *et al.* 2006; Chapter 8). Pounds *et al.* (2006) also hypothesized that global warming has produced microclimatic conditions conducive to the optimum growth and spread of disease-causing *Batrachochytrium*, even in remote highland areas.

9.7   Extinction proneness

Certain life history, behavioural, morphological and physiological characteristics make some species more susceptible than others to the types of processes now being imposed upon the natural world by human activities. In general, large-sized species with restricted distributions that demonstrate habitat specialization tend to be at greater risk of extinction from human agency than others within their taxa. This is especially true for processes that cause rapid habitat loss, such as deforestation, when there is insufficient time for evolutionary adaptation to these new selective pressures. In this section, we discuss the effects of various factors (e.g. biogeographical, morphological, and behavioural) on the extinction proneness of several tropical plants and animals.

9.7.1   Biogeography

Whether populations at the edges of species' distributions are more vulnerable to deforestation than their counterparts deep within their primary ranges remains unclear (Brooks 2000). One hypothesis is that populations are at their physiological and ecological limits at their distributional edge because they are more vulnerable to genetic (e.g. inbreeding depression) and environmental stressors (e.g. pesticides – Møller 1995) resulting from relatively lower population densities. For example, 19 of 29 bird species at the upper or lower limit of their elevational distributions were extirpated from the western Andes (Kattan *et al.* 1994), and similar observations have been reported elsewhere (Christiansen and Pitter 1997; Gillespie 2001). However, in the Colombian Andes, Renjifo (1999) found no evidence that bird species ranging beyond their usual elevational limits were disproportionately more likely to become extinct, suggesting perhaps that some species are resilient to putatively harsher conditions at their distributional limits.

Rare species may also be more prone to extinction than common species because there are generally fewer of them, unless they are locally rare but geographically widespread. But rarity itself is generally an expression of other characteristics such as body size, habitat specificity and geographic distribution (Kattan 1992; Goerck 1997). In Ecuador, some inherently rare species, such as the tiny hawk (*Accipiter superciliosus*) and crane hawk (*Geranospiza caerulescens*), were possibly extirpated (Leck 1979); this was attributed to their small initial population sizes, requirement for large territories, and diet specialization. Similarly, rare species were more likely to be extirpated than common species in the Colombian Andes (Renjifo 1999), and were more readily affected than common species by fragmentation in the tropical moist rain forests of eastern Queensland, Australia (Warburton 1997).

However, rare species are not always more extinction prone than common species. Karr (1982a) sampled the understorey fauna at a site near Barro Colorado Island (Panama) that was presumed to be the source of individuals colonizing the island. Species extirpated from the island were no less abundant at the source site (e.g. song wren, *Cyphorhinus phaeocephalus*), but the source populations of locally extinct species did exhibit higher annual population variability. Karr hypothesized that the latter type of species tracks variable food resources and requires larger areas for foraging. Similarly, apparently widespread and common species are sometimes extirpated (e.g. the brown-hooded parrot, *Pionopsitta haematotis*, and red-capped manakin, *Pipra mentalis*, in Ecuador) when habitat degradation is particularly acute (Leck 1979).

The area of a species' range can also dictate extinction proneness (Ribon *et al.* 2003). Small ranges may make species more vulnerable to local or regional stochastic perturbations, even if local abundance is high. For example, proportionally more passerines with relatively small geographic ranges in the Americas are threatened than their more widely distributed counterparts (Manne *et al.* 1999). A similar relationship is evident in tropical Australia, where the bird species most at threat of extinction are those occupying few biogeographic regions within Australia and, interestingly, those also resident in the fewest countries outside Australia (Garnett and Brook 2007). Such trends are worrisome because those species with shrinking ranges due to adverse human activities, such as 173 threatened mammal species, are predicted to have ever-increasing chances of extinction without some positive intervention (Ceballos and Ehrlich 2002). Range-restricted butterfly species seem to be particularly vulnerable to changes in their habitat (Hill *et al.* 1995, 2001; Hamer *et al.* 1997, 2003; Ghazoul 2002). In Vietnam (Tam Dao Mountains), endemic butterflies that are confined to mature montane forests (Spitzer *et al.* 1993) are more common in the understorey than in the canopy. This result illustrates the high conservation value of these types of forests for some invertebrate taxa and reinforces the notion that loss of understorey habitats will probably lead to pronounced increases in extinction rates of a wide range of taxa.

Deforestation can also reduce the habitat patch sizes necessary for species with large home range requirements. For example, habitat area was the main predictor of forest primate and carnivore persistence in 129 fragments in southern Brazil (Michalski and Peres 2005). Additionally, four canopy frugivores requiring large home ranges (e.g. the scaly-naped parrot, *Amazona mercenaria*) disappeared from San Antonio, Ecuador (Kattan *et al.* 1994). Assuming that home range scales positively with body mass, Beier *et al.* (2002) determined that threshold patch size in Ghana for a 24-g and a 920-g species was 10 ha and 8000 ha, respectively. Renjifo (1999) suggested that for some species, large home ranges may not be a constraint because such species could be adapted to track variable food resources, and thus be able to travel through the landscape matrix (areas surrounding their primary forest habitat).

9.7.2 Disadvantage of large body size

Larger-bodied vertebrates are considered to be more extinction prone than smaller-bodied ones [e.g. Kattan 1992; Gaston and Blackburn 1995; Brook *et al.* 2003a; but see Gotelli and Graves (1990) for island extinctions]. A common explanation for this trend is that body size is inversely correlated with population size (Pimm *et al.* 1988), making large-bodied animals less abundant and more vulnerable to environmental perturbations, despite being better buffered against short-term fluctuations (Brook and Bowman 2005). The extinction proneness of large-bodied animals is further enhanced because of other correlated traits such as their requirement for large area, greater food intake, high habitat specificity and lower reproductive rates (Terborgh 1974; Leck 1979; Kattan 1992; Sieving and Karr 1997).

In birds, large species are generally more vulnerable to human persecution such as hunting, whereas smaller species are generally more vulnerable to habitat loss (Owens and Bennett 2000). It is important, however, to be cautious when constructing generalized rules regarding the role of body size in the extinction process. Owing to a slower reproductive rate, larger parrots are more vulnerable than smaller finches, despite fewer of the former being captured for the pet trade (Beissinger 2000). However, some smaller species (e.g. Muscicapinid flycatchers) with small population sizes are also vulnerable to extinction owing to heavy harvest rates for the pet trade, suggesting that in certain cases factors other than body size may enhance extinction proneness, or trends expected on biological grounds can be reversed by human choice. Likewise, in the case of terrestrial mammals, Cardillo *et al.* (2005) found that smaller species (< 3 kg) are more affected by environmental factors than their larger counterparts with generally lower vital rates (e.g. lower reproductive rate). Indeed, threatened mammals are an order of magnitude heavier than non-threatened ones (Cardillo *et al.* 2005); just as in the prehistoric past (Brook and Bowman 2002).

Many studies report the disproportionate loss of large-bodied or heavier species from tropical communities (Leck 1979; Diamond *et al.* 1987; Renjifo 1999; Castelletta *et al.* 2000). Karr (1982a) found that heavier species were lost more readily from a tropical island population in Panama. However, there was no correlation between body mass and rarity, suggesting that persisting larger species may not always be rare. Likewise, species extirpated between 1911 and 1997 from the Colombian Andes were heavier than extant species (Renjifo 1999). However, this relationship was not manifested when the missing species from smaller fragments were accounted for, indicating resilience in some extant large-bodied species. On the same note, Castelletta *et al.* (2000) showed that larger species suffer more extirpations than smaller species initially, but that this is not the case in recent losses (past 50 years). Together these observations suggest that body size may be important in early extirpations, but other factors may become more critical as the extinction wave proceeds, after the most vulnerable taxa have been eliminated. Insects also demonstrate a similar pattern; butterfly species in the forest gaps in Sabah (Malaysian Borneo) have relatively broader and larger thoraxes, indicating that they have superior flying abilities (Hill *et al.* 2001). Forest gaps contain more widespread and mobile butterfly species than closed-canopy areas, so canopy openings might be more harmful to the smaller forest butterfly species with restricted distributions.

On the other hand, a few studies have failed to find evidence for a clear relationship between body size and extinction proneness. For instance, Newmark (1991), found that body mass was not correlated with the occurrence of persisting birds. However, that result should be carefully interpreted because of relatively small variation in the body masses (7–138 g) investigated and small sample sizes (for 22 out of 26 species there were fewer than 30 captures). Body size also did not predict abundance in the forests of Sumba and Buru, Indonesia (M.J. Jones *et al.* 2001).

There may also be an interaction between a species' foraging habits and its body size. Large canopy frugivores and large understorey insectivores represent the most extinction-prone avian guilds in the Neotropics (Willis 1979; Bierregaard and Lovejoy 1989). Further, extinct large-canopy frugivores were longer (body lengths) than extant species in San Antonio, Colombia (Kattan *et al.* 1994). These larger species depend on difficult to locate and highly variable fruits in isolated forests, whereas the ability to forage on other items such as insects and the use of an array of habitats may explain why large-bodied species like some tanagers (family Thraupidae) are unexpectedly resilient to habitat loss and extinction.

In addition to body size, other morphological variables can affect extinction proneness. Excessive investment in secondary sexual characteristics (e.g. display crests, large horns or combative rituals) may render highly dimorphic species less adaptable to a changing environment (McLain *et al.* 1999). However, sexual dimorphism was not one of the predictors of the forest bird abundance on the islands of Sumba and Buru (M.J. Jones *et al.* 2001), and Morrow and Fricke (2004) found no effect of sexual dimorphism on extinction risk in mammals. However, sexual dichromatism (variation in coloration) was one of the predictors of extinction in tropical butterflies (Koh *et al.* 2004b). Overall, though, the role of sexual dimorphism in extinction proneness remains equivocal (Sodhi *et al.* 2004a).

9.7.3 Modern curse of specialization

Evolution engenders the speciation of taxa that occupy all available niches, so that in relatively stable systems, both specialist and generalist species can co-exist (Futuyma and Moreno 1988). However, when an environment is altered abruptly or deterministically at a rate above normal background change, specialist species with narrow ecological niches often bear the brunt of progressively unfavourable conditions such as habitat loss, degradation and invasive competitors or predators (McKinney 1997; Purvis *et al.* 2000; Mooney and Cleland 2001). In Singapore, for example, larval host plant specificity was found to be the most important correlate of extinction risk for butterflies (Koh *et al.* 2004b). Previous studies from temperate regions have reported similar trends, with larval host plant specificity being an important factor in the survival and distribution of butterflies from these areas (Shahabuddin *et al.* 2000; Shapiro 2002; Thomas 1991; Thomas *et al.* 2001; Thomas and Morris 1994).

As predicted, highly specialized forest-dependent taxa have been shown to be acutely vulnerable to extinction following deforestation and forest fragmentation. Examining a wide range of taxonomic groups (vascular plants, phasmids, butterflies, decapods, freshwater fish, amphibians, reptiles, birds and mammals), Brook *et al.* (2003a) showed that Singapore's forest-dependent taxa (those residing predominantly in primary or secondary rain forest or mangrove forest interiors) lost 33% of their species following large-scale clearance of the island's forests, compared with a mere 7% loss in open-habitat or edge-tolerant taxa. A probable explanation is that forest-dependent species are more likely to decline in concert with the loss of forest cover (due to reductions in breeding and feeding sites, increased predation, elevated soil erosion and nutrient loss, dispersal limitation, enhanced edge effects, etc.), and thus suffer higher extinction rates than non-forest-dependent species that are better able to persist in disturbed landscapes.

Foraging specialization is one mechanism that can compromise a species' ability to persist in altered habitats. Many studies have shown that frugivores and insectivores are more extinction prone than other avian guilds (reviewed in Sodhi *et al.* 2004a), with the lack of year-round access to fruit plants in smaller forests being the likely culprit (Leck 1979). A number of mechanistic hypotheses are proposed for the disappearance of insectivorous birds from deforested or fragmented areas. First, deforestation may impoverish the insect fauna and reduce preferred insectivore microhabitats (e.g. dead leaves). Second, insectivores may be poor dispersers and have near-ground nesting habits where they become more vulnerable to nest predators penetrating smaller forest fragments. For example, the insectivorous bird guild was more depauperate in small fragments (4–5 ha) than in a large fragment (227 ha) in Costa Rica (Sekercioglu *et al.* 2002), with most large and specialized insectivorous bird species (e.g. the black-faced antthrush, *Formicarius analis*) probably absent from small fragments. However, invertebrate abundance, average length and dry biomass were similar among fragments, and fragment size had little effect on the diet composition, prey biomass and prey items per sample of most sampled bird species. These observations led Sekercioglu *et al.* (2002) to conclude that the absence of some insectivorous bird species from small fragments may not be related to food scarcity; rather, it may be due to their poorer dispersal abilities.

Other avian guilds may be more extinction prone in different areas. For instance in Java, proportionally more extirpations occurred among carnivorous birds than in birds of any other foraging guild (Diamond *et al.* 1987). Similarly, carnivorous bird species were found to be more vulnerable than other guilds in Nicaragua because of their lower population densities and higher susceptibility to hunting (Gillespie 2001). These data might imply that species with specialist foraging habits are more extinction prone. Nevertheless, in Singapore and Java, mono-diet (specialist) bird species were more vulnerable initially, but as the extinction event unfolded, multi-diet (generalist) species were also later extirpated. This suggests that with heavy forest loss, even generalist species are affected negatively (Castelletta *et al.* 2000; Sodhi *et al.* 2006a).

Behaviours other than specialized foraging habits can also affect extinction proneness. Species forming mixed-species flocks disappeared more frequently from smaller fragments (1 and 10 ha) in Manaus, Brazil (Stouffer and Bierregaard 1995a). Some of the mixed-species flock members (e.g. the cinereous antshrike, *Thamnomanes caesius*) may have high foraging success in mixed flocks, or have large territories and high territory fidelity (Munn and Terborgh 1979; G.V.N.

Powell 1985), aspects that are eroded in isolated fragments. In Peninsular Malaysia, mixed-species flocks in forests were more species diverse than those observed in urban habitats (T.M. Lee *et al.* 2005), and flock participation was influenced by environmental characteristics such as the amount of canopy cover. The flocking species sensitive to habitat disturbance were likely to be from the families Corvidae, Nectariniidae and Sylviidae, which had restricted altitudinal ranges and were exclusively dependent on primary forest and understorey microhabitat. That study showed that bird flocks of submontane mixed species tend to be more sensitive to habitat loss and urbanization. Undoubtedly, many such specialized behaviours are negatively affected by deforestation and the break-up of resources, but more empirical evidence is required.

Obligate army ant-following bird species restricted to the Neotropics also disappeared from smaller fragments, probably because of their large area requirements (Bierregaard and Lovejoy 1989; Bierregaard *et al.* 1992). These species (e.g. the white-plumed antbird, *Pithys albifrons*) may have home ranges between 1 and 5 km in diameter (Willis and Oniki 1978; Harper 1989; Bieregaard and Lovejoy 1989), an area not available in smaller fragments (< 10 ha). Harper (1989) demonstrated that obligate ant-followers (e.g. the rufous-throated antbird, *Gymnopithys rufigula*) disappeared from fragments not containing army ant colonies but were more likely to persist if such colonies were made available experimentally. In addition to mixed-species flocking and army ant-following birds, colonial nesting icterids may also be more vulnerable because they cannot form viable colonies in small forest fragments (Renjifo 1999).

Poor dispersal ability can also affect species persistence following disturbance (Terborgh 1974; Bierregaard *et al.* 1992). The ability to disperse may depend on morphological characteristics, such as wing loading in birds and insects, and physiological restrictions, such as an intolerance to sunlight when moving within non-forested matrix between fragments (Johns 1992). As such, poor dispersal ability may make certain bird species vulnerable to extinction because they cannot readily colonize new areas (Tilman *et al.* 1994; Lens *et al.* 2002). This is critical because indirect evidence shows that species with superior colonization ability, and those that can exploit secondary habitats, are less extinction prone (Stouffer and Bierregaard 1995a; Gascon *et al.* 1999; Sekercioglu *et al.* 2002). Similarly, small mammal species that rarely venture into the grassy vegetation surrounding forest patches were highly vulnerable to extinction owing to habitat degradation and extensive fragmentation of Brazil's Atlantic forest (de Castro and Fernandez 2004) and Australia's monsoon vine thickets (Brook *et al.* 2002). Habitat loss may also disrupt some behaviours such as mating, although associated data are generally lacking for tropical fauna.

9.7.4 Role of life history and genetics in tropical extinctions

Regardless of the variation in life history traits that could make a species more or less vulnerable to extinction by human agency, it is beyond question that habitat loss affects species negatively (Beissinger 2000; Purvis *et al.* 2000). Species with high fecundity, short generation time, low to moderate survival rate and small body size are predicted to be vulnerable because they are more susceptible to extreme environmental conditions more common in degraded and restricted habitats. Alternatively, species with low fecundity, long generation time, high survival and large body size are also susceptible because they recover slowly from reductions in population size resulting from abrupt or rapid habitat loss or overexploitation (Brook and Bowman 2005). However, attempts to understand the role of life history traits in the extinction process of tropical fauna are still rare. Exceptions include work done on terrestrial insectivorous birds in Tanzania between 1989 and 2004, which were severely affected by forest disturbance and had long recovery times (Newmark 2006).

Deforestation, and the concomitant reduction in population size and connectedness, can also be one of the major factors in the loss of genetic variability within populations (Heywood and Stuart 1992). Because of poor dispersal ability, patchy distributions and generally low population densities, genetic diversity for some tropical species may be difficult to maintain. However, empirical evidence for this hypothesis is lacking. Spielman *et al.* (2004) reported that heterozygosity was, on average, 35% lower in 170 threatened taxa than in taxonomically related unthreatened taxa, indicating that threatened species may have elevated extinction risk as a result of compromised reproductive fitness resulting from declining heterozygosity. This was evident in a taxonomically broad range of tropical species, including Lumholtz's tree kangaroo (*Dendrolagus lumholtzi*), Indian rhinoceros (*Rhinoceros unicornis*), Sumatran orang-utan (*Pongo abelii*), golden lion tamarin (*Leontopithecus rosalia*), Mauritius kestrel (*Falco punctatus*), Komodo dragon (*Varanus komodoensis*) and Sumatran pine (*Pinus merkusii*) (Spielman *et al.* 2004). Overall, however, the role of inbreeding in extinctions from isolated tropical populations requires more detailed and wide-ranging scientific attention.

Plants are obviously the first organisms impacted by deforestation and forest disturbance. For instance, Singapore has lost 99% of its primary forest and original mangroves since 1819 (I.M. Turner *et al.* 1994; Hilton and Manning 1995), which led to a consequent loss of 26% of its original 2277 native vascular plant species (I.M. Turner *et al.* 1994; Brook *et al.* 2003a). Coastal habitats (including mangroves) and inland forests lost 39% and 29% of their vascular plant species, respectively (I.M. Turner *et al.* 1997), 88% of the original 196 native orchid species vanished, and 62% of the original 297 epiphyte species have been lost. High losses in orchids and epiphytes may be due to factors such as the loss of preferred hosts (big trees), microclimate changes brought about by fragmentation and overexploitation for the horticultural trade (I.M. Turner *et al.* 1994). Because Singapore is a land-bridge island with low endemism, these losses represent local extirpations, not global extinctions; however, such a telling confirmation of the habitat loss–species loss relationship is still disturbing and is probably representative of many tropical systems.

Moreover, considering that some plants can live for centuries, many of currently extant species in Singapore may nevertheless be committed to extinction due to unviable population sizes associated with the decreasing ability to cross-pollinate (I.M. Turner *et al.* 1994). This has been illustrated by a study of plant extinctions from an isolated 4-ha fragment of lowland rain forest in Singapore (I.M. Turner *et al.* 1996). Although 49% of 448 vascular plant species recorded in the

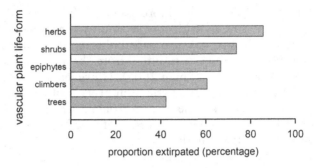

Figure 9.9 Proportion of different vascular plant life-forms that were extirpated from the Singapore Botanic Garden. (Data derived from I.M. Turner *et al.* 1996.)

1890s were extirpated by 1994, more shorter-lived shrubs, herbs, climbers and epiphytes were lost than long-living trees (Figure 9.9). However, the last group is still committed to extinction because over half of the tree species surveyed were represented by only one or two individuals (> 5 cm dbh; diameter at breast height) that would be incapable of perpetuating the population.

9.7.5   Role of taxonomy and phylogeny in tropical extinctions

It is well known that controlling for similarity due to shared evolutionary history (phylogenetic effects) is important when comparing behavioural or ecological characteristics among species because spurious relationships may otherwise arise (Harvey and Pagel 1991). Lockwood (1999) correctly pointed out that phylogenetic effects are commonly measured as taxonomic effects (i.e. families or genera are taken as monophyletic). Despite the practical ease of considering taxonomy (a proxy for phylogeny, assuming equal evolutionary branch lengths) and the non-random nature of extinctions with respect to phylogeny (P.M. Bennett and Owens 1997; McKinney and Lockwood 1999), many studies concerning extinction vulnerability of tropical biotas have not taken this problem into account. Further, most of the studies concerned with phylogenetic biases have been done on a global scale (P.M. Bennett and Owens 2002), so the practical usefulness of these studies for tropical conservation management decision-making in the tropics is not high.

On a related note, the act of saving species *per se* cannot be enough to preserve the variability of life on planet earth – it is necessary simultaneously to safeguard evolutionary history (Vazquez and Gittleman 1998). However, the phylogeny of most tropical taxa is not well enough understood to organize a conservation approach that could maintain genetic and phylogenetic lineages intact, resulting in the observation that poorly studied taxa are generally at a higher direct risk of extinction (McKinney 1999). Because tropical systems are much less studied than temperate ones, even for a comparatively well-known taxon such as birds, it can be predicted that tropical birds are even more at risk than is generally accepted owing to lower professional and public awareness of their plight (Stutchbury and Morton 2001).

Although some authors suggest that phylogenetics should be incorporated into conservation decisions (Heard and Mooers 1999), others argue that species richness can be used as an adequate surrogate of phylogenetic diversity (Rodrigues *et al.* 2004a). It is, however, unclear how far theoretical considerations can apply in practice, especially in the tropics, where decision-making and the implementation of even simple conservation management strategies are fraught with uncertainty. Practical management for vulnerable species in the tropics needs to be done carefully and with scientific rigour to test the efficacy of available models (Nee and May 1997; Heard and Mooers 1999). Results thus far have been mixed. For instance, Fjedeså and Lovett (1997) pointed out that more recently evolved species of African birds are more patchily distributed and, hence, relatively more vulnerable to extinction. However, Lockwood *et al.* (2000) found that the age of a lineage does not seem to affect community homogenization – a process whereby particular lineages are lost and the community becomes swamped by only a few invasive lineages.

Few specific tropical case studies have dealt with phylogenetic effects specifically. Studies using taxonomic family as a control show that this phylogenetic proxy had little impact on predictions of extinction proneness in butterflies and birds (Koh *et al.* 2004b; Sodhi *et al.* 2006a). Other studies provide conflicting conclusions; Thiollay (1997) suggested that the presence of sympatric congeners may lead to competitive pressure, narrower niches and lower abundance of at least some species. However, Terborgh and Winter (1980) and M.J. Jones *et al.* (2001) suggested that the presence of congeners did not affect the survival prospects of their relatives using bird data from Trinidad, Venezula, Sumba (Indonesia) and Buru (Indonesia). Terborgh and Winter (1980) also found that bird families containing disproportionately more susceptible species in Trinidad, Hainan (China), Sri Lanka and Tasmania (a temperate island of Australia) were the Bucerotidae, Cracidae, Falconidae, Phasianidae, Picidae, Timaliidae, Tinamidae and Ramphastidae, although patterns for other families were mixed (i.e. many contained a similar number of susceptible and resilient species). In the Colombian Andes, only the bird family Icteridae was more prone to extinctions (Renjifo 1999). In Costa Rica, Sekercioglu *et al.* (2002) found a positive correlation between the number of species of a bird family present in non-forested habitats relative to small fragments, suggesting that family-level dispersal characteristics and the ability to exploit deforested habitats may assist species persistence in small fragments.

Similarly, Hamer *et al.* (2003) reported that species' responses to selective logging may be taxonomically biased. Butterfly species of the families Satyrinae and Morphinae, which have higher shade preference and narrower geographical distributions, were affected adversely by selective logging, whereas the light-preferring and more cosmopolitan species of other families benefited. Hamer *et al.* (2003) argued that these trends resulted from the loss of habitat heterogeneity through logging. A mosaic of dense shade and open gaps in unlogged forests attracts species with opposing habitat requirements; therefore, the conservation value of logged areas can be enhanced by increasing habitat heterogeneity.

9.8   Extinction and the perturbation of ecological processes

The extinction of some 'keystone' species such as large predators and pollinators may have more devastating ecological consequences than the extinction of others (Terborgh 1992; Crooks and Soulé 1999; Chapter 2). Ironically, avian vulnerability to predation is often exacerbated when certain large predatory species become rarer in tropical communities. For instance, although large cats such as jaguars (*Panthera onca*) do not prey on birds directly, they exert a limiting force on smaller predators such as medium-sized and small mammals (meso-predators), which become more abundant following the top predators' decline. The corollary is that abundant mesopredators incur an above average predation rate on avian young and eggs (Terborgh 1992; Crooks and Soulé 1999; Figure 9.10). Although this 'mesopredator release' hypothesis has been applied largely to mammals (e.g. C.N. Johnson *et al.* 2007), the loss of large raptorial birds such as the harpy eagle (*Harpia harpyja*) may have similar ecosystem effects.

Similarly, the disappearance of a competitor can result in the niche expansion and higher densities of subordinate species. In the undergrowth of Barro Colorado Island, Panama, the disappearance of the black-faced antthrush, *Formicarius analis* (a large, superior competitor), appeared to permit the chestnut-backed antbird (*Myrmeciza exsul*) to become more densely and evenly distributed (Sieving and Karr 1997), signalling the likelihood of unbalanced predation rates on target invertebrate taxa. Extinction of forest species, coupled with habitat changes such as an increase in edge habitats, may result in edge-tolerant species thriving in isolated areas. Leck (1979) suggested that an increase in the amount of forest edge in Ecuador may have contributed to the increased abundance and diversity of brood parasites such as the shiny cowbird (*Molothrus bonariensis*) and olive-crowned yellowthroat (*Geothlypis semiflava*). More than two-thirds of the species that increased over time near Lagoa Santa, Brazil, live in the understorey (Christiansen and Pitter 1997) and target the abundant insect fauna that populate forest remnants with high edge–interior ratios. This sort of phenomenon has also been observed between unrelated taxa – the extinction of insectivorous birds from scrub forests of West Indian islands correlated with the subsequent higher biomass of competing anolis lizards (Wright 1981).

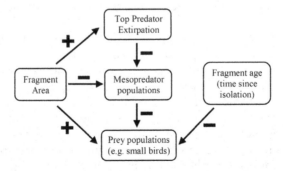

Figure 9.10 Effects of top predator extirpation on mesopredator and prey populations. (After Crooks and Soulé 1999. Copyright, Nature Publishing Group.)

Following deforestation, more non-forest or edge-adapted species can colonize the newly degraded areas (Renjifo 1999); however, this change is more likely to be due to the creation of optimal habitats for disturbance specialists than to any competitive release caused by the extinction of forest interior species. For example, the loss of insectivorous bird species from experimentally isolated fragments in Manaus did not facilitate the increase of non-forest or previously uncommon species, even after 9 years of isolation (Stouffer and Bierregaard 1995a). Hence, the role of direct inter-specific competition in the extinction process remains unclear, at least over short time scales. Similarly, how competitive release affects the abundance and reproductive success of extant species still needs to be demonstrated conclusively.

That said, it would be simplistic to assume that the demise of one species would have no wider ecological ramifications for ecosystems (Koh *et al.* 2004a), which by definition are suites of interacting species. Tropical rain forests are characterized by highly diverse and complex ecological communities in which many species are inextricably co-dependent. However, conservation biologists have tended to focus on the study of the declines or extinctions of individual species independently, largely ignoring the possible cascading effects of species co-extinctions (e.g. hosts and their parasites – Koh *et al.* 2004a; Stork and Lyal 1993). This is despite Diamond (1984) famously listing chains of extinctions as one of his 'evil quartet' (the four principal causes of extinction), along with habitat destruction, overexploitation and invasive species. Tropical examples are, however, beginning to emerge. For instance, a recent study showed that the decline and loss of butterfly species in Singapore was positively associated with the decline and loss of their specific larval host plants (Koh *et al.* 2004b). It is likely that many similar co-extinctions between other interdependent taxa have occurred throughout the tropics, but most have gone unnoticed in these relatively under-studied systems. The dynamics and incidence of host–parasite interactions may also be affected by extinctions. The brood parasitic brush cuckoo (*Cacomantis variolosus*) disappeared from the Bogor Botanical Garden (Java), most likely because of the decline in abundance of its hosts, the pied fantail (*Rhipidura javanica*) and hill blue flycatcher (*Cyornis banyumas*). In contrast, the parasitic plaintive cuckoo (*C. merulinus*) survived in this forested isolate because of the high abundance of its primary nest host, the ashy tailorbird (*Orthotomus ruficeps* – Diamond *et al.* 1987). In modified tropical habitats, important pollinators such as bees and wasps suffer increased parasitoid attacks (Tylianakis *et al.* 2007). This may have implications for pollination by bees and wasps in agroecosystems. Clearly, many more studies are urgently needed to investigate the phenomenon of species co-extinctions because they are likely to have important ecological and conservation implications such as an improved understanding of species distribution patterns and a better estimation of extinction rates (Koh *et al.* 2004a).

The disruption of ecological processes by extinction or declines in abundance may also lead to cascading and catastrophic co-extinctions. Frugivory, a key interaction linking plant reproduction and dispersal with animal nutrition, is placed in jeopardy by habitat degradation causing compounded species loss. Many tropical trees produce large, lipid-rich fruits adapted for animal dispersal (H.F.

Howe 1984), so the demise of bird frugivores may have serious consequences for forest regeneration, even if the initial drivers of habitat loss and degradation are controlled. For example, one of the dominant fruiting trees in Puerto Rico, Tabonuco (*Dacryodes excelsa*), failed to re-establish in areas it previously occupied because of the extinction and decline of frugivores that disseminated its seeds (Brash 1987). On the island of Negros (Philippines), while early-successional tree species were visited by a wide spectrum of frugivores, mid- and late-successional trees were largely visited only by specialist hornbills and fruit pigeons (Hamann and Curio 1999). Clearly, should these avian frugivores be eliminated by overhunting, mid- and late-successional trees will lose their primary dispersers, fail to replace themselves, and eventually cause the collapse of mature forest stands that support thousands of other specialist plant and animal species.

Research to determine whether frugivorous bird assemblages were adversely affected by forest disturbance has examined the composition of species visiting figs (*Ficus* spp.) in primary dipterocarp and recently logged forest in Sabah, Malaysian Borneo (Zakaria and Nordin 1998). Those investigators determined that the species composition was similar between the forests, with only one species unique to each site – the jambu fruit dove (*Ptilinopus jambu*) in primary forest and the olive-winged bulbul (*Pycnonotus plumosus*) in logged forest. However, fig trees in logged forests had a lower visitation by obligate frugivores such as the fairy bluebird (*Irena puella*). Consequently, Zakaria and Nordin (1998) suggested that there may be less food available for frugivores in logged forests, and that altered microclimatic conditions such as greater light penetration make logged forests less attractive to the species upholding vital ecosystem functions. A similar predicament has been found for African trees that rely heavily on birds for seed dispersal (Cordeiro and Howe 2001, 2003). In Tanzania, an endemic tree, *Leptonychia usambarensis*, relies on birds to disperse its seeds (Cordeiro and Howe 2003), so in fragmented forests where the birds were absent or rare, trees experienced lower seed dispersal (Figure 9.11). Thus, concern has been raised that the declining availability of fruits in disturbed tropical forests caused by the loss of avian frugivores in the canopy may also result in the extirpation of understorey and forest floor frugivores that have relatively small ranges and are generally unable to tolerate extreme microclimates in disturbed habitats (Lambert 1991; Zakaria and Nordin 1998).

Essential ecosystem functions provided by forest invertebrates are also highly susceptible to species loss resulting from habitat degradation. Figs, a keystone resource in Southeast Asian rain forests (Lambert 1991), rely on tiny (1–2 mm) species-specific symbiotic wasps for pollination. Some fig wasps may have limited dispersal ability (Harrison *et al.* 2003), suggesting that forest fragmentation causing individual fig trees to be more widely scattered could have negative influences on these wasps and the figs that they pollinate, creating a self-reinforcing feedback loop. Indeed, Koh *et al.* (2004a) showed that at least 3% of fig wasp species in this region are liable to become extinct followed by the extinction of their hosts. Similarly, dung beetles are critically important components of tropical biodiversity. In Venezuela, heavier dung beetles were found to be more extinction prone than lighter species (Larsen *et al.* 2005), which spells particularly dire

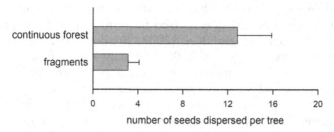

Figure 9.11 Differences in seed removal by birds between trees in continuous forest and small forest fragments. Error bars represent one standard error. (After Cordeiro and Howe 2003. Copyright, National Academy of Sciences, USA.)

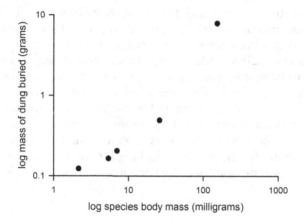

Figure 9.12 Positive correlations between the mass of dung buried and beetle mass. (After Larsen *et al.* 2005. Copyright, Blackwell Publishing Limited.)

predictions of ecosystem functional loss given the former species' capacity to dispose of more dung (Figure 9.12) (Larsen *et al.* 2005).

Mammals can also play a critical role in the regulation of plant populations and communities. For example, the wild pig (*Sus scrofa*) exhibits several behaviours, such as soil rooting, nest building and seed predation, that can influence the structuring of understorey vegetation (Lacki and Lancia 1983). Although soil rooting can kill seedlings and facilitate the spread of invasive weeds (Aplet *et al.* 1991), it can also promote plant growth (Lacki and Lancia 1983). Wild pigs were superabundant (27–47 pigs/km^2) in Pasoh Forest Reserve (Peninsular Malaysia) because of the demise of its natural predators, such as tigers (*Panthera tigris*), and the year-round food supply of African oil palm fruits (*Elaeis guineenis*) (Ickes and Thomas 2003). At such high densities, pigs reduced overall plant recruitment by as much as 67%; however, the growth of trees between 1 and 7 m tall increased by 52% (Ickes 2001). Therefore, pigs clearly influence plant dynamics in the understorey, and their demise (or superabundance caused by the demise of their predators) can strongly affect forest composition and structure.

Clearly, therefore, deforestation has ramifications beyond the direct effects of biomass removal and degradation. Almost all flowering plants in tropical rain forests are pollinated by animals (Bawa 1990), and an estimated one-third of the human diet in tropical countries is derived from insect-pollinated plants (Crane and Walker 1983). Sekercioglu *et al.* (2004) predicted that up to 14% of all bird species will be extinct by year 2100, and many others functionally extinct, with the obvious result of compromised seed dispersal and pollination (Table 9.1). Therefore, the sad irony is that a decline of forest-dwelling pollinators may impede plant reproduction not only in forests, but also in neighbouring agricultural areas. Lowland coffee (*Coffea canephora*) is an important tropical cash crop, and it depends on bees for cross-pollination. In central Sulawesi, Indonesia, the number of social bees species visiting coffee fields decreased with increasing distance from adjacent forests (Klein *et al.* 2003a) (see Figure 2.4; see also Chapter 2).

These studies and others suggest that the relative susceptibility of tropical species to habitat disturbance may be related to particular ecological traits (Koh *et al.* 2004b), as has been shown for a range of taxa in other environmental settings (Purvis *et al.* 2000). Identifying the ecological correlates of extinction proneness due to habitat loss is therefore key for defining effective conservation priorities, especially in the tropics, where long-term and broad-scale monitoring programmes are almost non-existent and traditional surrogate methods of threat assessment, such as the International Union for the Conservation of Nature and Natural resources (now World Conservation Union, IUCN) Red List criteria (Randrianasolo *et al.* 2002), are either lacking for tropical species owing to data deficiencies or do not provide adequate resolution of relative endangerment (Koh *et al.* 2004a).

9.9   Biotic resilience

Despite our focus on tropical extinctions, it is important here to re-emphasize that not all tropical species are equally prone to extinction due to human activities. The species–area relationship leads to a rule of thumb that a 90% loss in habitat leads to approximately 25–50% loss of species (Simberloff 1992a). However, the predictive power of this relationship may be weak because it does not account for habitat diversity or fragmentation (Simberloff 1992b). The species–area predictions are also probabilistic, meaning that some fraction of the biota will, by chance or fortuity, persist even in the face of extreme habitat loss (e.g. Brook *et al.* 2003a). That said, it remains a useful tool for making crude predictions of the extent and magnitude of biotic extinctions in the absence of more complete mechanistic knowledge (see Spotlight 9: Stuart Pimm).

Observed extinction rates from some locations may be lower than those predicted by the species–area equation (Brook *et al.* 2002), at least over the short term, indicating that some species have temporary or permanent resilience or lagged responses. Brook *et al.* (2006a) illustrated the effects of this extinction debt for Southeast Asia endemic species in predicting that the number of extinctions by 2030 should deforestation begin to slow immediately, and cease completely within 25 years, will be only 4% fewer than if clearing were to continue at its current alarming rate into the future (Figure 9.13). In addition to preservation,

Table 9.1 Ecological and economical contributions of avian functional groups. (After Sekercioglu et al. 2004. Copyright, National Academy of Sciences, USA)

Functional group	Ecological process	Ecosystem service and economical benefits	Negative consequences of loss of functional group
Frugivores	Seed dispersal	Removal of seeds from parent tree; escape from seed predators; improved germination; increased economical yield; increased gene flow; recolonization and restoration of disturbed ecosystems	Disruption of dispersal mutualisms; reduced seed removal; clumping of seeds under parent tree; increased seed predation; reduced recruitment; reduced gene flow and germination; reduction or extinction of dependent species
Nectarivores	Pollination	Outbreeding of dependent and/or economically important species	Pollinator limitation; inbreeding and reduced fruit yield; evolutionary consequences; extinction
Scavengers	Consumption of carrion	Removal of carcasses; leading other scavengers to carcasses; nutrient recycling sanitation	Slower decomposition; increases in carcasses; increases in undesirable species; disease outbreaks; changes in cultural practices
Insectivores	Predation on invertebrates	Control of insect populations; reduced plant damage; alternative to pesticides	Loss of natural pest control; pest outbreaks; crop losses; trophic cascades
Piscivores	Predation on fishes and invertebrates and production of guano	Controlling unwanted species; nutrient deposition around rookeries; soil formation in polar environments; indicators of fish stocks; environmental monitors	Loss of guano and associated nutrients; impoverishment of associated communities; loss of socioeconomic resources and environmental monitors; trophic cascades
Raptors	Predation on vertebrates	Regulation of rodent populations; secondary dispersal	Rodent pest outbreaks; trophic cascades; indirect effects
All species	Miscellaneous	Environmental monitoring; indirect effects; bird watching tourism; reduction of agricultural residue; cultural and economic uses	Losses of socioeconomic resources and environmental monitors; unpredictable consequences

Spotlight 9: Stuart L. Pimm

Biography

I am the Doris Duke Professor of Conservation Ecology at the School of the Environment at Duke University and have a secondary appointment of Extraordinary Professor at the Conservation Ecology Research Unit at the University of Pretoria, South Africa. My interests are endangered species conservation, biodiversity, species extinction and habitat loss. I am the author of over 200 scientific publications, many of them in *Nature* and *Science*, and have written four books, the most recent being the critically acclaimed *World According to Pimm: a Scientist Audits the Earth*. In 2006, Prince Willem-Alexander presented me with the Dr. A.H. Heineken Prize for Environmental Sciences on behalf of the Royal Netherlands Academy of Arts and Sciences. How did all this happen?

Like many of my peers, I started out as a naturalist during my adolescence, read zoology at university, and then did a PhD in ecology. Unlike others, I worked in Hawai'i soon afterwards, where I was deeply shaken by the total absence of many of the birds I expected to see – they were either extinct or close to it. I was curious about why some species succumbed while other survived. Importantly, these losses were an outrage. Scientists, I realized, could help prevent extinctions. Vitally, they had an obligation to do so. Thereafter, my research group has sought out the species and ecosystems that are in most urgent need of protection. That work takes us to the Everglades, the Amazon and the coastal forests of Brazil, to southern Africa and to Madagascar. We work with local organizations and governments to provide the best possible advice to solve conservation problems. We emphasize solving problems – and that means we develop whatever skills are need in their solution. We have always had good quantitative skills, but in addition, my group members all use geographic information systems and analyses of satellite imagery – skills we developed only in the last decade. And, yes, some of the solutions come from sharing our knowledge with politicians and advising on policy issues.

Major publications

Pimm, S. L. (1991) *The Balance of Nature? Ecological Issues in the Conservation of Species and Communities.* University of Chicago Press, Chicago.

Pimm, S. L. (2001) *The World According to Pimm: A Scientist Audits the Earth*. McGraw Hill, New York.

Pimm, S. L. and Jenkins, C. (2005) Sustaining the variety of life. *Scientific American* **293**, 66–73.

Pimm, S. L., Russell, G. J., Gittleman, J. L. and Brooks, T. M. (1995) The future of biodiversity. *Science* **269**, 347–350.

Pimm S., Raven, P., Peterson, A., Sekercioglu, C. H. and Ehrlich, P. R. (2006) Human impacts on the rates of recent, present, and future bird extinctions. *Proceedings of the National Academy of Sciences of the USA* **103**, 10941–10946.

Questions and answers

The current biodiversity crisis has been termed the 'sixth extinction', an allusion to the five largest mass die-offs in earth's past. Is this comparison justified?

In the previous five die-offs – the last killed off the dinosaurs – more than half the variety of life disappeared. It took roughly 10 million years to recover the former numbers of species. Human actions in the last thousand years have probably wiped out about 10% of species, while actions in the last century have threatened at least 10% of the remainder. By threatened, I mean that expert opinion judges that these species will become extinct in the next few decades if we do nothing to protect them. It gets worse. Tropical forests hold perhaps two-thirds of all species on land and tropical oceans, especially coral reefs, the great majority of marine species. If current trends continue, human actions will so massively reduce these ecosystems that a third or more of the remaining species will be on a path to extinction within a few decades.

How reliable are biogeographic proxies such as the species-area relationship for inferring extinction rates?

Our ability to predict future trends on land comes from the species–area relationship. It's one of the great ecological laws – that is, a commonly observed pattern across different species groups in different areas. An oceanic island, half the size of larger island, will have about 15% fewer species according to this law. Imagine we convert what was once a continuous forest – say, eastern North America – into islands of about half the forest cover. There are about 30 species of bird endemic to the forests of the region, so we'd expect to lose 4.5 species. And, indeed, four species of bird became extinct as eastern North America lost its forests in the four centuries since European colonization – and another species is threatened with extinction! Detailed calculations like this one have now been done on many areas of tropical forest, which often contain hundreds of endemic species. The numbers of species the model predicts will become extinct and those that have done so are very similar (or are presently in danger of doing so, for extinctions take time to happen). These excellent calibrations of the law mean that we can predict how many species will become extinct if we reduce tropical forests further.

How can scientists most effectively engage the often pseudo-scientific arguments posed by environmental 'sceptics', who claim global hazards, such as the large-scale death of species and climate change, are illusory or inconsequential?

The most effective strategy is not to engage the sceptics. I'm for honest scientific debate – it's what I do every day. The evidence for global change and massive loss of species is unassailable, however. Sceptics ignore the evidence, usually, in my experience, because they are paid to do so. There is nothing honest in the debate; indeed, it usually isn't a debate. Would you debate someone who thought the world was flat? If you were so foolish as to do so, what would happen? You would present all the familiar observations – earth's shadow on the Moon, for example, – and demolish your opponent. Would he continue with his foolish ideas? You bet! He would loudly trumpet that he'd debated a competent, thoughtful scientist at the University of Somewhere. To outsiders, his pathetically ignorant ideas would gain credibility and his sponsors would continue to pay him. Lots of good people work hard to address ways to reverse global change and reduce species loss. Get on with solving problems and don't waste time with fools.

What is the future of tropical biodiversity … 'according to Pimm'?

The important message is that we can stem the loss of tropical biodiversity – its future is not yet written. We can slow the rate of deforestation and we know enough about the patterns of where the most vulnerable species live to make their protection a priority.

other management efforts such as reintroductions and supplementation of numbers or food supplies may be needed in deforested landscapes to offset these trends.

Since European colonization of the South American continent about 500 years ago, the Atlantic coastal rain forest of Brazil has been reduced to 12% of its approximate original size, with no recently reported avian extirpations (K.S.J. Brown and Brown 1992). This observation probably reflects genuine resilience in some of the species, but more generally indicates that many undocumented extinctions of unknown species have already occurred, and that many more will disappear with time regardless of the future suppression of habitat loss and degradation (Balmford 1996; Brooks and Balmford 1996; Brooks *et al.* 1999c). Likewise, in a 15-km² remnant forest of Cebu Island, Philippines, 7 of 14 forest

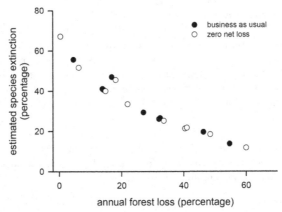

Figure 9.13 Negative relationships of endemic species extinction (%) with rate of forest loss (%), assuming that rate of forest loss is constant ('business as usual'– filled circles), or is on a linear decline to point of zero net deforestation by 2030 ('zero net loss' – unfilled circles). (Data derived from Brook *et al.* 2006a.)

endemic bird species/subspecies still persist (Magsalay *et al.* 1995), which is expected given that a fragmented landscape containing mature forest can retain as much as 96% of the original avifauna several decades after isolation (Renjifo 1999). Many other examples show that a large proportion of the forest biota can be found in disturbed habitats (e.g. Daily *et al.* 2003; Mayfield *et al.* 2005; Peh *et al.* 2005); however, few data are available that demonstrate whether persisting species in degraded areas are reproducing and surviving as well as their counterparts in pristine areas.

In Sumatra, birds that do not exploit agroforests (e.g. rubber tree plantations) were predominantly forest-dependent species that included large insectivores and frugivores (Thiollay 1995). In Peninsular Malaysia, Peh *et al.* (2005) found that, of the 159 extant primary forest bird species, only 28–32% existed in disturbed areas such as plantations and farmland. The types of microhabitats on which these Malaysian species depend also seemed to influence their vulnerability to disturbance. For instance, ground-dwelling species tended not to occur in disturbed areas, presumably because most of the ground-dwelling species also nest on or close to the ground. Many have observed higher predation of artificial eggs in disturbed over undisturbed forests (Wong *et al.* 1998; Sodhi *et al.* 2003), indicating that ground-dwelling primary forest birds may suffer higher nest predation in disturbed areas. These results emphasize the susceptibility of still-persisting taxa to the large-scale conversion of forests for agriculture that is now occurring throughout much of the tropics.

9.10   The future of tropical biodiversity

Many studies relying on the species–area equation have projected the probable magnitude and extent of tropical extinctions that will result from deforestation

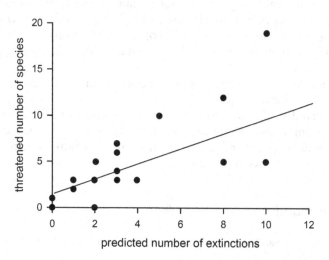

Figure 9.14   Relationship between the predicted extinctions of bird species following deforestation and the number of bird species currently threatened in insular Southeast Asia. (After Brooks *et al.* 1997. Copyright, Blackwell Publishing Limited.)

in the near future (J.B. Hughes *et al.* 1997; Brooks *et al.* 1999b; Grelle 2005). For example, Brooks *et al.* (1997) predicted that the number of bird extinctions in Southeast Asia arising from deforestation closely matches the number of threatened species currently listed (Figure 9.14). This relationship suggests that a large proportion of threatened bird species in Southeast Asia are likely to become extinct should the current rate of deforestation continue. In our opinion, this is a likely scenario considering that the deforestation rate in this region is accelerating at an unprecedented rate. Similarly, Brook *et al.* (2003a) analysed reported and inferred extirpations from Singapore and used these to project future biotic losses from Southeast Asia (Figure 9.4) with much the same dire predictions. Analogous environmental scenarios to those that occurred in the microcosm of Singapore over the last two centuries are already unfolding on a much larger scale in other Southeast Asian countries such as Indonesia (Jepson *et al.* 2001). Grelle (2005) predicted a two- to threefold increase in the number of mammals committed to extinction in the Brazilian Amazon, and a high likelihood of future extinction cascades. Taking a slightly different perspective, Cardillo *et al.* (2006) identified areas where terrestrial mammals will be threatened in the future using predictions based on their biological traits (e.g. body size). They argued that these 'hotspots of latent extinction risk' should not be ignored even though they are not currently under the same level of threat as classic 'hotspot' areas of high endemism (Myers 1988; Chapter 1). Thirteen of these 20 (65%) latent-risk hotspots were in the tropics, and should be added to the other known high-endemism hotspots as areas in dire need of conservation intervention.

9.11 Summary

1 Extinctions are a normal part of evolution, but human modifications to earth's ecosystems from local to regional to global scales have greatly accelerated the rate of species losses relative to the long-term background extinction rate.
2 Habitat loss is the main driver of tropical extinctions but it may act synergistically with other drivers such as overharvesting, alien species invasion and climate change.
3 Large-bodied, rare species and habitat specialists are particularly prone to extinction due to human agency.
4 Extinctions can disrupt vital ecological processes such as pollination and seed dispersal, leading to ecosystem collapse and a higher overall extinction rate.
5 Extinction rates are predicted to accelerate in the near future due to extensive and ongoing habitat loss in the tropics.

9.12 Further reading

Brook, B.W., Sodhi, N.S. and Ng, P.K.L. (2003) Catastrophic extinctions follow deforestation in Singapore. *Nature* **424**, 420–423.
Dirzo, R. and Raven, P.J. (2003) Global state of biodiversity and loss. *Annual Review of Environment and Resources* **28**, 137–167.
Pimm, S.L. and Raven, P. (2000) Extinction by numbers. Nature **403**, 843–845.

10

Lights at the End of the Tunnel: Conservation Options and Challenges

Our coverage of the conservation crises within the tropical realm has revealed, perhaps intuitively, that the major challenges to overcome the present and imminent threats to biodiversity are primarily socioeconomic in origin. These include extensive poverty, chronic shortage of conservation resources (e.g. expertise and funding) and weak governance from impotent national institutions. A particular aspect of these problems characteristic of the tropics is the inevitable marginalization of environmental issues as developing societies strive to match the living standards of developed nations. Despite these challenges, there are still conceivable ways to conserve at least some of the rich natural resources within the tropics for future generations. Given that many of the drivers of biodiversity loss (e.g. international demand for rain forest timber, climate change, commercial fisheries) are issues that transcend national boundaries, any realistic solution will need to involve a multinational and multidisciplinary strategy. Such strategies must include political, socioeconomic and scientific input in which all major stakeholders (governmental, non-governmental, national and international organizations, and the general public) must partake (see Spotlight 10: Peter Raven). The objective of this chapter is to make recommendations for the adequate conservation of extant tropical biodiversity. We suggest ways to improve the governance of tropical resources and discuss the importance of preserved areas for biodiversity conservation by highlighting examples of successes in the tropics. We also stress the need to integrate the social issues (e.g. human hunger) in order to achieve tangible conservation outcomes.

10.1 Protected areas are critical for tropical conservation

Ideally, protected areas (reserves or national and regional parks) should shelter the full range of species and ecosystem diversity to safeguard important ecosystem functions such as the preservation of catchments and the regulation of climate (see Chapters 2 and 8). Globally, 10–27% of threatened vertebrates do not fall

Spotlight 10: Peter H. Raven

Biography

When I was about 8 years old, I was inspired by the beauty and interest of the insects and plants of California. I didn't know what kind of a career might be possible, but as I went through high school and college, inspired by my participation in the Student Section of the California Academy of Sciences in San Francisco, I gradually came to understand that it was possible to spend my life studying the magnificent biosphere that is all around us. I then began to realize that we were facing a huge problem: with a rapidly increasing human population, increasing consumption and inappropriate technologies, we were driving a majority of the species that share this earth with us to extinction. The opportunities we have now will never be as rich in the future, and I do what I can to take advantage of them.

Major publications

Dirzo, R. and Raven, P. H. (2003) Global state of biodiversity and loss. *Annual Review of Environment and Resources* **28**, 137–167.

Raven, P. H. (1980) *Research Priorities in Tropical Biology*. Committee on Research Priorities in Tropical Biology, National Research Council. US National Academy of Sciences, Washington, DC.

Raven, P. H. (1988) *We're Killing Our World, The Global Ecosystem in Crisis*. MacArthur Foundation Occasional Paper, December 1987.

Raven, P. H. (2002) Science, sustainability, and the human prospect. *Science* **297**, 954–958.

Raven, P. H. and Berg, L. R. (2007) *Environment*, 6th edn. John Wiley & Sons, Inc., Hoboken, NJ.

Questions and answers

Could you briefly describe some conservation initiatives undertaken by the Missouri Botanical Garden in tropical countries and their positive outcomes to date?

In the California-sized island of Madagascar, the Missouri Botanical Gardens employs about 50 citizens, who, along with two foreign science advisors, work out conservation futures for protection and the sustainable use of the biosphere. Over 90% of the species of plants and animals there are found nowhere else, and our skill and diligence in dealing with them will determine how many survive for our use, or enjoyment, or simply because we have no right to destroy them. We are making a significant difference there and in many countries for the lives of people.

How can we, as conservation scientists, work to ensure that protected areas in tropical regions function as more than just 'paper parks'?

Mainly by understanding and living with the people there, to ensure that their priorities and ours are the same. Information is very important in managing parks properly, and in conserving biodiversity between parks and protected areas, but it must have demonstrated meaning to the people who live there. We must all work hard to adopt an international attitude based on caring among people everywhere.

Is it possible to strike a balance between development and preservation in developing nations?

It is not only possible, it is necessary, and the opportunities for striking that balance are greater now than they will ever be in the future. People must be able to improve their lives, but the resources available must be used sustainably or there will not be a satisfactory and sustainable future for anyone else. We need new ways of thinking in order to develop this balance properly, and there are relatively few models for us.

Is cross-disciplinary research the future for conservation science and, if so, how might it be achievable in practice?

We need to know a great deal more than we do at present about the functioning of ecosystems, and of the complexes of species that make them up. Productive agriculture in suitable places is part of the key, and sustaining biodiversity between protected areas, which must contribute to the sustainability of the whole region, is likewise of great importance. The alleviation of poverty must be undertaken not only as a matter of social justice, but because without it there can be no sustainable future.

under the umbrella of any legally protected areas (Rodrigues *et al.* 2004b), and this proportion is woefully elevated in the tropics. For example, today there is formalized protection for only 28% of the remaining forests in the tropics (Figure 10.1), a worrying figure considering that protected areas may be the only hope for retaining a reasonable proportion of tropical biodiversity still remaining (Bruner *et al.* 2001; Rodrigues *et al.* 2004b). Perhaps even more distressing is the fact that the situation is much worse for marine protected areas (Table 10.1 and Chapter 7).

Much of the remaining native forests in the tropics are not only unprotected, they are situated in countries that do not have the resources to protect them adequately, such as Myanmar [annual gross domestic product (GDP) per capita US$225], Niger (US$228), Eritrea (US$218) and Sierra Leone (US$202). Currently, less than 10% of the costs needed for effectively maintaining protected areas are being met in the tropics, and of an estimated total US$6 billion spent each year on managing protected areas, less than 12% is spent in developing countries (Balmford *et al.* 2003; Figure 10.2). A chronic lack of funding is hampering the success of protected areas, as exemplified in many African countries (Struhsaker *et al.* 2005). In addition to adequate funds, factors such as high public support, effective law enforcement and low human population densities are conducive to

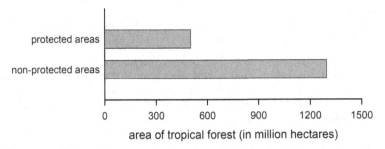

Figure 10.1 Comparison of areas of tropical forests designated for the conservation of biological diversity and for other uses. (Data derived from FAO 2005.)

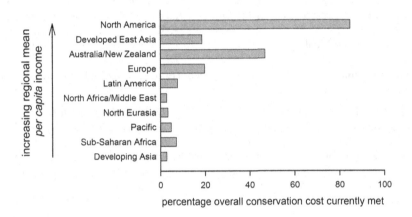

Figure 10.2 Regional variation in the percentage of the overall cost of effective reserve networks that are met. (After Balmford *et al.* 2003. Copyright, National Academy of Sciences, USA.)

the success of protected areas (Struhsaker *et al.* 2005). The current trend in at least some tropical areas suggests that forest loss will probably increase along with human population density (Chapter 1). For example, human population density around reserves in West Africa influenced the projected extinction likelihood of large mammals (Brashares *et al.* 2001; Figure 10.3).

Forested areas with high conservation value (e.g. 'hotspots' rich in endemic species – Myers *et al.* 2000; Chapter 1) could be purchased and protected in a competent manner by international agencies through collaborations with national organizations (A.N. James *et al.* 1999; T. Whitten *et al.* 2001). This strategy, although expensive, would not only provide immediate protection for native biodiversity, it would also establish a mechanism to sustain the effort. A reasonable proportion of the funding required for such endeavours could come from developed countries, both in the region and beyond (Balmford and Whitten 2003). Involvement of private-sector institutions and private philanthropists in biodiversity conservation (e.g. purchase of land for biodiversity reserves) needs to be explored more aggressively (Daily and Walker 2000; Balmford and Whitten

Table 10.1 Length of coastline and the ratio of marine/littoral protected areas to length of coastlines in tropical countries (data from the World Resources Institute). Countries arranged in decreasing ratio of protected area relative to length of coastline.

Country	Length of coastline (km)	Marine and Littoral protected areas (km²)	Ratio of protected area relative to length of coastline
Namibia	1754	74020	42.20
Angola	2252	29100	12.92
Mauritania	1268	14960	11.80
Jamaica	895	8190	9.15
Zaire	177	1000	5.65
Dominican Republic	1612	8550	5.30
Mexico	23761	82060	3.45
Venezuela	6762	21340	3.16
Panama	5637	17480	3.10
Mozambique	6942	21170	3.05
Kenya	1586	3550	2.24
Suriname	620	1350	2.18
Equatorial Guinea	603	1310	2.17
Cuba	14519	31410	2.16
Cameroon	1799	3890	2.16
Ecuador	4597	9920	2.16
Martinique	369	760	2.06
Costa Rica	2069	3930	1.90
Gabon	2019	2980	1.48
Malaysia	9323	13520	1.45
Colombia	5875	8500	1.45
Indonesia	95181	135590	1.42
Honduras	1878	2610	1.39
Brazil	33379	42990	1.29
Guatemala	445	560	1.26
Peru	3362	3390	1.01
Belize	1996	1870	0.94
India	17181	15930	0.93
Somalia	3898	3340	0.86
Sri Lanka	2825	2320	0.82
Thailand	7066	5780	0.82
Nicaragua	1915	1300	0.68
Brunei Darussalam	269	180	0.67
Liberia	842	550	0.65
Senegal	1327	850	0.64
Seychelles	747	410	0.55
Philippines	33900	16630	0.49

Continued overleaf

Table 10.1 *Continued.*

Country	Length of coastline (km)	Marine and Littoral protected areas (km²)	Ratio of protected area relative to length of coastline
Guadeloupe	581	260	0.45
Gambia	503	220	0.44
Côte d'Ivoire	797	330	0.41
Netherlands Antilles	361	120	0.33
United Republic of Tanzania	3461	1070	0.31
Antigua and Barbuda	289	70	0.24
Samoa	463	110	0.24
Puerto Rico	1094	230	0.21
Vietnam	11409	2160	0.19
Togo	53	10	0.19
Cambodia	1127	210	0.19
French Guiana	763	140	0.18
Mauritius	496	90	0.18
Papua New Guinea	20197	3520	0.17
New Caledonia	3624	570	0.16
Saint Vincent and the Grenadines	264	40	0.15
Sudan	2245	260	0.12
Trinidad and Tobago	704	80	0.11
El Salvador	756	70	0.09
Dominica	152	10	0.07
Saint Lucia	166	10	0.06
Réunion	219	10	0.05
Singapore	268	10	0.04
Myanmar	14708	480	0.03
French Polynesia	5830	180	0.03
Guinea	1615	40	0.02
Madagascar	9935	160	0.02
Fiji	4638	50	0.01
Solomon Islands	9880	80	0.01
Barbados	97	0	0.00
Djibouti	443	0	0.00
Cape Verde	1121	0	0.00
Micronesia (Federated States of)	1295	0	0.00
Maldives	2002	0	0.00
Vanuatu	3132	0	0.00
Guinea-Bissau	3176	0	0.00

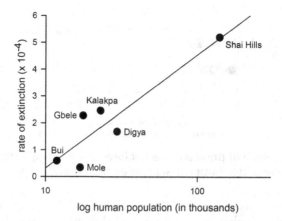

Figure 10.3   Rate of local extinction of large mammals in Ghanaian reserves in relation to total human population within 50 km of the reserves. (After Brashares *et al*. 2001. Copyright, The Royal Society London.)

2003). Further, protected areas should also encompass poorly studied and unexplored natural habitats, as well as different habitat types (e.g. mangroves, swamps) and elevations (montane areas).

About 23% of existing tropical humid forests are officially protected worldwide (Chape *et al*. 2003). However, it is not known if these reserves receive adequate protection against activities such as illegal logging and poaching. Some so-called 'protected' forests in the tropics have become isolated, degraded or deforested. Between the early 1980s and 2001, 25–70% of 198 protected areas containing tropical forests lost forest area within their boundaries surrounding buffer zones (DeFries *et al*. 2005; Figure 10.4). Buffer zones are important because they augment the function of protected areas by safeguarding them from catastrophic events such as fires (Gascon *et al*. 2000). Other studies illustrate similar problems faced by specific protected areas throughout tropics (e.g. Sánchez-Azofeifa *et al*. 1998; Curran *et al*. 2004). For example, protected lowland forests of the hyperbiodiverse region of Kalimantan (Indonesia) declined by 56% between 1985 and 2001, primarily as a result of intensive logging (Curran *et al*. 2004). This forest decline was not restricted to the parks and also occurred within the buffer zones. Similarly, owing to expansion of agriculture and cattle ranching in the Amazon, it is predicted that by 2050, 40% of the forests in its protected areas could be logged (Soares-Filho *et al*. 2006).

As a generality, many protected forests in the tropics suffer from four main threats: (1) illegal logging, (2) encroachment by shifting cultivators, (3) fires and (4) overharvest of high-value resources such as bush meat. For example, in Pulau Kaget Nature Reserve (Kalimantan, Indonesia), excessive infringement by expanding farms has resulted in the loss of habitat for the threatened proboscis monkey (*Nasalis larvatus*). Translocation of these monkeys resulted in their ultimate demise from the reserve, and the action failed to establish new populations elsewhere because of poor planning (e.g. translocation to unprotected areas) and ineffective execution (e.g. 13 of 84 monkeys died during capture) (Meijaard and Nijman 2000).

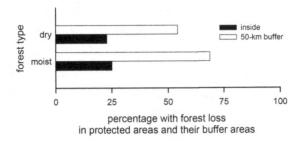

Figure 10.4   Percentage of tropical protected forest areas losing forest within their boundaries and surrounding buffer. (Data derived from DeFries *et al*. 2005.)

Likewise, a large proportion of the protected areas of Myanmar suffer from excessive human disturbance such as establishment of agricultural fields (Rao *et al*. 2002; Figure 10.5). Insufficient numbers of poorly trained staff are left to manage the parks of Myanmar (e.g. to prevent poaching), but cannot do so adequately (Rao *et al*. 2002). For tropical protected areas, the density of guards charged with the task of enforcing reserves correlates positively with the effectiveness of park protection (Bruner *et al*. 2001). Additionally, data from Sulawesi (Indonesia) demonstrate that resource extraction from protected areas is less common among people who are supportive of conservation endeavours than among those opposed to protected areas (T.M. Lee *et al*. 2007; Figure 10.6).

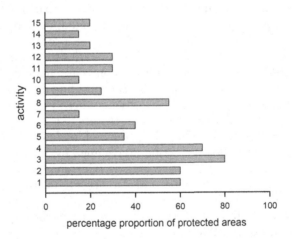

Figure 10.5   Proportion of protected areas in Myanmar with human activities. 1, hunting for subsistence and wildlife trade; 2, fuel wood collection; 3, extractions of non-timber forest products (orchids, palm leaves, grass, rattan, honey, mushrooms, bamboo, resin from dipterocarps and medicinal plants); 4, grazing by domestic cattle, sheep and horses; 5, fishing (crabs, prawns and fish); 6, shifting cultivation; 7, mining; 8, permanent human settlements; 9, roads and railway lines; 10, plantations of sugar cane, rubber and oil palm; 11, military or insurgents camps, indicating availability of firearms; 12, permanent cultivation; 13, tourism; 14, breeding centres for ducks, fish and other animals; and 15, extraction of timber species such as teak. (Data derived from Rao *et al*. 2002.)

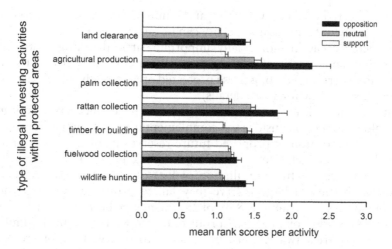

Figure 10.6 The extent of resource harvesting activities in the protected areas (PAs) of Sulawesi (Indonesia) in relation to the three levels of support for establishment of PAs. All error bars represent standard error. (After T.M. Lee *et al.* 2007. Copyright, Cambridge University Press.)

Protected areas remain a major hope for biodiversity conservation in the tropics. In their analysis of 93 protected areas in 22 tropical countries, Bruner *et al.* (2001) found that protected area status may be succeeding in preventing land clearing; 43% of them had no net land clearance since their establishment. They recommended that management actions such as genuine law enforcement, boundary demarcation and direct compensation to local communities can all assist in better protection of parks and thus of tropical biodiversity.

Establishment of appropriate protected areas is also critical in halting future extinctions. About 5% of savannas are converted for human use every year in the Serengeti (Tanzania) (Sinclair *et al.* 2002). Considering that only 28% of savanna birds occur in agricultural areas, their expansion may jeopardize this taxon. Likewise, Brook *et al.* (2003a) showed that the already alarming number of biotic extinctions in Singapore will further increase by 66% with the loss of its few remaining protected areas.

It is important that new protected areas are identified correctly for optimal conservation outcomes. Irreplaceability and vulnerability should take precedence in conservation planning (Brooks *et al.* 2006). For example, most, if not all, areas of high bird endemism should be protected (K. Norris and Harper 2004). Threatened endemic bird species are mostly habitat specialists and as such are particularly vulnerable to habitat disturbance; the taxon is clearly a high priority for conservation (K. Norris and Harper 2004). In the tropics, the extent of current protected areas does not correspond particularly well with the number of threatened species (Kerr and Burkey 2002). Effective forest reserves should also be of sufficient area to maintain viable populations of their resident threatened species, as mere presence does not guarantee long-term survival. The monitoring of ecological data such as population size and demographic rates will be needed to evaluate adequately the population viability of at least the keystone species within a protected area (see Caughley and Gunn 1996). It is logical to conclude that the security of tropical biodiversity can be augmented effectively by creating new protected areas in biologically important sites, diverting more funds to improve the management of existing protected areas, and perhaps by creating links (habitat corridors) among those protected areas already established (Bowles *et al.* 1998; see below for more details).

Because of the typically patchy distribution of many tropical taxonomic groups, care should be taken to consider the complementarity and comprehensiveness of the species composition of proposed future reserves planned as part of a larger protected areas network (Diamond 1980; Van Balen *et al.* 1999). An important part of this process is the consultation of lists of threatened species and their distributions because a large reserve may fail to protect all or even the most vulnerable species. For example, 83% of 29 endangered butterfly taxa do not occur in any of 14 existing or proposed protected terrestrial areas in the Philippines (Danielsen and Treadaway 2004). In addition to reserve location, reserve size is critical. Small reserves may (1) not contain sufficient volume or diversity of resources, (2) permit the infiltration and spread of parasites and invasive species from the surrounding matrix and (3) contain only small populations of many species, leaving them vulnerable to the perils of environmental stochasticity (e.g. localized wildfires, hurricanes). Ten carnivore species including tiger (*Panthera tigris*) were more likely to disappear from smaller than larger reserves (Woodroffe and Ginsberg 1998). Wide-ranging species were more at risk because of their higher probability of exposure and persecution by humans than those with smaller home ranges (Woodroffe and Ginsberg 1998; see also Chapter 6).

Although large protected areas are clearly often required, there is no consensus on the minimum reserve sizes needed in tropical systems. Leck (1979) recommended that tropical reserves should be at least as large as 20–30 km², but suggested that those > 100 km² would be more effective. Only large fragments (≥ 100 km²) are probably capable of supporting intact populations of many vertebrate fauna with large home range requirements (Curran *et al.* 2004). Other authors have also suggested that reserves of several thousand square kilometres may be needed in the tropics to halt or diminish the chances of mass extinctions (Terborgh 1974, 1992; Whitmore 1980; Myers 1986; Thiollay 1989). Such large reserves may be possible in some areas of Borneo, Sulawesi and New Guinea, where extensive undisturbed forests still exist (W.F. Laurance 1999). In areas where large forested tracts are unavailable, forests around the existing reserves could be restored (Whitmore and Sayer 1992; see below). Ecosystems and countries with higher species diversity and endemism need to designate a larger proportion of their areas as protected (Rodrigues and Gaston 2001).

The abundance of potential predators should not be unnaturally high in reserves (Terborgh *et al.* 1997). Mature and high-quality forests should ideally be protected. Sodhi (2002) found that reserves composed of poor-quality mature forest ('unhealthy' forest) contained fewer rare bird species than high-quality, but otherwise similar-sized, mature forests. This suggests that some rare species may need high-quality forest with a greater diversity of ecological niches to persist.

Long-term viability of protected areas can only be assured when the balance between wildlife and human needs are considered at the design stage. For example, Kremen *et al.* (1999) showed that by allowing sustainable harvest of timber by locals in the buffer zones, it was far easier to gazette Masoala National Park, the largest lowland reserve in Madagascar. The ecosystem services that biodiversity provides to humankind (Chapter 2) should be highlighted to people to gain support for protected areas (Brooks *et al.* 2004). Greater attention to local conditions and institutions is needed for sustainable conservation outcomes (T.M. Hayes 2006).

10.2 Poor governance as a threat to tropical biodiversity

For effective management of tropical resources, good governance is essential. However, political corruption is rife in many tropical countries (Talbott and Brown 1998; W.F. Laurance 2004; Norgrove and Hulme 2006), which can hinder conservation by reducing effective funding and distorting priorities through the overexploitation of forests, wildlife, fisheries and other natural resources. Corruption is also rampant in the tropical timber industry. Pervasive corruption has been shown to promote rampant illegal logging in Kalimantan (Indonesian Borneo) (J. Smith *et al.* 2003) and the Philippines (Kummer and Turner 1994). In Borneo, timber entrepreneurs apparently bribed local authorities to issue excessive numbers of short-term logging contracts. Local tribal leaders, military, police and provincial forestry officials were also bribed (J. Smith *et al.* 2003). To facilitate the export of wood to Malaysian Borneo, border patrols and Malaysian officials were also paid to remain silent about the illegal activity. This has resulted in a catastrophic loss of much of the island's forest. This example dictates that efforts to reduce or eradicate the effects of corruption are urgently required for achieving effective conservation. Corruption also reduces revenues – Indonesian government losses are estimated at US$4 billion annually due to illegal logging (W.F. Laurance 2004).

In response to demand for ivory and rhinoceros horns, the African elephant (*Loxodonta africana*) and black rhinoceros (*Diceros bicornis*) are fully protected through the Convention on International Trade in Endangered Species (CITES). The populations of both species have also recovered in countries with good governance (R.J. Smith *et al.* 2003; Figure 10.7). Good governance of tropical resources is thus a requisite for the successful outcome of conservation projects (Sodhi *et al.* 2006b). A large problem in this regard though is the fact that countries containing biodiversity hotspots, Endemic Birds Areas and Terrestrial Ecoregions have poorer governance than countries without these important conservation targets (R.J. Smith *et al.* 2003; Figure 10.8). Soares-Filho *et al.* (2006) also show that the establishment of good governance (i.e. implementation of all environmental legislation) by 2050 could eliminate deforestation from protected areas in the Amazon and reduce it by 35% in unprotected forests. In addition to benefits to biodiversity, such an achievement would also result in important reductions in carbon emissions (see Chapter 2).

A large proportion of funds needed for conservation will realistically come from developed nations, and W.F. Laurance (2004) suggested that developing nations should be assisted by the governments of developed countries in their

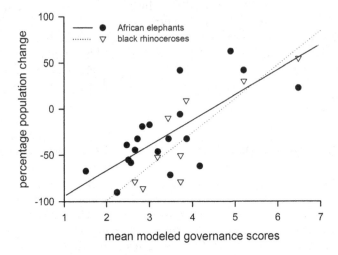

Figure 10.7 Mean modelled governance scores (based on a system that assigns a maximum value of 10 to the least corrupt countries) and changes in national populations of African elephants and black rhinoceroses from 1987 to 1994. (Data derived from R.J. Smith *et al.* 2003.)

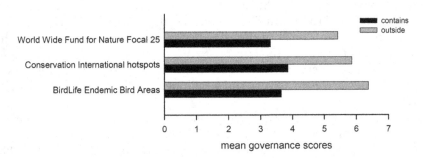

Figure 10.8 Governance scores (based on a system that assigns a maximum value of 10 to the least corrupt countries) and priority areas for conservation. (Data from R.J. Smith *et al.* 2003.)

efforts to curb corruption. In addition, richer nations can assist developing countries in training resource managers and in improving governance. It has been recommended that all developing tropical countries should be assisted by the developed world in establishing better environmental practices (even though we concede that many, if not all, developed nations need to improve their own environmental performance simultaneously). For example, Wetlands International and Wildlife Habitat Canada are working with local people to prevent fires in the peat swamps of Borneo and Sumatra (Aldhous 2004). This includes building dams using local materials and supplies so that peat swamps have enough moisture to resist drying out and becoming highly flammable during environmental events such as super El Niños (see Chapters 4 and 8).

The fostering of stronger collaborations among national, regional and international stakeholders is also a good step towards maximizing the persistence of good environmental governance. Some partnerships already exist between organizations such as ADB (Asian Development Bank), World Bank and UNEP (United Nations Environmental Program). Existing partnerships should be strengthened, and newly forged relationships developed to maximize international assistance for poorer countries to improve their environmental performance. Environmental indicators (e.g. the state of protected areas) for each country should be collected rigorously across all countries in the region (www.adb.org). If possible, nations should be subjected to periodic environmental audits, conducted under the auspices of a relatively impartial international body such as the United Nations, and mediated by an international panel of experts.

Local and national non-governmental organizations (NGOs) can also be used to evaluate the degradation of habitats and the adequacy of biodiversity protection. Environmental audits should aim to assess objectively and quantitatively the level and efficacy of protection of each country's natural habitats and biodiversity. The results of such audits should preferably be tied to appreciable consequences. Countries with poor conservation track records could also be penalized by being refused monetary assistance with their economic development (excluding humanitarian aid), while countries that have become more environmentally conscious through time could be rewarded. International development loans and grants would ideally be tied to the soundness and sustainability of environmental practices in much the same way that the current and proposed carbon credits programme has been developed (Kremen *et al.* 2000; Robertson and van Schaik 2001; see also Chapter 8). Only with such measures in place will there be the political will to better protect biodiversity. We admit that this approach may initially result in some environmental destruction and international ill-feeling, but we can see few other avenues for realizing tangible biodiversity protection in tropical regions over the long term. Alternatives such as a blanket ban on national exports of timber from nations that allow unsustainable forest extraction would probably be ineffective, difficult to enforce, create black markets and stymie initiatives that have the potential to foster conservation management (Jepson *et al.* 2001). However, promoting consumer boycotts of 'environmentally harmful' products should also be explored (see below).

Subsidies that have adverse effects on the environment as well as on society need to be removed (Millennium Ecosystem Assessment 2005). The promotion of good environmental governance can also be stimulated by international market forces, especially with respect to the mitigation of two main drivers of deforestation – cattle ranching and soy (*Glycine max*) farming (Soares-Filho *et al.* 2006). For example, such positive international pressures resulted in the expansion of protected areas in the Amazon (Soares-Filho *et al.* 2006).

A large part of the illegal logging money is laundered through banks. Realizing this problem, in 2003 Indonesia took the step of imposing extremely heavy penalties for laundering the proceeds of forestry and environmental crimes, with a maximum penalty of 15 years jail and a fine of US$1.7 billion [Centre for International Forestry (CIFOR) Research Annual Report 2003]. Stockmarkets also need to be vigilant so that money is not invested into companies guilty of unsustainable timber practices, enforced via the requisite use of certificates of sustainability for timber products (Barr 2001).

10.3 Improving logging practices

Poor logging practices still prevail in the tropics (Bowles *et al.* 1998; W.F. Laurance *et al.* 2000; Putz *et al.* 2000). It has been estimated that only a fraction of forest allocated for logging activities (production forests) is being harvested sustainably (International Tropical Timber Organization 2006; Figure 10.9). One of the reasons for this unfortunate situation is that 'unsustainable' logging can yield higher short-term profits than sustainable activities, with as much as 450% more money to be made through the former (Bowles *et al.* 1998; Putz *et al.* 2000). Sadly, there seems to be neither the institutional capacity nor political will to counter such a large financial disparity (R.J. Smith *et al.* 2003). Certainly, international assistance through environmentally friendly policies such as carbon-credit offset funds and other such means may be needed to promote sustainable logging in the tropics (Kremen *et al.* 2000; Putz *et al.* 2000; Chapter 8). Unsustainable logging may benefit certain individuals, but if one accounts for losses to humanity (e.g. flood protection – Bradshaw *et al.* 2007a – and carbon stocks – Chapter 2), it results in an average of 14% economic loss compared to sustainable logging (Balmford *et al.* 2002). For logging to become sustainable, it therefore needs the support of markets, development assistance agencies, NGOs and the public (ITTO 2006). Other hurdles preventing sustainable harvesting include the lack of long-term commitment to land and resource management, and the chronic shortage of staff and equipment (e.g. vehicles) to manage forests sustainably (ITTO 2006).

In addition to clear-felling, selective logging of rain forests is widely practised in the tropics, but the way in which such operations are done can often be improved substantially to strike a better balance with biodiversity conservation (Table 10.2). In Malaysia, for example, forestry guidelines dictate that all commercial non-dipterocarp species with a diameter at breast height (dbh) ≥ 4 cm and dipterocarp species with a dbh ≥ 50 cm can be harvested, provided sufficient stems remain in

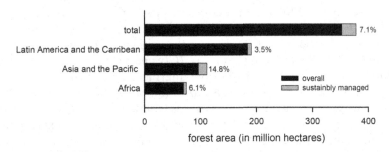

Figure 10.9 Global summary of the proportion of forested areas in tropical PFEs (protected forest estates) that are managed sustainably. [Data derived from International Tropical Timber Organization (ITTO) 2006.]

Table 10.2 Actions needed to make forest harvesting biodiversity-friendly. (After Sist *et al.* 2003. Copyright, Elsevier.)

1	When selective logging is practised, reduce the impact by minimizing skid trails and damage to the residual vegetation
2	Determine the maximum number of trees that can be harvested sustainably. For dipterocarp forests, harvest of eight trees per ha or less is recommended
3	Logging cycles should be more than 80 years
4	Assess minimum-diameter cutting limits based on the structure, density and diameter at reproduction of the target species
5	Minimize the size (e.g. $< 600\,m^2$) and connectivity of gaps
6	Refrain from understorey clearance
7	Set aside pristine areas of appropriate sizes (e.g. $> 1000\,ha$) with adequate connections to other such areas within logging concessions
8	Identify the species of greatest conservation concern (e.g. endangered/threatened species) and ensure that their needs are catered for adequately during logging activities
9	Convert logged rather than pristine areas into plantations for commercially valuable trees
10	Prevent excessive human invasion into logged areas. Attempts should also be made to deny access to the poachers of such areas

the logging concession for a repeat harvest in about 30 years (Appanah 1998). Similarly, in Indonesia, all commercial trees with dbh of $> 60\,cm$ can be felled over a 35-year cutting (logging) cycle (Sist *et al.* 2003). Under this scheme, understorey vegetation is removed two, four and six years after logging to control woody climbers and non-commercial saplings. This treatment is intended to facilitate the growth of commercial tree species, but can be detrimental to those (often species-rich) biotas that depend on the presence of a healthy understorey. More importantly, destruction of the forest floor vegetation can disrupt vital ecological processes such as pollination by decreasing habitat suitability for bees, and reduce the efficiency of nutrient recycling, thereby damaging a forest's capacity to regenerate (Chapter 2).

Trunk diameter limits to harvesting described above are usually designed to accommodate the market demands and processing technologies, rather than being made with biological considerations in mind. Sist *et al.* (2003) recommended that minimum diameter cuts should be limited to $> 60\,cm$ for dipterocarp trees to increase the probability of successful reproduction prior to harvesting. Dipterocarps, a key and iconic element of the Southeast Asian lowland forest flora, are renowned for their interspecific mass fruiting events, usually in cycles of 3–5 years (Appanah 1993; Curran *et al.* 1999). This mass fruiting is thought to have evolved as a mechanism to satiate seed predators such as pigs (*Sus* spp.) and thus allow a larger proportion of fruits to survive and germinate (Curran and Leighton 2000). Therefore, mass fruiting is vital to forest regeneration. Logging practices that reduce the capacity for forest stands to produce mass fruiting increase the likelihood of stand extinction. Similar ecological considerations should be given when deciding minimum-diameter cutting limits for other commercial tree species (Table 10.2).

A simulation study by Huth and Ditzer (2001) showed that the standard 40-year logging cycles may be too frequent for the forests of Sabah (Malaysia). They recommended logging cycles of > 80 years for the greatest benefit to biodiversity conservation and in minimizing the risks of erosion and nutrient loss, whilst still being viable commercially. Such logging cycles should be combined with careful logging practices, including minimizing the creation of logging access roads and careless felling of untargeted stems during selective logging. Lengthening the logging cycles will certainly result in lower economic returns (Sist *et al.* 2003), meaning that government subsidies may be needed to offset some economic losses. Selective logging can damage > 50% of a targeted forest stand (Sist *et al.* 2003). Therefore, reduced-impact logging practices (e.g. carefully controlled felling and skidding, and reduced damage to soil and residual trees) have been proposed as an alternative. Reduced impact logging (RIL) also reduces tree injury and death by 18% compared with conventional selective logging in Kalimantan (Bertault and Sist 1997). RIL is also less harmful to wildlife than conventional logging (Meijaard *et al.* 2005). Clearly, RIL can be effective in achieving more sustainable forestry and should be practised more widely (Kremen *et al.* 2000).

Lindenmayer *et al.* (2006) suggest that conservation approaches during a logging operation should include (1) long logging rotation cycles, (2) adequate fire management and buffers for aquatic areas, (3) retention of some structural features such as dead trees and (4) more reliance on plantations. Others have also suggested that to dampen or even halt the deforestation rates in tropical areas, a prudent measure would be to rely more on plantations for timber and timber products (Durst *et al.* 2004). These plantations can be encouraged, especially in already deforested areas, and may serve to provide on-going employment opportunities for local communities (see also below). There is also a need to curtail illegal logging that can operate clandestinely deep within protected areas (T. Whitten *et al.* 2001). It is worrisome that illegal loggers sometimes hold prominent political posts and that resident military have been observed supporting such activities (Kinnaird and O'Brien 2001; T. Whitten *et al.* 2001). A network of illegal loggers operating not only within Indonesia, which has received most publicity, but also working in close collaboration with industries from Singapore and Hong Kong, is thriving (Robertson and van Schaik 2001). There have been reports of illegal tropical wood smuggling from Sumatra to Singapore and Malaysia (Anonymous 2004b). Owing to the absence of witness protection programmes, activists and forestry officials receive death threats, and may even be killed (Robertson and van Schaik 2001; www.ecologyasia.com). Similarly, in Myanmar and Cambodia, illegal logging has been widely used to fund military operations and may go hand in hand with drug trafficking as well (Talbott and Brown 1998).

Timber certification can be used as means to restrict illegal logging and promote sustainable practices. However, such certification should also ideally consider the impacts of logging on the biota (E.L. Bennett 2000). For example, logging companies can be encouraged to retain pristine areas within their concessions via timber certification regulations and enforce trade bans on the commercial exploitation of wildlife and wildlife products in their own forestry concessions (J.G. Robinson *et al.* 1999). One of the good outcomes of timber certification has been the establishment of local organizations such as the Indonesian Ecolabelling Institute to curb illegal logging through certification. In summary, more sustainable forestry practices in the tropics are urgently needed and logging companies can be encouraged to conserve biodiversity through financial incentives by governments (Goldstein *et al.* 2006).

10.4   Livelihoods and conservation

Protected areas in the tropics will inevitably cause some disruption to the livelihood of indigenous communities. For example, the creation of nature reserves may be detrimental to local inhabitants who are reliant on the forest for sustenance (e.g. bush meat and localized traditional agriculture), and have few available substitutes at acceptable prices (Milner-Gulland and Bennett 2003). Continued conflicts around certain tropical protected areas indicate that social issues need better consideration for sustainable conservation (Bawa *et al.* 2004a; Sodhi *et al.* 2006b). Nature conservation is a luxury that the short-term requirements induced by hunger usually override, so poorer rural communities generally need to be compensated when their livelihoods are compromised by conservation actions. Suggestions have been made to give direct payments to rural communities to safeguard biodiversity (du Toit *et al.* 2004). Costa Rica, for example, began a Payment for Environment Services programme in 1997 in which private landowners are paid for preserving forests. Such forest preservation is considered to provide the following benefits: (1) biodiversity conservation, (2) carbon sequestration, (3) hydrological services and (4) preservation of scenic beauty that attracts ecotourists (Pagiola 2002). Direct payments may work in some instances but, wherever possible, efforts should be made to find long-term and sustainable solutions that involve the development of compatible employment opportunities (e.g. resource management) or the introduction of alternative food sources for the affected communities (Chapter 6). Further, direct payments to conserve biodiversity can work only if they result in durable conservation practices. The due consideration of such socioeconomic forces is paramount to the success of any conservation exercise (Adams *et al.* 2004). Outside assistance, however, may be critical to alleviate poverty around protected areas. For example, money from carbon credit funds (Chapters 8) can be channelled to developing alternatives for rural communities so that they reduce forest cutting (W.F. Laurance 2006a).

Tools used by various governmental and non-governmental organizations to accommodate rural communities as forest stakeholders must include among their criteria relative legislative power (e.g. government authorities) and the capacity to alleviate the loss of livelihood and settlement to enhance the likelihood of sustainable conservation outcomes. Such an approach has been successfully adopted in Tubbataha National Park (the Philippines), where coral cover and commercial fish biomass increased by 50% and 100%, respectively, between 1999 and 2004 as a result of successful partnerships among all stakeholders (Sodhi *et al.* 2006b; see also Chapter 7). Furthermore, similar creative interdisciplinary efforts are urgently required to design and implement effective conservation strategies.

Although at times controversial, the idea of community-based conservation needs to be explored seriously (Berkes 2004). Community-based ecotourism can assist local communities economically, but such projects may require long-term funding commitments and do not always succeed (Kiss 2004). Naidoo and Adamowicz (2005) showed that money generated by ecotourism can be distributed to communities so that they have the incentive to protect forest biodiversity adequately. Similarly, about half of the entrance fees to the Parc National de Ranamofana in Madagascar (amounting to US$50 000) are spent on economic development projects in the surrounding communities (Lovejoy 2006).

In some cases, it may be possible for rural communities to extract non-timber products sustainably from protected areas. Shanker *et al.* (2005) show that an indigenous community can sustainably harvest nelli fruits (*Phyllanthus* spp. – an edible fruit with additional value as a traditional Indian medicine and ingredient in cosmetics) from the Biligiri Rangaswamy Temple Wildlife Sanctuary in India's Western Ghats if they use non-destructive methods (i.e. cutting off branches rather than killing small trees). Some non-forest products are used for traditional medicines and are essential for certain remote rural communities with little access to, or desire for, Western medicines (Mendelsohn and Balick 1995). For example, Belsky and Siebert (1995) explored whether it is possible for local people to extract rattan (*Calamus exilis*) from the Kerinci-Seblat National Park (Sumatra, Indonesia). Rattan is a coppicing cane used in local handicrafts and basketry. If rattan extraction can provide rural communities with a viable means of earning their livelihood, then there would be less incentive for them to convert the Park to agriculture. Belsky and Siebert (1995) proposed that rattan can be extracted sustainably at 4-year intervals from designated areas of the park. Examples such as this provide good illustrations of the practicability of the long-term coexistence potential of protected areas and human communities, and should be considered as feasible real-world options for shared land use across many parts of the tropics.

In unprotected areas, the negative impacts of human development on native biodiversity could be minimized by discouraging the large-scale and often homogeneous conversion of land for agriculture. For example, some agricultural subsidies could be removed and replaced by other, more environmentally friendly, incentives (e.g. alternative employment opportunities) to compensate affected communities (A.N. James *et al.* 1999; O'Brien and Kinnaird 2003). Funds saved through elimination of perverse subsidies can be channelled instead towards better biodiversity conservation (A. James *et al.* 2000). Furthermore, appropriate land use decisions (e.g. retaining sufficiently large and connected parcels of native vegetation) in degraded habitats (e.g. countryside currently partially converted for crops) could substantially increase the long-term conservation value of such areas (Sodhi *et al.* 2005d).

10.5   Conservation education and advocacy

It is probably fair to say that given the global conservation crisis, conservation biologists have failed to convey effectively the importance and complexity of healthy, functioning ecosystems to the general public. Studies show that people who have a personal connection with natural areas are more motivated to protect them (Schultz 2000). Thus, reconnecting people with nature should be a challenge tackled immediately by conservation biologists (Balmford and Cowling 2006). Ecologically literate people will also be able to mitigate better the conflicts over natural resources (C.D. Saunders *et al.* 2006). Better conservation education can be achieved only through inter-disciplinary approaches involving psychologists, educators, economists, anthropologists, social scientists, architects and marketing scientists (Balmford *et al.* 2006). Conservation education should also target all sectors of society, from children to adults (Feinsinger *et al.* 1997).

Conservation-friendly decisions are a must for the everyday operations of private and public sectors (Balmford and Cowling 2006). At the local level, scientists, non-governmental organizations and government agencies need to recognize the possibility and importance of attempting to sustain economic development without obliterating their natural resources. In the tropics in particular, there seems to be a general ignorance or apathy about environmental issues (T. Whitten *et al.* 2001). For instance, only 27% of 74 people surveyed in Indonesia thought that conservation of species was of major concern (Jepson 2001). More worryingly, this percentage declined to 15% when asked about the necessity of preserving native forests and other wild places. It is imperative to convince more of the populace and a higher proportion of elected government officials in the region that biodiversity matters to secure long-term prosperity in their societies (Kinnaird and O'Brien 2001; see Chapter 1).

Embracing conservation advocacy is controversial among scientists because many believe that taking a particular stand might compromise scientific objectivity. However, others think that conservation biologists have a responsibility to enlighten and educate the public (Noss 1996; Matsuda 1997). We suggest that through activities such as intensive public education programmes, general public awareness should be raised about the severity and the possible short- and long-term ramifications of the looming disaster facing tropical biodiversity. The media could be, and in many cases already has been, engaged creatively for this purpose. For example, local radio and television could be used as communication vehicles for spreading this message to the older generations through local languages and dialects and to future generations through episodes targeting children. This media campaign could also be extended internationally to canvass for much needed conservation funds from foreign sources. Pictorial billboards depicting the importance of nature are also potentially useful media (Figure 10.10). The lack of a truly free press in some tropical countries may hinder public awareness campaigns, but this problem can be countered using global communication media such as the internet.

We emphasize that it would be wise to include environmental awareness into the education curricula of children. Children who play in wild areas show a greater affinity to and appreciation of them later in life (Bixler *et al.* 2002). Early experiences in wild areas also result in forming a baseline that people can compare and advocate should environmental destruction threaten them later in their life (J.R. Miller 2005). Thus, children should be given ample opportunities to visit intact natural areas and observe wilderness (Feinsinger *et al.* 1997).

Figure 10.10 Billboards depicting the importance of nature. (Photo by M. K. Pandit.)

Without such measures, there is a real danger of developing a general alienation towards environmental issues in youths and adults, as has been noted in heavily urbanized Singapore (Kong *et al.* 1997, 1999). Worse still, they could become environmental destroyers (i.e. 'lost generation'), rather than protectors early in life. Young people are also key stakeholders because they will experience the future consequences of environmental decisions made today (Millennium Ecosystem Assessment 2005).

One excellent example of a positive public education campaign is provided by E.L. Bennett *et al.* (2000). The Sarawak Forest Department (Malaysian Borneo) has been attempting to educate local people about environmental issues for a number of years. Innovative methods such as dance and role playing are being used to educate people about wildlife conservation in a stimulating way. In addition, people are being instructed in how to run small businesses, attract tourism dollars and communicate in English in the hope that they can obtain jobs in the tourism industry and other related enterprises. Becker *et al.* (2005) showed that by involving local people in environmental and bird monitoring in Ecuador, they became more aware of ecosystem services, learned about local birds and their conservation status, and became more environmentally literate to include conservation and sustainable development planning in their community's actions. Conservation education resulted in biodiversity being included in children's art and songs. This project also facilitated collaboration among local, regional, national and international agencies. In India, some NGOs organize school children to protest against deforestation and teach them the importance of reforestation (Figure 10.11). Deforestation in Amazonia is threatening medicinal plants: five of nine top-selling medicinal plants with no apparent botanical substitute are now being harvested for timber (Shanley and Luz 2003). Material from a book (*Frutíferas e Plantas Úteis na Vida Amazônica*) is now being used in workshops and schools and on billboards to educate the rural communities

Figure 10.11 In India, some NGOs organize school children to protest against deforestation and teach them the importance of reforestation. (Photo by M. K. Pandit.)

and policy makers about the plight of medicinal plants in the Amazon (Bruce Campbell, personal communication). We feel that these are superb, albeit rare, examples of what can be achieved, and they should be implemented by relevant national and international organizations in other tropical countries.

T. Whitten *et al.* (2001) posited that local NGOs remain a valuable, but often underutilized, hope for biodiversity conservation in many tropical countries. The local people working in these organizations can also play important roles in compiling biodiversity inventories and assisting in the development of sound environmental impact assessments (Janzen 2004; Sheil and Lawrence 2004). However, organizations should be aware that some NGOs may be infiltrated by people with their own economic and political agendas (see R.L. Bryant 2002). Additionally, NGO staff often lack biological training (e.g. in proper scientific sampling methods, data analyses and interpretation), and efforts should be made to rectify this lack of expertise and avoid duplication of effort (Mace *et al.* 2000). Further, training of students in conservation biology and forestry guards against improper management practices, and will certainly help to facilitate good conservation outcomes (Kinnaird and O'Brien 2001; Bawa *et al.* 2004b). It has also been suggested that national and local NGOs should work in close collaboration with international NGOs to ensure that the perpetrators of environmental destruction are prosecuted (J.M.Y. Robertson and van Schaik 2001). It is all too often the case that offenders walk free due to a lack of funds to pursue legal proceedings.

Governance of natural resources and protected areas should involve civil society, improve the quality of life and sustain livelihoods. The formation of 'village conservation organizations' by the Lore Lindu National Park (Sulawesi, Indonesia) Authority and The Nature Conservancy (TNC) exemplifies institutional change addressing those issues (Sodhi *et al.* 2006b). Such organizations provide conservation education and monitor protected areas in collaboration with customary councils as adjudicators imposing sanctions for infractions of park regulations. These hybrid arrangements combine the intrinsic value of saving imperilled biota, maintaining vital ecosystem services for sustaining rural livelihoods and re-engaging and empowering community members. Traditional ecological knowledge is gained over generations by humans acting with and in nature and can be useful in modern conservation programmes. Traditional knowledge has been used for managing fisheries in I-Kiribati (Drew 2005; also see Johannes 2002).

10.6   Better technologies

Greater effort is needed to develop environmentally friendly technologies to reduce the damage humans are causing to the biosphere. For example, agriculture is a main driver of land use change in the tropics (Chapter 1), with a 34% expansion of agricultural lands predicted for developing countries by year 2030 (www.fao.org). Worldwide, agriculture accounts for 70% of freshwater withdrawals and results in severe water shortages such as those seen in some parts of China (www.fao.org). Clearly, there is an urgent need to develop agricultural technologies that increase the production of food per unit area without a concomitant increase of water consumption, nutrient loading and pesticide use (Millennium Ecosystem Assessment 2005). However, care must be taken that these new agricultural technologies do not promote further land clearing (Angelsen and Kaimowitz 2001).

One potential avenue of development yielding promising results is the use of genetic experiments to boost crop yield (Gur and Zamir 2004). However, a potential disadvantage is that genetically modified crops may result in population declines of some birds in certain agro-ecosystems (Watkinson *et al.* 2000). Organic farming that excludes the use of fertilizers and pesticides should be encouraged and made more economically feasible. Similarly, technological acceptance and advances to reduce greenhouse gas emissions are needed. These can be achieved through fuel switching (from coal/oil dependency to gas), increased efficiency of power plants and the more efficient use of energy in general (Millennium Ecosystem Assessment 2005; Chapter 8). Buildings and transportation need to become more energy efficient; for example, heavily urbanized Singapore has introduced a number of good measures such as the use of solar panels to reduce energy consumption (www.mewr.gov.sg).

10.7   Good examples of tropical conservation

Here we highlight some examples to show how the efforts of conservation biologists combined with activities aimed at convincing stakeholders about the value of nature can result in successful conservation outcomes.

Assessments by local and international economists have concluded that ongoing upland deforestation in Madagascar promotes siltation and reduced water flow for lowland farmers (Carrett and Loyer 2003). To counter these trends, it was concluded that conserving forest makes sound economic sense because of the hydrological and ecotourism benefits that ensue. These results were considered instrumental in the decision by the Malagasy government to triple the size of protected forests in 2003 (Balmford and Bond 2005). Another example comes from the hyper-biodiverse cloud forests in western Ecuador. Despite difficulties, a collaboration with NGOs and villagers resulted in joint data collection which showed that forests are important in fog interception – an essential process bringing moisture to the forests, which are otherwise devoid of major water input. It was estimated that deforestation of the cloud forests costs US$640 per household annually – roughly equivalent to half the average annual family income (Becker 1999; Becker *et al.* 2005; Figure 10.12). This realization resulted in the preservation of about 3000 ha of cloud forest in an ecological reserve (Becker 1999; Becker *et al.* 2005).

In Thailand, Poonswad *et al.* (2005) attempted to integrate 28 known hornbill poachers into hornbill monitoring programmes using mostly locally generated funds. Individuals donated US$120 each to adopt a hornbill family – 68% of funds were generated from within Thailand alone. Over 3 years, their efforts to offer alternative livelihoods to poachers, thus alleviating nest poaching, increased the number of nests with fledglings by 39%. It is our hope that such locally initiated conservation successes become a norm in the tropics.

10.8 Organizations assisting with tropical conservation

There are a number of organizations with the principal aim of protecting biodiversity and ecosystems, and we list some of the prominent ones here. Conservation International's (CI) mission is to save biodiversity and natural heritage (www.conservation.org). CI has projects around the globe and is working

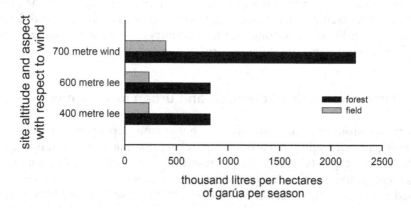

Figure 10.12 Seasonal interception of garúa fog from intact cloud forest and areas cleared for pasture or fields at three different combinations of altitude and aspect at Loma Alta, Ecuador. (After Balmford and Bond 2005. Copyright, Blackwell Publishing Limited.)

to save terrestrial biodiversity hotspots and key marine regions. TNC aims to preserve both terrestrial and aquatic biotas (www.nature.org) and identifies landscapes and seascapes that deserve conservation priority. TNC has projects in major biomes throughout the world and also runs ecotours to generate funds. The Wildlife Conservation Society (WCS) aims to save wildlife and wild lands through careful science and education (www.wcs.org). With projects operating worldwide, WCS works towards terrestrial and marine conservation. The World Wide Fund for Nature (WWF) has a global network active in over 90 countries with an objective to slow climate change, reduce environmental toxicity, protect aquatic ecosystems, halt deforestation and save species from extinction (www. panda.org). Worldwide, WWF has recognized over 200 ecoregions in need of immediate conservation measures (see Chapter 1).

The World Conservation Union (IUCN) is mandated to support cutting-edge conservation science and link research to policy (www.iucn.org). The most famous role of IUCN has been to maintain its celebrated *Red List of Threatened Species* (www.iucnredlist.org). The Convention on International Trade in Endangered Species of Wild Fauna and Flora (CITES) is an international agreement among governments that endeavours to prevent wildlife trade from threatening the survival of species. CITES accords protection to over 30 000 species by listing them on one of its three appendices: Appendix I lists species that should not be traded except in exceptional cases; Appendix II lists species whose trade should be restricted; and Appendix III lists species that are threatened in at least one country and for which some trade restrictions apply (e.g. they may be traded only when accompanied by a certificate of origin).

The above-mentioned agencies have worked towards successful conservation of tropical biodiversity. For example, in 2006, both TNC and WWF worked towards gazetting 1.2 million ha of marine protected areas in Berau (Indonesia) – an area holding the highest level of coral biodiversity in the world (www.nature. org). This protected area was established with governmental and local support. Local fishers agreed to implement the sustainable harvesting of marine resources from some parts of this area. Other examples include TNC and WWF arranging a debt pardon in exchange for protecting rain forest reserves in Peru and Panama and encouraging Brazil to become the first country to have a management plan for freshwater resources (www.nature.org).

10.9   Restoration, reintroductions and urban management

Conservation biologists can also assist in enhancing depauperate habitats and biodiversity, and we highlight a few examples here. Owing to heavy deforestation of tropical landscapes (Chapter 1), reforestation is considered to be one of the management activities that can lead to solid conservation outcomes. Some degraded areas have the potential to recover through succession; however, the chances and rate of succession can be improved in recently deforested sites when some residual trees, seedling banks and soil seed stores are left after logging (Lamb *et al.* 2005). Regenerating forests usually lack large-fruited plant species because of an absence of seed dispersers (Neilan *et al.* 2006), so the biodiversity

Table 10.3 Different forms of reforestation used when secondary forests are present or when planting is needed. Combinations of various techniques could be used in degraded landscapes depending on ecological circumstances and on the goals of the land managers. [After Lamb et al. 2005. Copyright, American Association for the Advancement of Science (AAAS).]

| Natural secondary forests | Plantings and plantations | |
	To restore biodiversity	To supply goods and ecological services
Protect and manage natural regrowth: Potentially able to supply a variety of goods and services depending on the age and condition of the forest	*Restoration plantings using a small number of short-lived nurse trees*: Acquisition of further diversity dependent on colonization from nearby forest remnants. Primary benefit is ecological services, although can supply some goods depending on species present	*Tree plantation monoculture of exotic species*: An efficient method of timber or food production for (mainly) industrial users; in most circumstances, it is less successful in supplying many services
Protect and manage natural regrowth plus enrichment with key species: Enrichment with commercially, socially or ecologically useful species can improve the value of these forests to local communities or industry	*Restoration plantings using large number of species from later successional stages*: Higher initial diversity that will also be supplemented by colonization from nearby forest remnants. Primary benefit is ecological services although can supply some goods depending on species used	*Tree plantation used as a nurse crop with underplantings of native species not otherwise able to establish at the site*: An initial fast-growing nurse crop supplying commercially useful timbers or other goods can facilitate (e.g. via nitrogen fixation and microclimate alterations) the subsequent establishment of more species-rich forests that supply a wider range of goods and services
	Direct seeding: The number of species that can be established by direct seeding is limited by seed supply but the establishment cost can be lower. Direct seeding can be used to initiate reforestation in open fields under appropriate conditions but it may be most useful when used to enhance diversity once some tree cover is already present	*Tree plantation mixtures of native species*: Mixed-species plantations can potentially supply a wider range of goods and services than monocultures. Biodiversity gains are greater than in plantation monocultures but are mostly still modest (usually fewer than five planted species)

value of such regenerating areas can be improved by planting species that would otherwise be unable to re-colonize themselves (Lamb *et al.* 2005; Table 10.3).

Most reforestation activities have been done using monocultures of *Pinus*, *Eucalyptus* and *Acacia* trees. Such species generally have low biodiversity benefits as monocultures, so mixed-species plantations of native trees are preferable for reforestation activities (Lamb *et al.* 2005; see Table 10.3). Mixed-species native plantations may also have more economic value and be able to withstand high insect damage (Piotto *et al.* 2003, 2004). Financial incentives for rural people to reforest degraded areas is also an important step in the right direction. For example, some Costa Rican farmers are being paid for reforestation and conservation activities (Zbinden and Lee 2005), but where this is not possible farmers can be encouraged to plant shade-tolerant cash crops [e.g. coffee, cocoa (*Theobroma cacao*) and cardamom (*Elettaria cardamomum*)] under reforested trees (Lamb *et al.* 2005).

Forest restoration should ideally be done at a landscape level for enhanced ecological benefits (e.g. improved hydrology). Denuded hills can be reforested to reduce the chance of landslides and floods, and preservation and restoring biodiversity in metropolitan areas should also be a high priority. The presence of seminatural ecosystems in or near urban areas helps to reconnect people to nature and contributes to their well-being (for example, exposure to nature relieves stress in humans – Kaplan and Kaplan 1989). However, conservation areas in urban areas requires partnerships among scientists, urban planners and architects (G.H. Miller 2005). Similarly, mangrove restoration is also being envisaged in various tropical countries because of their ecological benefits and their current rates of rapid loss (Ellison 2000). As with the forests, it is suggested that monoculture plantations of mangrove trees for restoration projects should be avoided because these generally yield low ecological benefits (Walters 2000).

Habitat links (e.g. fence rows and windbreaks) between forest patches or reserves may also facilitate dispersal and buffer against local extinctions and population declines in fragmented landscapes (Lovejoy *et al.* 1997; Sekercioglu *et al.* 2002; Chapter 3). Data on efficacy and adequacy of habitat corridors in the tropics are limited; however, it is known that high vegetation cover (both native and non-native) can enhance the attractiveness of corridors for forest species, thereby enhancing the likelihood of movement through these areas and the recolonization and use of edge habitats (Castellon and Sieving 2006). For example, the maintenance of safe corridors may be essential for migrating species such as the African elephant (*Loxodonta africana*). Castelletta *et al.* (2005) also recommended that patches isolated about 10 years ago in Singapore should be reconnected to facilitate bird movements. On the negative side, corridors can be expensive to create and maintain, may not serve well in heavily fragmented landscapes and may be counterproductive by facilitating the spread of predators and diseases (Simberloff and Cox 1987; Simberloff 1992a). Nonetheless, we conclude that the available evidence suggests that more conservation advantages than disadvantages are derived in the creation of habitat linking corridors.

Reintroductions are defined as the introduction of a species to previously occupied areas. The objective of reintroductions is to establish viable wild population of a species that has become globally or locally extinct. Reintroduced individuals can come from captive breeding programmes or can be translocated from wild populations. Reintroduction programmes can be costly and should include a feasibility study, preparation time, release and follow-up monitoring. A feasibility study should include estimation of carrying capacity and the potential for negative impacts on extant flora and fauna. For example, a feasibility study for the release of agile gibbons (*Hylobates agilis albibarbis*) on an island in Kalimantan (Indonesia) included estimation of food abundance (fruit abundance and fruit productivity) and concluded that between 3 and 19 gibbons could be sustained on the island (Cheyne 2006).

It is also wise to restore original habitat if it has been heavily reduced and disturbed prior to reintroduction. In addition, it is essential that the underlying causes of the original extirpation be removed or suppressed prior to reintroduction for the probability of establishment success and persistence to be high. For example, if overharvesting was the main cause of the original extirpation, efforts should be put in place to minimize hunting or restrict it to highly managed parcels of land away from the introduction site. In some cases, population augmentation may be needed after the first release. There are some high-profile examples of reintroductions that have successfully improved the conservation status of a threatened species. The Seychelles warbler (*Acrocephalus sechellensis*) was once a highly threatened single-island species with only 26 individuals remaining at its nadir. Conservation efforts resulted in population recovery to over 300 individuals, with successful translocations to two neighbouring islands carried out to reduce the chance of total species extinction (Komdeur 1994).

10.10 Role of sound biological science in tropical conservation

Sadly, our overall understanding of earth's biodiversity is superficial (Lovejoy 2006), and especially in the tropics, large-scale, coordinated research efforts are needed urgently to remedy the paucity of biodiversity studies. Pereira and Cooper (2005) suggested that to review and modify conservation strategies, global monitoring of biodiversity and ecosystems is needed. They suggested that plants and birds should be used as indicator taxa and that they must be surveyed in key areas over long time scales. Conservation efforts including both strategic and on-ground management will undoubtedly benefit from a better understanding of the biology of native biotas.

A comprehensive programme of environmental and biodiversity mapping must be done to reverse the sad state of affairs now confronting tropical biodiversity. Pereira and Cooper (2006) also suggested that efforts should be made to develop a global land-cover map to monitor ecosystem change. More and better-directed research will also help train citizens of local communities to protect the natural resources upon which their livelihoods and health depend in a more strategic and efficient manner. Scientists must collect more data showing whether applied conservation measures, such as the establishment of reserves and the adoption of alterative land use methods in the tropics, are effective at abating extinctions and arresting population declines. More information is also needed on the value of biodiversity in maintaining vital ecological processes in addition to providing ecosystem services to humanity (e.g. flood and disease protection; Chapter 2).

Balmford and Bond (2005) highlighted four areas in conservation science where more or better data are needed: (1) rate of habitat and population changes; (2) predictions of the future state of biodiversity and natural ecosystems; (3) understanding the drivers of biodiversity and habitat change; and (4) understanding the link between nature, its ecosystem services and human well-being. Bawa *et al.* (2004b) recommended that more basic research is needed on the description, maintenance and functioning of tropical ecosystems and how these are affected by humans. They also suggested more interdisciplinary research involving social and natural scientists, and a better linkage between science and policy. Along the same lines, Sodhi *et al.* (2006b) argue that any weak link among biological, social and governance issues could further endanger tropical biodiversity and human well-being. They suggested that academic institutions and funding agencies can also be critical facilitators in tropical resource governance by recognizing academic participation in conservation and livelihood projects, and increasing financial support for multidisciplinary conservation research and discourse. Scientists in biodiversity-related research should be better integrated, and they need to communicate regularly on biodiversity issues to policy makers, governments, NGOs and the public (Loreau *et al.* 2006). If we are to avert the massive impending tropical biodiversity crisis now already under way, issues relating to politico-economic drivers (e.g. corruption), disciplinary divisions, resource governance and rural empowerment must be addressed together and urgently.

10.11 Summary

1 There are feasible ways to conserve at least some tropical natural resources.
2 Key solutions should include enhancing public environmental awareness (which will also help to change political decision-making), delineating adequately protected reserves and providing economic incentives for conservation.
3 Good governance, political accountability, independent and fair judiciary, and a freer press all facilitate biodiversity conservation and should be encouraged.
4 Given that many of the drivers of biodiversity loss (e.g. international demand for rain forest timber, global warming associated with elevation in CO_2 concentration) are issues that transcend national boundaries, any realistic solution will need to involve a multinational and multidisciplinary strategy, including political, socioeconomic and biophysical input in which all major stakeholders (governmental, non-governmental, national and international organizations) must partake.
5 Technological advances are needed to increase agricultural yield efficiently and to curb the emission of greenhouse gases and energy use.
6 Large-scale reforestation activities should be carried out in degraded tropical landscapes.
7 More biological data are needed to understand tropical biodiversity.

10.12 Further reading

Balmford, A. and Whitten, T. (2003) Who should pay for tropical conservation, and how could the costs be met? *Oryx* 37, 238–250.

Balmford, A., Gaston, K. J., Blyth, S., James, A. and Kapos, V. (2003) Global variation in terrestrial conservation costs, conservation benefits, and unmet conservation needs. *Proceedings of the National Academy of Sciences of the USA* 100, 1046–1050.

Bawa, K. S., Kress, W. J., Nadkarni, N. M. and Lele, S. (2004) Beyond paradise – meeting the challenges in tropical biology in the 21st century. *Biotropica* 36: 437–446.

References

Abal, E. G. and Dennison, W. C. (1996) Seagrass depth range and water quality in southern Moreton Bay, Queensland, Australia. *Marine and Freshwater Research* **47**, 763–771.

Abdullah, A. R., Bajet, C. M., Matin, M. A., Nhan, D. D. and Sulaiman, A. H. (1997) Ecotoxicology of pesticides in the tropical paddy field ecosystem. *Environmental Toxicology and Chemistry* **16**, 59–70.

Abercrombie, D. L., Clarke, S. C. and Shivji, M. S. (2005) Global-scale genetic identification of hammerhead sharks: Application to assessment of the international fin trade and law enforcement. *Conservation Genetics* **6**, 775–788.

Achard, F., Eva, H. D., Stibig, H.-J., Mayaux, P., Gallego, J., Richards, T. and Malingreau, J.-P. (2002) Determination of deforestation rates of the world's humid tropical forests. *Science* **297**, 999–1002.

Adams, W. M., Aveling, R., Brockington, D., Dickson, B., Elliott, J., Hutton, J., Roe, D., Vira, B. and Wolmer, W. (2004) Biodiversity conservation and the eradication of poverty. *Science* **306**, 1146–1149.

Adeel, Z. and Pomeroy, R. (2002) Assessment and management of mangrove ecosystems in developing countries. *Trees – Structure and Function* **16**, 235–238.

Adeney, J. M., Ginsberg, J. R., Russell, G. J. and Kinnaird, M. F. (2006) Effects of an ENSO-related fire on birds of a lowland tropical forest in Sumatra. *Animal Conservation* **9**, 292–301.

Agrios, G. N. (1988) *Plant Pathology*. Academic Press, San Diego.

Aiken, S. R. and Leigh, C. H. (1992) *Vanishing Rain Forests*. Clarendon Press, Oxford.

Aizen, M. A. and Feinsinger, P. (1994) Habitat fragmentation, native insect pollinators, and feral honey bees in Argentine Chaco Serrano. *Ecological Applications* **4**, 378–392.

Albaret, J. J. and Lae, R. (2003) Impact of fishing on fish assemblages in tropical lagoons: The example of the Ebrie lagoon, West Africa. *Aquatic Living Resources* **16**, 1–9.

Albott, K. L. (2004) Alien ant invasion on Christmas Island, Indian Ocean: The role of ant-scale associations in the dynamics of supercolonies of the yellow crazy ant, *Anoploepis gracilipes*. PhD thesis, Monash University, Melbourne.

Aldhous, P. (2004) Land remediation: Borneo is burning. *Nature* **432**, 144–146.

Aldhous, P. (2006) Drugs, crime and a conservation crisis. *New Scientist* **191**, 6-8

Aldrich, P. R. and Hamrick, J. L. (1998) Reproductive dominance of pasture trees in a fragmented tropical forest mosaic. *Science* **281**, 103–105.

Alho, C. J. R. and Vieira, L. M. (1997) Fish and wildlife resources in the Pantanal wetlands of Brazil and potential disturbances from the release of environmental contaminants. *Environmental Toxicology and Chemistry* **16**, 71–74.

Ali, A. (1996) Vulnerability of Bangladesh to climate change and sea level rise through tropical cyclones and storm surges. *Water Air and Soil Pollution* **92**, 171–179.

Allan, J. D. and Flecker, A. S. (1993) Biodiversity conservation in running waters. *BioScience* **43**, 32–43.

Allendorf, F. W. and Lundquist, L. L. (2003) Introduction: Population biology, evolution, and control of invasive species. *Conservation Biology* 17, 24–30.

Altizer, S., Dobson, A., Hosseini, P., Hudson, P., Pascual, M. and Rohani, P. (2006) Seasonality and the dynamics of infectious diseases. *Ecology Letters* 9, 467–484.

Alvard, M. S. (2000) The potential for sustainable harvests by traditional Wana hunters in Morowali Nature Reserve, Central Sulawesi, Indonesia. *Human Organization* 59, 428–440.

Alvard, M. S. and Winarni, N. L. (1999) Avian biodiversity in Morowali Nature Reserve, Central Sulawesi, Indonesia and the impact of human subsistence activities. *Tropical Biodiversity* 6, 59–74.

Alverson, D. L. and Hughes, S. E. (1996) Bycatch: From emotion to effective natural resource management. *Reviews in Fish Biology and Fisheries* 6, 443–462.

Alverson, D. L., Freeberg, M. H., Murawski, S. A. and Pope, J. G. (1994) A global assessment of fisheries by-catch and discards. *United Nations Food and Agriculture Organization FAO Fisheries Technical Paper 339*. FAO, Rome.

Andersen, A. N., Braithwaite, R. W., Cook, G. D., Corbett, L. K., Williams, R. J., Douglas, M. M., Gill, A. M., Setterfield, S. A. and Muller, W. J. (1998) Fire research for conservation management in tropical savannas: Introducing the Kapalga fire experiment. *Australian Journal of Ecology* 23, 95–110.

Anderson, P. K., Cunningham, A. A., Patel, N. G., Morales, F. J., Epstein, P. R. and Daszak, P. (2004) Emerging infectious diseases of plants: Pathogen pollution, climate change and agrotechnology drivers. *Trends in Ecology and Evolution* 19, 535–544.

Andreae, M. O. (1991) *Global Biomass Burning: Atmospheric, Climatic, and Biospheric Implications.* MIT Press, Cambridge, MA.

Andresen, E. (2003) Effect of forest fragmentation on dung beetle communities and functional consequences for plant regeneration. *Ecography* 26, 87–97.

Andrew, N. L. and Pepperell, J. G. (1992) The by-catch of shrimp trawl fisheries. *Oceanography and Marine Biology Annual Review* 30, 527–565.

Angelsen, A. and Kaimowitz, D. (2001) *Agricultural Technologies and Tropical Deforestation.* CAB International Publishing and Centre for International Forestry Research (CIFOR), New York.

Anonymous (2004a) Graft and poverty at root of deforestation. *The Straits Times*, 4 December 2004.

Anonymous (2004b) We've 'proof' of wood-smuggling: Greenpeace. *The Straits Times*, 7 October 2004.

Apaza, L., Wilkie, D., Byron, E., Huanca, T., Leonard, W., Perez, E., Reyes-Garcia, V., Vadez, V. and Godoy, R. (2002) Meat prices influence the consumption of wildlife by the Tsimane Amerindians of Bolivia. *Oryx* 36, 382–388.

Aplet, G. H., Anderson, S. J. and Stone, C. P. (1991) Association between feral pig disturbance and the composition of some alien plant assemblages in Hawaii Volcanoes National Park. *Vegetatio* 95, 55–62.

Appanah, S. (1993) Mass flowering of dipterocarp forests in the aseasonal tropics. *Journal of BioSciences* 18, 457–474.

Appanah, S. (1998) Management of natural forests. In: *A Review of Dipterocarps, Taxonomy, Ecology and Silviculture* (eds. S. Appanah and J. M. Turnbull), pp. 133–149. CIFOR, Bogor, Indonesia.

Archibald, S., Bond, W. J., Stock, W. D. and Fairbanks, D. H. K. (2005) Shaping the landscape: Fire–grazer interactions in an African savanna. *Ecological Applications* 15, 96–109.

Argeloo, M. and Dekker, R. W. R. J. (1996) Exploitation of megapode eggs in Indonesia: The role of traditional methods in the conservation of megapodes. *Oryx* 30, 59–64.

Atkinson, C. T., Dusek, R. J., Woods, K. L. and Iko, W. M. (2000) Pathogenicity of avian malaria in experimentally-infected Hawaii amakihi. *Journal of Wildlife Diseases* 36, 197–204.

Atkinson, C. T., Lease, J. K., Dusek, R. J. and Samuel, M. D. (2005) Prevalence of pox-like lesions and malaria in forest bird communities on leeward Mauna Loa Volcano, Hawaii. *Condor* 107, 537–546.

Atkinson, C. T., Woods, K. L., Dusek, R. J., Sileo, L. S. and Iko, W. M. (1995) Wildlife disease and conservation in Hawaii: Pathogenicity of avian malaria (*Plasmodium relictum*) in experimentally infected Iiwi (*Vestiaria coccinea*). *Parasitology* 111, S59–S69.

Attwood, D. K., Hendee, J. C. and Mendez, A. (1992) An assessment of global warming stress on Caribbean coral reef ecosytems. *Bulletin of Marine Science* 51, 118–130.

Bailey, K. M. and Houde, E. D. (1989) Predation on eggs and larvae of marine fishes and the recruitment problem. *Advances in Marine Biology*, 24, 1–83.

Baird, I. G. and Beasley, I. L. (2005) Irrawaddy dolphin *Orcaella brevirostris* in the Cambodian Mekong River: An initial survey. *Oryx* **39**, 301–310.

Baker, C. S., Lento, G. M., Cipriano, F. and Palumbi, S. R. (2000) Predicted decline of protected whales based on molecular genetic monitoring of Japanese and Korean markets. *Proceedings of the Royal Society B: Biological Sciences* **267**, 1191–1199.

Baker, P. S., Khan, A., Mohyuddin, A. I. and Waage, J. K. (1992) Overview of biological control of Lepidoptera in the Caribbean. *Florida Entomologist* **75**, 477–483.

Balazs, G. H. and Chaloupka, M. (2004) Thirty-year recovery trend in the once depleted Hawaiian green sea turtle stock. *Biological Conservation* **117**, 491–498.

Balirwa, J. S., Chapman, C. A., Chapman, L. J., Cowx, I. G., Geheb, K., Kaufman, L., Lowe-McConnell, R. H., Seehausen, O., Wanink, J. H., Welcomme, R. L. and Witte, F. (2003) Biodiversity and fishery sustainability in the Lake Victoria Basin: An unexpected marriage? *BioScience* **53**, 703–715.

Balmford, A. (1996) Extinction filters and current resilience: The significance of past selection pressures for conservation biology. *Trends in Ecology and Evolution* **11**, 193–196.

Balmford, A. and Bond, W. (2005) Trends in the state of nature and their implications for human well-being. *Ecology Letters* **8**, 1218–1234.

Balmford, A. and Cowling, R. M. (2006) Fusion or failure? The future of conservation biology. *Conservation Biology* **20**, 692–695.

Balmford, A. and Whitten, T. (2003) Who should pay for tropical conservation, and how could the costs be met? *Oryx* **37**, 238–250.

Balmford, A., Bruner, A., Cooper, P., Costanza, R., Farber, S., Green, R. E., Jenkins, M., Jefferiss, P., Jessamy, V., Madden, J., Munro, K., Myers, N., Naeem, S., Paavola, J., Rayment, M., Rosendo, S., Roughgarden, J., Trumper, K. and Turner, R. K. (2002) Economic reasons for conserving wild nature. *Science* **297**, 950–953.

Balmford, A., Gaston, K. J., Blyth, S., James, A. and Kapos, V. (2003) Global variation in terrestrial conservation costs, conservation benefits, and unmet conservation needs. *Proceedings of the National Academy of Sciences of the USA* **100**, 1046–1050.

Baran, E. and Hambrey, J. (1998) Mangrove conservation and coastal management in southeast Asia: What impact on fishery resources? *Marine Pollution Bulletin* **37**, 431–440.

Barber, R. T. and Chavez, F. P. (1983) Biological consequences of El-Niño. *Science* **222**, 1203–1210.

Barbier, E. B. (1993) Economic aspects of tropical deforestation in Southeast Asia. *Global Ecology and Biogeography Letters* **3**, 215–234.

Barker, M. J. and Schluessel, V. (2005) Managing global shark fisheries: Suggestions for prioritizing management strategies. *Aquatic Conservation-Marine and Freshwater Ecosystems* **15**, 325–347.

Barker, N. H. L. and Roberts, C. M. (2004) Scuba diver behaviour and the management of diving impacts on coral reefs. *Biological Conservation* **120**, 481–489.

Barlow, J. and Peres, C. A. (2006) Effects of single and recurrent wildfires on fruit production and large vertebrate abundance in a central Amazonian forest. *Biodiversity and Conservation* **15**, 985–1012.

Barlow, J., Haugaasen, T. and Peres, C. A. (2002) Effects of ground fires on understorey bird assemblages in Amazonian forests. *Biological Conservation* **105**, 157–169.

Barlow, J., Peres, C. A., Lagan, B. O. and Haugaasen, T. (2003) Large tree mortality and the decline of forest biomass following Amazonian wildfires. *Ecology Letters* **6**, 6–8.

Barlow, J., Peres, C. A., Henriques, L. M. P., Stouffer, P. C. and Wunderle, J. M. (2006) The responses of understorey birds to forest fragmentation, logging and wildfires: An Amazonian synthesis. *Biological Conservation* **128**, 182–192.

Barnes, R. F. W. (2002) The bushmeat boom and bust in West and Central Africa. *Oryx* **36**, 236–242.

Barnosky, A. D., Koch, P. L., Feranec, R. S., Wing, S. L. and Shabel, A. B. (2004) Assessing the causes of Late Pleistocene extinctions on the continents. *Science* **306**, 70–75.

Barr, C. (2001) *Banking on Sustainability: Structural Adjustment and Forestry Reform in Post-Suharto Indonesia.* WWF Macroeconomics Program Office and CIFOR, Washington, DC.

Barton, B. A., Morgan, J. D. and Vijayan, M. M. (2002) Physiological and condition-related indicators of environmental stress in fish. In: *Biological Indicators of Ecosystem Stress* (ed. S. M. Adams), pp. 111–148. American Fisheries Society, Bethesda, MD, USA.

Bascompte, J., Melian, C. J. and Sala, E. (2005) Interaction strength combinations and the overfishing of a marine food web. *Proceedings of the National Academy of Sciences of the USA* **102**, 5443–5447.

Baum, J. K. and Myers, R. A. (2004) Shifting baselines and the decline of pelagic sharks in the Gulf of Mexico. *Ecology Letters* 7, 135–145.

Baum, J. K. and Vincent, A. C. J. (2005) Magnitude and inferred impacts of the seahorse trade in Latin America. *Environmental Conservation* 32, 305–319.

Baum, J. K., Meeuwig, J. J. and Vincent, A. C. J. (2003a) Bycatch of seahorse (*Hippocampus erectus*) in a Gulf of Mexico shrimp trawl fishery. *Fishery Bulletin* 101, 721–731.

Baum, J. K., Myers, R. A., Kehler, D. G., Worm, B., Harley, S. J. and Doherty, P. A. (2003b) Collapse and conservation of shark populations in the Northwest Atlantic. *Science* 299, 389–392.

Bawa, K. S. (1990) Plant–pollinator interactions in tropical rain forests. *Annual Review of Ecology and Systematics* 21, 399–422.

Bawa, K. S. and Dayanandan, S. (1997) Socioeconomic factors and tropical deforestation. *Nature* 386, 562–563.

Bawa, K. S., Kress, W. J., Nadkarni, N. M. and Lele, S. (2004a) Beyond paradise – meeting the challenges in tropical biology in the 21st century. *Biotropica* 36, 437–446.

Bawa, K. S., Kress, W. J., Nadkarni, N. M., Lele, S., Raven, P. H., Janzen, D. H., Lugo, A. E., Ashton, P. S. and Lovejoy, T. E. (2004b) Tropical ecosystems into the 21st century. *Science* 306, 227–228.

Bax, N., Williamson, A., Aguero, M., Gonzalez, E. and Geeves, W. (2003) Marine invasive alien species: A threat to global biodiversity. *Marine Policy* 27, 313–323.

Bayliss, P. and Yeomans, K. M. (1989) Distribution and abundance of feral livestock in the 'Top End' of the Northern Territory (1985–86), and their relation to population control. *Australian Wildlife Research* 16, 651–676.

Beasley, I., Robertson, K. M. and Arnold, P. (2005) Description of a new dolphin, the Australian snubfin dolphin *Orcaella heinsohni* sp. n. (Cetacea, Delphinidae). *Marine Mammal Science* 21, 365–400.

Becker, C. D. (1999) Protecting a Garua forest in Ecuador: The role of institutions and ecosystem valuation. *Ambio* 28, 156–161.

Becker, C. D., Agreda, A., Astudillo, E., Costantino, M. and Torres, P. (2005) Community-based monitoring of fog capture and biodiversity at Loma Alta, Ecuador, enhance social capital and institutional cooperation. *Biodiversity and Conservation* 14, 2695–2707.

Beddington, J. R. and Kirkwood, G. P. (2005) The estimation of potential yield and stock status using life-history parameters. *Philosophical Transactions of the Royal Society B: Biological Sciences* 360, 163–170.

Beebee, T. J. C. (1992) Amphibian decline. *Nature* 355, 120.

Beier, P., Van Drielen, M. and Kankam, B. O. (2002) Avifaunal collapse in West African forest fragments. *Conservation Biology* 16, 1097–1111.

Beissinger, S. R. (2000) Ecological mechanisms of extinction. *Proceedings of the National Academy of Sciences of the USA* 97, 11688–11689.

Beissinger, S. R. (2001) Trade of live birds: Potential, principles and practices of sustainable use. In: *Conservation of exploited species* (eds. J. D. Reynolds, G. M. Mace and J. G. Robinson), pp. 183–202. Cambridge University Press, Cambridge.

Bell, D., Roberton, S. and Hunter, P. R. (2004) Animal origins of SARS coronavirus: Possible links with the international trade in small carnivores. *Philosophical Transactions of the Royal Society B: Biological Sciences* 359, 1107–1114.

Bellwood, D. R., Hoey, A. S., Ackerman, J. L. and Depczynski, M. (2006) Coral bleaching, reef fish community phase shifts and the resilience of coral reefs. *Global Change Biology* 12, 1587–1594.

Belsky, J. M. and Siebert, S. F. (1995) Managing rattan harvesting for local livelihoods and forest conservation in Kerinci-Seblat National Park, Sumatra. *Selbyana* 16, 212–222.

Benitez-Malvido, J. (1998) Impact of forest fragmentation on seedling abundance in a tropical rain forest. *Conservation Biology* 12, 380–389.

Benitez-Malvido, J. and Martinez-Ramos, M. (2003) Impact of forest fragmentation on understory plant species richness in Amazonia. *Conservation Biology* 17, 389–400.

Bennett, E. L. (2000) Timber certification: Where is the voice of the biologist? *Conservation Biology* 14, 921–923.

Bennett, E. L. (2002) Is there a link between wild meat and food security? *Conservation Biology* 16, 590–592.

Bennett, E. L. and Caldecott, J. O. (1981) Unexpected abundance: The trees and wildlife of the Lima Belas Estate forest reserve, near Slim River, Perak. *The Planter* 57, 516–519.

Bennett, E. L., Nyaoi, A. J. and Sompud, J. (2000) Saving Borneo's bacon: The sustainability of hunting in Sarawak and Sabah. In: *Hunting for Sustainability in Tropical Forests* (eds. J. G. Robinson and E. L. Bennett), pp. 305–324. Columbia University Press, New York.

Bennett, K. D. (1990) Milankovitch cycles and their effects on species in ecological and evolutionary time. *Paleobiology* 16, 11–21.

Bennett, P. M. and Owens, I. P. F. (1997) Variation in extinction risk among birds: Chance or evolutionary predisposition? *Proceedings of the Royal Society B: Biological Sciences* 264, 401–408.

Bennett, P. M. and Owens, I. P. F. (2002) *Evolutionary Ecology of Birds*. Oxford University Press, Oxford.

Benning, T. L., LaPointe, D., Atkinson, C. T. and Vitousek, P. M. (2002) Interactions of climate change with biological invasions and land use in the Hawaiian Islands: Modeling the fate of endemic birds using a geographic information system. *Proceedings of the National Academy of Sciences of the USA* 99, 14246–14249.

Benstead, J. P., March, J. G., Pringle, C. M. and Scatena, F. N. (1999) Effects of a low-head dam and water abstraction on migratory tropical stream biota. *Ecological Applications* 9, 656–668.

Benstead, J. P., De Rham, P. H., Gattolliat, J. L., Gibon, F. M., Loiselle, P. V., Sartori, M., Sparks, J. S. and Stiassny, M. L. J. (2003) Conserving Madagascar's freshwater biodiversity. *BioScience* 53, 1101–1111.

Berbert, M. L. C. and Costa, M. H. (2003) Climate change after tropical deforestation: Seasonal variability of surface albedo and its effects on precipitation change. *Journal of Climate* 16, 2099–2104.

Berkes, F. (2004) Rethinking community-based conservation. *Conservation Biology* 18, 621–630.

Bermudez, A., Oliveira-Miranda, M. A. and Velazquez, D. (2005) Ethnobotanical Brazil, research on medicinal plants: A review of its goals and current approaches. *Interciencia* 30, 453.

Berrio, J. C., Hooghiemstra, H., van Geel, B. and Ludlow-Wiechers, B. (2006) Environmental history of the dry forest biome of Guerrero, Mexico, and human impact during the last c. 2700 years. *Holocene* 16, 63–80.

Bertault, J. G. and Sist, P. (1997) An experimental comparison of different harvesting intensities with reduced-impact and conventional logging in East Kalimantan, Indonesia. *Forest Ecology and Management* 94, 209–218.

Bhushan, R., Thapar, S. and Mathur, R. P. (1997) Accumulation pattern of pesticides in tropical freshwaters. *Biomedical Chromatography* 11, 143–150.

Bierregaard, R. O. and Lovejoy, T. E. (1989) Effects of forest fragmentation on Amazonian understorey bird communities. *Acta Amazonica* 19, 215–241.

Bierregaard, R. O., Lovejoy, T. E., Kapos, V., Dossantos, A. A. and Hutchings, R. W. (1992) The biological dynamics of tropical rain-forest fragments. *BioScience* 42, 859–866.

BirdLife International (2000) *Threatened Birds of the World* (eds. A. J. Stattersfield and D. R. Capper). Lynx Edicions, Barcelona, and BirdLife International, Cambridge.

Bixler, R. D., Floyd, M. E. and Hammutt, W. E. (2002) Environmental socialization: Qualitative tests of the childhood play hypothesis. *Environment and Behaviour* 34, 795–818.

Blaber, S. J. M. (2002) 'Fish in hot water': The challenges facing fish and fisheries research in tropical estuaries. *Journal of Fish Biology* 61, 1–20.

Blackburn, T. M., Cassey, P., Duncan, R. P., Evans, K. L. and Gaston, K. J. (2004) Avian extinction and mammalian introductions on oceanic islands. *Science* 305, 1955–1958.

Blaustein, A. R. and Kiesecker, J. M. (2002) Complexity in conservation: Lessons from the global decline of amphibian populations. *Ecology Letters* 5, 597–608.

Bohnsack, J. A. (1998) Application of marine reserves to reef fisheries management. *Australian Journal of Ecology* 23, 298–304.

Bojsen, B. H. and Barriga, R. (2002) Effects of deforestation on fish community structure in Ecuadorian Amazon streams. *Freshwater Biology* 47, 2246–2260.

Bolido, E. (2004) Living proof: Apo Island's journey from ruin to modest riches. *Ecos* 121, 8–9.

Bolle, H. J., Seiler, W. and Bolin, B. (1986) Other greenhouse gases and aerosols: Assessing their roles for atmospheric radiative transfer. In: *The Greenhouse Effect, Climate Change and Ecosystems* (eds. B. Bolin, B. Warrick and D. Jager), pp. 157–203. John Wiley and Sons, New York.

Bond, W. J., Midgley, G. F. and Woodward, F. I. (2003) The importance of low atmospheric CO_2 and fire in promoting the spread of grasslands and savannas. *Global Change Biology* 9, 973–982.

Boot, R. G. A. and Gullison, R. E. (1995) Approaches to developing sustainable extraction systems for tropical forest products. *Ecological Applications* 5, 896–903.

Booth, D. J. and Beretta, G. A. (2002) Changes in a fish assemblage after a coral bleaching event. *Marine Ecology Progress Series* 245, 205–212.

Borbor-Cordova, M. J., Boyer, E. W., McDowell, W. H. and Hall, C. A. (2006) Nitrogen and phosphorus budgets for a tropical watershed impacted by agricultural land use: Guayas, Ecuador. *Biogeochemistry* 79, 135–161.

Botsford, L. W., Fiorenza, M. and Hastings, A. (2003) Principles for the design of marine reserves. *Ecological Applications* 13, S25–S31.

Boulé, M. E. (1994) An early history of wetland ecology. In: *Global Wetlands: Old World and New* (ed. W. J. Mitsch), pp. 57–74. Elsevier, Amsterdam.

Bourke, L., Selig, E. and Spalding, M. (2002) *Reefs at Risk in Southeast Asia*. World Resources Institute, Cambridge.

Bowen-Jones, E. and Entwistle, A. (2002) Identifying appropriate flagship species: The importance of culture and local contexts. *Oryx* 36, 189–195.

Bowles, I. A., Rice, R. E., Mittermeier, R. A. and da Fonseca, G. A. B. (1998) Logging and tropical forest conservation. *Science* 280, 1899–1900.

Bowman, D. M. J. S. (1998) Tansley Review No. 101 – the impact of Aboriginal landscape burning on the Australian biota. *New Phytologist* 140, 385–410.

Bowman, D. (2000) Tropical rain forests. *Progress in Physical Geography* 24, 103–109.

Bowman, D. (2005) Understanding a flammable planet – climate, fire and global vegetation patterns. *New Phytologist* 165, 341–345.

Bowman, D., Walsh, A. and Milne, D. J. (2001) Forest expansion and grassland contraction within a Eucalyptus savanna matrix between 1941 and 1994 at Litchfield National Park in the Australian monsoon tropics. *Global Ecology and Biogeography* 10, 535–548.

Bowman, D., Walsh, A. and Prior, L. D. (2004) Landscape analysis of Aboriginal fire management in Central Arnhem Land, north Australia. *Journal of Biogeography* 31, 207–223.

Bowman, D. M. J. S., Franklin, D. C., Price, O. F. and Brook, B. W. (2007) Land management affects grass biomass in the *Eucalyptus tetrodonta* savannas of monsoonal Australia. *Austral Ecology* 32, 446–452.

Bradshaw, C. J. A. and White, W. W. (2006) Rapid development of cleaning behaviour by Torresian crows *Corvus orru* on non-native banteng *Bos javanicus* in northern Australia. *Journal of Avian Biology* 37, 409–411.

Bradshaw, C. J. A., Isagi, Y., Kaneko, S., Bowman, D. and Brook, B. W. (2006a) Conservation value of non-native banteng in northern Australia. *Conservation Biology* 20, 1306–1311.

Bradshaw, C. J. A., Fukuda, Y., Letnic, M. I. and Brook, B. W. (2006b) Incorporating known sources of uncertainty to determine precautionary harvests of saltwater crocodiles. *Ecological Applications* 16, 1436–1448.

Bradshaw, C. J. A., Sodhi, N. S., Peh, K. S. H. and Brook, B. W. (2007a) Global evidence that deforestation amplifies floods in the developing world. *Global Change Biology* (in press).

Bradshaw, C. J. A., Mollet, H. F. and Meekan, M. G. (2007b) Inferring population trends for the world's largest fish from mark-recapture estimates of survival. *Journal of Animal Ecology* 76, 480–489.

Brainard, R. E. and McLain, D. R. (1987) Seasonal and interannual subsurface temperature variability off Peru, 1952 to 1984. In: *The Peruvian Anchoveta and its Upwelling Ecosystem: Three Decades of Change* (ed. D. P. Tsukayama), pp. 15–45. International Centre for Living Aquatic Resources Management (ICLARM) Studies and Reviews 15.

Brash, A. R. (1987) The history of avian extinction and forest conversion on Puerto Rico. *Biological Conservation* 39, 97–111.

Brashares, J. S., Arcese, P. and Sam, M. K. (2001) Human demography and reserve size predict wildlife extinction in West Africa. *Proceedings of the Royal Society B: Biological Sciences* 268, 2473–2478.

Brashares, J. S., Arcese, P., Sam, M. K., Coppolillo, P. B., Sinclair, A. R. E. and Balmford, A. (2004) Bushmeat hunting, wildlife declines, and fish supply in West Africa. *Science* 306, 1180–1183.

Briani, D. C., Palma, A. R. T., Vieira, E. M. and Henriques, R. P. B. (2004) Post-fire succession of small mammals in the Cerrado of central Brazil. *Biodiversity and Conservation* 13, 1023–1037.

Brodie, J., Fabricius, K., De'Ath, G. and Okaji, K. (2005) Are increased nutrient inputs responsible for more outbreaks of crown-of-thorns starfish? An appraisal of the evidence. *Marine Pollution Bulletin* 51, 266–278.

Brook, B. W. and Bowman, D. M. J. S. (2002) Explaining the Pleistocene megafaunal extinctions: Models, chronologies, and assumptions. *Proceedings of the National Academy of Sciences of the USA* 99, 14624–14627.

Brook, B. W. and Bowman, D. M. J. S. (2004) The uncertain blitzkrieg of Pleistocene megafauna. *Journal of Biogeography* 31, 517–523.

Brook, B. W. and Bowman, D. M. J. S. (2005) One equation fits overkill: Why allometry underpins both prehistoric and modern body size-biased extinctions. *Population Ecology* 47, 137–141.

Brook, B. W. and Sodhi, N. S. (2006) Rarity bites. *Nature.* 444, 555–557.

Brook, B. W. and Whitehead, P. J. (2005) Sustainable harvest regimes for magpie geese (*Anseranas semipalmata*) under spatial and temporal heterogeneity. *Wildlife Research* 32, 459–464.

Brook, B. W., Griffiths, A. D. and Puckey, H. L. (2002) Modelling strategies for the management of the critically endangered Carpentarian rock-rat (*Zyzomys palatalis*) of northern Australia. *Journal of Environmental Management* 65, 355–368.

Brook, B. W., Sodhi, N. S. and Ng, P. K. L. (2003a) Catastrophic extinctions follow deforestation in Singapore. *Nature* 424, 420–423.

Brook, B. W., Sodhi, N. S., Soh, M. C. K. and Lim, H. C. (2003b) Abundance and projected control of invasive house crows in Singapore. *Journal of Wildlife Management* 67, 808–817.

Brook, B. W., Bowman, D. M. J. S. and Bradshaw, C. J. A. (2005) Mapping the future: Spatial predictions of decadal-scale landscape change in northern Australia. In: *MODSIM 2005. Proceedings of the International Congress on Modelling and Simulation. Advances and Applications for Management and Decision Making. Modelling and Simulation Society of Australia and New Zealand, December 2005.* http://www.mssanz.org.au/modsim05/proceedings/papers/brook.pdf

Brook, B. W., Bradshaw, C. J. A., Koh, L. P. and Sodhi, N. S. (2006a) Momentum drives the crash: Mass extinction in the tropics. *Biotropica* 38, 302–305.

Brook, B. W., Traill, L. W. and Bradshaw, C. J. A. (2006b) Minimum viable population sizes and global extinction risk are unrelated. *Ecology Letters* 9, 375–382.

Brooks, T. M. (2000) Living on the edge. *Nature* 403, 27–29.

Brooks, T. M. and Balmford, A. (1996) Atlantic forest extinctions. *Nature* 380, 115.

Brooks, T. M., Pimm, S. L. and Collar, N. J. (1997) Deforestation predicts the number of threatened birds in insular southeast Asia. *Conservation Biology* 11, 382–394.

Brooks, T. M., Pimm, S. L., Kapos, V. and Ravilious, C. (1999a) Threat from deforestation to montane and lowland birds and mammals in insular South-east Asia. *Journal of Animal Ecology* 68, 1061–1078.

Brooks, T. M., Pimm, S. L. and Oyugi, J. O. (1999b) Time lag between deforestation and bird extinction in tropical forest fragments. *Conservation Biology* 13, 1140–1150.

Brooks, T. M., Tobias, J. and Balmford, A. (1999c) Deforestation and bird extinctions in the Atlantic forests. *Animal Conservation* 2, 211–222.

Brooks, T. M., Bakarr, M. I., Boucher, T., Da Fonseca, G. A. B., Hilton-Taylor, C., Hoekstra, J. M., Moritz, T., Olivier, S., Parrish, J., Pressey, R. L., Rodrigues, A. S. L., Sechrest, W., Stattersfield, A., Strahm, W. and Stuart, S. N. (2004) Coverage provided by the global protected-area system: Is it enough? *BioScience* 54, 1081–1091.

Brooks, T. M., Mittermeier, R. A., da Fonseca, G. A. B., Gerlach, J., Hoffmann, M., Lamoreux, J. F., Mittermeier, C. G., Pilgrim, J. D. and Rodrigues, A. S. L. (2006) Global biodiversity conservation priorities. *Science* 313, 58–61.

Broomhall, S. D. (2004) Egg temperature modifies predator avoidance and the effects of the insecticide endosulfan on tadpoles of an Australian frog. *Journal of Applied Ecology* 41, 105–113.

Brown, B. E. and Suharsano (1990) Damage and recovery of coral reefs affected by El Niño-related seawater warming in Thousand Islands, Indonesia. *Coral Reefs* 8, 163–170.

Brown, J. H. and Kodric-Brown, A. (1977) Turnover rates in insular biogeography: Effect of immigration on extinction. *Ecology* 58, 445–449.

Brown, K. A. and Gurevitch, J. (2004) Long-term impacts of logging on forest diversity in Madagascar. *Proceedings of the National Academy of Sciences of the USA* 101, 6045–6049.

Brown, K. S. J. and Brown, G. G. (1992) Habitat alteration and species loss in Brazilian forests. In:

Tropical Deforestation and Species Extinction (eds. T. C. Whitmore and J. A. Sayer), pp. 119–142. Chapman & Hall, London.

Bruijnzeel, L. A. (2004) Hydrological functions of tropical forests: Not seeing the soil for the trees? *Agriculture Ecosystems and Environment* 104, 185–228.

Bruna, E. M. (1999) Seed germination in rainforest fragments. *Nature* 402, 139–139.

Bruner, A. G., Gullison, R. E., Rice, R. E. and da Fonseca, G. A. B. (2001) Effectiveness of parks in protecting tropical biodiversity. *Science* 291, 125–128.

Bryant, D. G., Burke, L., McManus, J. and Spalding, M. (1998) *Reefs at Risk: A Map-based Indicator of Threats to the World's Coral Reefs.* World Resources Institute, Washington, DC.

Bryant, R. L. (2002) False prophets? Mutant NGOs and Philippine environmentalism. *Society and Natural Resources* 15, 629–639.

Bryant, R. L., Rigg, J. and Stott, P. (1993) Forest transformations and political ecology in Southeast Asia. *Global Ecology and Biogeography Letters* 3, 101–111.

Buckley, Y. M., Anderson, S., Catterall, C. P., Corlett, R. T., Engel, T., Gosper, C. R., Nathan, R., Richardson, D. M., Setter, M., Spiegel, O., Vivian-Smith, G., Voigt, F. A., Weir, J. E. S. and Westcott, D. A. (2006) Management of plant invasions mediated by frugivore interactions. *Journal of Applied Ecology* 43, 848–857.

Bulte, E. H. and Horan, R. D. (2002) Does human population growth increase wildlife harvesting? An economic assessment. *Journal of Wildlife Management* 66, 574–580.

Bunker, D. E., DeClerck, F., Bradford, J. C., Colwell, R. K., Perfecto, I., Phillips, O. L., Sankaran, M. and Naeem, S. (2005) Species loss and aboveground carbon storage in a tropical forest. *Science* 310, 1029–1031.

Bunn, S. E. and Arthington, A. H. (2002) Basic principles and ecological consequences of altered flow regimes for aquatic biodiversity. *Environmental Management* 30, 492–507.

Burkey, T. V. (1993) Edge effects in seed and egg predation at two Neotropical rain forest sites. *Biological Conservation* 66, 139–143.

Burnett, S. (1997) Colonizing cane toads cause population declines in native predators: Reliable anecdotal information and management implications. *Pacific Conservation Biology* 3, 65–72.

Burney, D. A. and Flannery, T. F. (2005) Fifty millennia of catastrophic extinctions after human contact. *Trends in Ecology and Evolution* 20, 395–401.

Buss, D. F., Baptista, D. F., Silveira, M. P., Nessimian, J. L. and Dorville, L. F. M. (2002) Influence of water chemistry and environmental degradation on macroinvertebrate assemblages in a river basin in south-east Brazil. *Hydrobiologia* 481, 125–136.

Butchart, S. H. M. and Baker, G. C. (2000) Priority sites for conservation of maleos (*Macrocephalon maleo*) in central Sulawesi. *Biological Conservation* 94, 79–91.

Byron, N. and Waugh, G. (1988) Forestry and fisheries in the Asian-Pacific Region: Issues in natural resource management. *Asian-Pacific Economic Literature* 2, 46–80.

Caddy, J. F. (1986) Stock assessment in data-limited situations – the experience in tropical fisheries and its possible relevance to evaluation of invertebrate fisheries. *Canadian Journal of Fisheries and Aquatic Sciences* 92, 379–392.

Caddy, J. F. and Garibaldi, L. (2000) Apparent changes in the trophic composition of world marine harvests: The perspective from the FAO capture database. *Ocean and Coastal Management* 43, 615–655.

Cahill, A. J. and Walker, J. S. (2000) The effects of forest fire on the nesting success of the red-knobbed hornbill *Aceros cassidix*. *Bird Conservation International* 10, 109–114.

Cahoon, D. R., Stocks, B. J., Levine, J. S., Cofer, W. R. and Oneill, K. P. (1992) Seasonal distribution of African savanna fires. *Nature* 359, 812–815.

Calaby, J. H. (1975) Introduction of Bali cattle to northern Australia. *Australian Veterinary Journal* 51, 108.

Camhi, M. (1995) Industrial fisheries threaten ecological integrity of the Galápagos Islands. *Conservation Biology* 9, 715–719.

Camhi, M. and Cook, S. (1994) Sharks in Galápagos in peril. *Shark News* 2, 1–3.

Campbell, L. M. (1998) Use them or lose them? Conservation and the consumptive use of marine turtle eggs at Ostional, Costa Rica. *Environmental Conservation* 25, 305–319.

Cane, M. A. (2005) The evolution of El Niño, past and future. *Earth and Planetary Science Letters* 230, 227–240.

Caniago, I. and Siebert, S. F. (1998) Medicinal plant ecology, knowledge and conservation in Kalimantan, Indonesia. *Economic Botany* 52, 229–250.

Canonico, G. C., Arthington, A., McCrary, J. K. and Thieme, M. L. (2005) The effects of introduced tilapias on native biodiversity. *Aquatic Conservation – Marine and Freshwater Ecosystems* 15, 463–483.

Cardillo, M., Purvis, A., Sechrest, W., Gittleman, J. L., Bielby, J. and Mace, G. M. (2004) Human population density and extinction risk in the world's carnivores. *PLoS Biology* 2, e909.

Cardillo, M., Mace, G. M., Jones, K. E., Bielby, J., Bininda-Emonds, O. R. P., Sechrest, W., Orme, C. D. L. and Purvis, A. (2005) Multiple causes of high extinction risk in large mammal species. *Science* 309, 1239–1241.

Cardillo, M., Mace, G. M., Gittleman, J. L. and Purvis, A. (2006) Latent extinction risk and the future battlegrounds of mammal conservation. *Proceedings of the National Academy of Sciences of the USA* 103, 4157–4161.

Carlton, J. T. (1999) The scale and ecological consequences of biological invasions in the world's oceans. In: *Invasive Species and Biodiversity Management*. (eds. O. T. Sandlund, P. J. Schei and A. Viken), pp. 195–212. Kluwer Academic Publishers, Dordrecht.

Carpaneto, G. M. and Fusari, A. (2000) Subsistence hunting and bushmeat exploitation in central-western Tanzania. *Biodiversity and Conservation* 9, 1571–1585.

Carpenter, A. I., Rowcliffe, J. M. and Watkinson, A. R. (2004) The dynamics of the global trade in chameleons. *Biological Conservation* 120, 291–301.

Carret, J.-C. and Loyer, D. (2003) *Madagascar Protected Area Network Sustainable Financing: Economic Analysis Perspective*. World Bank, Washington DC, and Agence Française de Développement, Paris.

Casey, J. M. and Myers, R. A. (1998) Near extinction of a large, widely distributed fish. *Science* 281, 690–692.

Castelletta, M., Sodhi, N. S. and Subaraj, R. (2000) Heavy extinctions of forest avifauna in Singapore: Lessons for biodiversity conservation in Southeast Asia. *Conservation Biology* 14, 1870–1880.

Castelletta, M., Thiollay, J. M. and Sodhi, N. S. (2005) The effects of extreme forest fragmentation on the bird community of Singapore Island. *Biological Conservation* 121, 135–155.

Castellón, T. D. and Sieving, K. E. (2006) An experimental test of matrix permeability and corridor use by an endemic understory bird. *Conservation Biology* 20, 135–145.

Castillo, L. E., delaCruz, E. and Ruepert, C. (1997) Ecotoxicology and pesticides in tropical aquatic ecosystems of Central America. *Environmental Toxicology and Chemistry* 16, 41–51.

Catling, P. C., Hertog, A., Burt, R. J., Wombey, J. C. and Forrester, R. I. (1999) The short-term effect of cane toads (*Bufo marinus*) on native fauna in the Gulf Country of the Northern Territory. *Wildlife Research* 26, 161–185.

Caughley, G. and Gunn, A. (1986) *Conservation Biology in Theory and Practice*. Blackwell Science, Cambridge, MA.

Ceballos, G. and Ehrlich, P. R. (2002) Mammal population losses and the extinction crisis. *Science* 296, 904–907.

Chakraborty, S., Tioedemann, A. V. and Teng, P. S. (2000) Climate change: Potential impact on plant diseases. *Environmental Pollution* 108, 317–326.

Chaloupka, M. and Limpus, C. (2001) Trends in the abundance of sea turtles resident in southern Great Barrier Reef waters. *Biological Conservation* 102, 235–249.

Chape, S., Fish, L., Fox, P. and Spalding, M. (2003) *United Nations List of Protected Areas*. IUCN/UNEP, Gland/Cambridge.

Charrette, N. A., Cleary, D. F. R. and Mooers, A. O. (2006) Range-restricted, specialist Bornean butterflies are less likely to recover from ENSO-induced disturbance. *Ecology* 87, 2330–2337.

Chateau-Degat, M. L., Chinain, M., Cerf, N., Gingras, S., Hubert, B. and Dewailly, E. (2005) Seawater temperature, *Gambierdiscus* spp. variability and incidence of ciguatera poisoning in French Polynesia. *Harmful Algae* 4, 1053–1062.

Chazdon, R. L. (2003) Tropical forest recovery: Legacies of human impact and natural disturbances. *Perspectives in Plant Ecology Evolution and Systematics* 6, 51–71.

Chesher, R. H. (1969) Destruction of Pacific corals by the sea star *Acanthaster planci*. *Science* 18, 280–283.

Chey, V. K. (2000) Moth diversity in the tropical rain forest of Lanjak-Entimau, Sarawak, Malaysia. *Malayan Nature Journal* 54, 305–318.

Cheyne, S. M. (2006) Unusual behaviour of captive-raised gibbons: Implications for welfare. *Primates* 47, 322–326.

Chin, S. C. (1977) The limestone hill flora of Malaya I. *Garden's Bulletin Singapore* 30, 166–219.

Chivian, E. (2002) *Biodiversity: Its Importance to Human Health.* Centre for Health and the Global Environment, Harvard Medical School.

Choat, J. H. and Robertson, D. R. (2002) Age-based studies. In: *Coral Reef Fishes. Dynamics and Diversity in a Complex Ecosystem* (ed. P. F. Sale), pp. 57–80. Academic Press, San Diego.

Choquenot, D. (1995) Assessing visibility bias associated with helicopter counts of feral pigs in Australia's semiarid rangelands. *Wildlife Research* 22, 569–578.

Christensen, V. (1996) Managing fisheries involving top predator and prey species components. *Reviews in Fish Biology and Fisheries* 6, 417–442.

Christian, C. E. (2001) Consequences of a biological invasion reveal the importance of mutualism for plant communities. *Nature* 413, 635–639.

Christiansen, M. B. and Pitter, E. (1997) Species loss in a forest bird community near Lagoa Santa in southeastern Brazil. *Biological Conservation* 80, 23–32.

Christy, M. (2002) Sulawesi's disappearing flagship bird. *Species* 38, 8–9.

Chua, K. B., Bellini, W. J., Rota, P. A., Harcourt, B. H., Tamin, A., Lam, S. K., Ksiazek, T. G., Rollin, P. E., Zaki, S. R., Shieh, W. J., Goldsmith, C. S., Gubler, D. J., Roehrig, J. T., Eaton, B., Gould, A. R., Olson, J., Field, H., Daniels, P., Ling, A. E., Peters, C. J., Anderson, L. J. and Mahy, B. W. J. (2000) Nipah virus: A recently emergent deadly paramyxovirus. *Science* 288, 1432–1435.

Chung, F. J. (1996) Interests and policies of the state of Sarawak, Malaysia regarding intellectual property rights for plant derived drugs. *Journal of Ethnopharmacology* 51, 201–204.

CIFOR (2003) *Forests and people: Research that makes a difference.* CIFOR Annual Report 2003, CIFOR, Bogor, Indonesia.

CIFOR (2004) *Forest for People and the Environment.* CIFOR Annual Report 2004, CIFOR, Bogor, Indonesia.

Cincotta, R. P., Wisnewski, J. and Engelman, R. (2000) Human population in the biodiversity hotspots. *Nature* 404, 990–992.

CITES (2002) CITES Appendix II Nomination of the Whale Shark, *Rhincodon typus.* Proposal 12.35 (CITES Resolutions of the conference of the parties in effect after the 12th Meeting, Santiago, Chile, 2002).

Civeyrel, L. and Simberloff, D. (1996) A tale of two snails: Is the cure worse than the disease? *Biodiversity and Conservation* 5, 1231–1252.

Clark, C. (1987) Deforestation and floods. *Environmental Conservation* 14, 67–69.

Clarke, S. (2004) Understanding pressures on fishery resources through trade statistics: A pilot study of four products in the Chinese dried seafood market. *Fish and Fisheries* 5, 53–74.

Clarke, S., McAllister, M. and Michielsens, C. (2004) Estimates of shark species composition and numbers associated with the shark fin trade based on Hong Kong auction data. *Journal of Northwest Atlantic Fishery Science* 35, 453–465.

Clarke, S. C., Magnussen, J. E., Abercrombie, D. L., McAllister, M. K. and Shivji, M. S. (2006a) Identification of shark species composition and proportion in the Hong Kong shark fin market based on molecular genetics and trade records. *Conservation Biology* 20, 201–211.

Clarke, S. C., McAllister, M. K., Milner-Gulland, E. J., Kirkwood, G. P., Michielsens, C. G. J., Agnew, D. J., Pikitch, E. K., Nakano, H. and Shivji, M. S. (2006b) Global estimates of shark catches using trade records from commercial markets. *Ecology Letters* 9, 1115–1126.

Claudi, R. and Leach, J. H. (1999) *Nonindigenous Freshwater Organisms: Vectors, Biology and Impacts.* CRC Press, Boca Raton, FL.

Clavero, M. and Garcia-Berthou, E. (2005) Invasive species are a leading cause of animal extinctions. *Trends in Ecology and Evolution* 20, 110.

Clay, J. W. (1997) The use of a keystone species for conservation and development. In: *Harvesting Wild Species – Implications for Biodiversity and Conservation* (ed. C. H. Freese), pp. 246–282. Johns Hopkins University Press, Baltimore, MD.

Clayton, D. H. and Milner-Gulland, E. J. (2000) The trade in wildlife in north Sulawesi, Indonesia. In: *Hunting for Sustainability in Tropical Forests* (eds. J. G. Robinson and E. L. Bennett), pp. 473–498. Columbia University Press, New York.

Clayton, D. H., Keeling, M. and Milner-Gulland, E. J. (1997) Bringing home the bacon: A spatial model of wild pig hunting in Sulawesi, Indonesia. *Ecological Applications* 7, 642–652.

Cleary, D. F. R. and Genner, M. J. (2004) Changes in rain forest butterfly diversity following major ENSO-induced fires in Borneo. *Global Ecology and Biogeography* 13, 129–140.

Clements, R., Sodhi, N. S., Schilthuizen, M. and Ng, P. K. L. (2006) Limestone karsts of Southeast Asia: Imperiled arks of biodiversity. *BioScience* 56, 733–742.

Cochrane, M. A. (2003) Fire science for rainforests. *Nature* 421, 913–919.

Cochrane, M. A. and Schulze, M. D. (1999) Fire as a recurrent event in tropical forests of the eastern Amazon: Effects on forest structure, biomass, and species composition. *Biotropica* 31, 2–16.

Cochrane, M. A., Alencar, A., Schulze, M. D., Souza, C. M., Jr., Nepstad, D. C., Lefebvre, P. and Davidson, E. A. (1999) Positive feedbacks in the fire dynamic of closed canopy tropical forests. *Science* 284, 1832–1835.

Colburn, T., Dumanoski, D. and Myers, J. P. (1996) *Our Stolen Future*. Dutton, New York.

Coles, S. L., DeFelice, R. C., Eldredge, L. G. and Carlton, J. T. (1999) Historical and recent introductions of non-indigenous marine species into Pearl Harbour, Oahu, Hawaiian Islands. *Marine Biology* 135, 147–158.

Coley, P. D. (1998) Possible effects of climate change on plant/herbivore interactions in moist tropical forests. *Climate Change* 39, 455–472.

Colinvaux, P. A., De Olivera, P. E. and Bush, M. B. (2000) Amazonian and neotropical plant communities on glacial time-scales: The failure of the aridity and refuge hypotheses. *Quaternary Science Reviews* 19, 141–169.

Collins, J. P. and Storfer, A. (2003) Global amphibian declines: Sorting the hypotheses. *Diversity and Distributions* 9, 89–98.

Collins, M. (2005) El Niño- or La Niña-like climate change? *Climate Dynamics* 24, 89–104.

Colman, J. G. (1997) A review of the biology and ecology of the whale shark. *Journal of Fish Biology* 51, 1219–1234.

Conover, D. O. and Munch, S. B. (2002) Sustaining fisheries yields over evolutionary time scales. *Science* 297, 94–96.

Cool, J. C. (1980) *Stability and Survival – The Himalayan Challenge*. Ford Foundation, New York.

Copper, P. (1988) Ecological succession in Phanerozoic reef ecosystems: Is it real? *Palaios* 3, 136–152.

Cordeiro, N. J. and Howe, H. F. (2001) Low recruitment of trees dispersed by animals in African forest fragments. *Conservation Biology* 15, 1733–1741.

Cordeiro, N. J. and Howe, H. F. (2003) Forest fragmentation severs mutualism between seed dispersers and an endemic African tree. *Proceedings of the National Academy of Sciences of the USA* 100, 14052–14056.

Corlett, R. T. (1988) The naturalized flora of Singapore. *Journal of Biogeography* 15, 657–663.

Corlett, R. T. (1992) The ecological transformation of Singapore, 1819–1990. *Journal of Biogeography* 19, 411–420.

Corlett, R. T. (2000) Environmental heterogeneity and species survival in degraded tropical landscapes. In: *The Ecological Consequences of Environmental Heterogeneity* (eds. M. J. Hutchings, E. A. John and A. Stewart), pp. 333–355. Blackwell Science, Oxford.

Corlett, R. T. (2004) Vegetation. In: *The Physical Geography of Southeast Asia* (ed. A. Gupta), pp. 105–119. Oxford University Press, Oxford.

Corlett, R. T. (2007) The impacts of hunting on the mammalian fauna of tropical Asian forests. *Biotropica* 39, 292–303.

Corrigan, P. J. (1992) *Investigation of the Southern Thailand Zebra Dove Industry*. TRAFFIC Southeast Asia Field Report No. 1, Selangor, Malaysia.

Cortés, E. (2002) Incorporating uncertainty into demographic modeling: Application to shark populations and their conservation. *Conservation Biology* 16, 1048–1062.

Cortés, J. N. and Risk, M. J. (1985) A reef under siltation stress: Cahuita, Costa Rica. *Bulletin of Marine Science* 36, 339–356.

Costa, M. H., Botta, A. and Cardille, J. A. (2003) Effects of large-scale changes in land cover on the discharge of the Tocantins River, Southeastern Amazonia. *Journal of Hydrology* 283, 206–217.

Cotton, P. A. (2003) Avian migration phenology and global climate change. *Proceedings of the National Academy of Sciences of the USA* 100, 12219–12222.

Council, N. R. (1990) *Decline of the Sea Turtles: Causes and Prevention.* National Academy Press, Washington, DC.

Courchamp, F., Angulo, E., Rivalan, P., Hall, R. J., Signoret, L., Bull, L. and Meinard, Y. (2006) Rarity value and species extinction: The anthropogenic Allee effect. *PLoS Biology* 4, e415.

Cowie, I. D. and Werner, P. A. (1993) Alien plant species invasive in Kakadu National Park, tropical northern Australia. *Biological Conservation* 63, 127–135.

Cox, P. A. and Elmqvist, T. (2000) Pollinator extinction in the Pacific Islands. *Conservation Biology* 14, 1237–1239.

Crane, E. and Walker, P. (1983) *The Impact of Pest Management on Bees and Pollination.* Tropical Development and Research Institute, London.

Cronin, T. M. and Schneider, C. E. (1990) Climatic influences on species – evidence from the fossil record. *Trends in Ecology and Evolution* 5, 275–279.

Cronk, Q. C. B. and Fuller, J. (1995) *Plant Invaders: The Threat to Natural Ecosystems.* Chapman & Hall, London.

Crooks, K. R. and Soulé, M. E. (1999) Mesopredator release and avifaunal extinctions in a fragmented system. *Nature* 400, 563–566.

Cross, P. C., Lloyd-Smith, J. O., Bowers, J. A., Hay, C. T., Hofmeyr, M. and Getz, W. M. (2004) Integrating association data and disease dynamics in a social ungulate: Bovine tuberculosis in African buffalo in the Kruger National Park. *Annales Zoologici Fennici* 41, 879–892.

Crossland, M. R. (2000) Direct and indirect effects of the introduced toad *Bufo marinus* (Anura: Bufonidae) on populations of native anuran larvae in Australia. *Ecography* 23, 283–290.

Crutzen, P. J. and Andreae, M. O. (1990) Biomass burning in the tropics – impact on atmospheric chemistry and biogeochemical cycles. *Science* 250, 1669–1678.

Cruz, C. and Segarra, A. (1990) Recent biological control experiences in Puerto Rico. *Caribbean Meetings on Biological Control*, 5–7 November 1990, Guadeloupe.

Cumming, G. S. and Van Vuuren, D. P. (2006) Will climate change affect ectoparasite species ranges? *Global Ecology and Biogeography* 15, 486–497.

Curran, L. M. and Leighton, M. (2000) Vertebrate responses to spatiotemporal variation in seed production of mast-fruiting Dipterocarpaceae. *Ecological Monographs* 70, 101–128.

Curran, L. M., Caniago, I., Paoli, G. D., Astianti, D., Kusneti, M., Leighton, M., Nirarita, C. E. and Haeruman, H. (1999) Impact of El Niño and logging on canopy tree recruitment in Borneo. *Science* 286, 2184–2188.

Curran, L. M., Trigg, S. N., McDonald, A. K., Astiani, D., Hardiono, Y. M., Siregar, P., Caniago, I. and Kasischke, E. (2004) Lowland forest loss in protected areas of Indonesian Borneo. *Science* 303, 1000–1003.

Cyranoski, D. (2006) Calls to conserve biodiversity hotspots. *Nature* 439, 774.

Dadzie, K. Y., Remme, J., Rolland, A. and Thylefors, B. (1989) Ocular onchocerciasis and intensity of infection in the community. 2. West African rain forest foci of the vector *Simulium yahense*. *Tropical Medicine and Parasitology* 40, 348–354.

D'Agrosa, C., Lennert-Cody, C. E. and Vidal, O. (2000) Vaquita bycatch in Mexico's artisanal gillnet fisheries: Driving a small population to extinction. *Conservation Biology* 14, 1110–1119.

Dahdouh-Guebas, F., Jayatissa, L. P., Di Nitto, D., Bosire, J. O., Lo Seen, D. and Koedam, N. (2005) How effective were mangroves as a defence against the recent tsunami? *Current Biology* 15, R443–R447.

Daily, G. C. and Ehrlich, P. R. (1996) Impacts of development and global change on the epidemiological environment. *Environment and Development Economics* 1, 311–346.

Daily, G. C. and Walker, B. H. (2000) Seeking the great transition. *Nature* 403, 243–245.

Daily, G. C., Ceballos, G., Pacheco, J., Suzan, G. and Sanchez-Azofeifa, A. (2003) Countryside biogeography of neotropical mammals: Conservation opportunities in agricultural landscapes of Costa Rica. *Conservation Biology* 17, 1814–1826.

Danielsen, F. and Treadaway, C. G. (2004) Priority conservation areas for butterflies (Lepidoptera: Rhopalocera) in the Philippine islands. *Animal Conservation* 7, 79–92.

Danielsen, F., Sørensen, M. K., Olwig, M. F., Selvam, V., Parish, F., Burgess, N. D., Hiraishi, T., Karunagaran, V. M., Rasmussen, M. S. and Hansen, L. B. (2005) The Asian tsunami: A protective role for coastal vegetation. *Science* 310, 643.

Darrigran, G. and de Drago, I. E. (2000) Invasion of the exotic freshwater mussel *Limnoperna fortunei* (Dunker, 1857) (Bivalvia: Mytilidae) in South America. *Nautilus* **114**, 69–73.

Daskalov, G. M. (2002) Overfishing drives a trophic cascade in the Black Sea. *Marine Ecology Progress Series* **225**, 53–63.

Daszak, P., Cunningham, A. A. and Hyatt, A. D. (2003) Infectious disease and amphibian population declines. *Diversity and Distributions* **9**, 141–150.

Davis, G. (2002) Bushmeat and international development. *Conservation Biology* **16**, 587–589.

Davis, J. R. and Garcia, R. (1989) Malaria mosquito in Brazil. In: *Eradication of Exotic Pests* (eds. D. L. Dahlsten and R. Garcia), pp. 274–283. Yale University Press, New Haven, CT.

Davis, M. B. and Shaw, R. G. (2001) Range shifts and adaptive responses to Quaternary climate change. *Science* **292**, 673–679.

Dayanandan, S., Dole, J., Bawa, K. and Kesseli, R. (1999) Population structure delineated with microsatellite markers in fragmented populations of a tropical tree, *Carapa guianensis* (Meliaceae). *Molecular Ecology* **8**, 1585–1592.

de Castro, E. B. V. and Fernandez, F. A. S. (2004) Determinants of differential extinction vulnerabilities of small mammals in Atlantic forest fragments in Brazil. *Biological Conservation* **119**, 73–80.

de Lima, M. G. and Gascon, C. (1999) The conservation value of linear forest remnants in central Amazonia. *Biological Conservation* **91**, 241–247.

de Thoisy, B., Spiegelberger, T., Rousseau, S., Talvy, G., Vogel, I. and Vie, J. C. (2003) Distribution, habitat, and conservation status of the West Indian manatee *Trichechus manatus* in French Guiana. *Oryx* **37**, 431–436.

DeFries, R. S., Houghton, R. A., Hansen, M. C., Field, C. B., Skole, D. and Townshend, J. (2002) Carbon emissions from tropical deforestation and regrowth based on satellite observations for the 1980s and 1990s. *Proceedings of the National Academy of Sciences of the USA* **99**, 14256–14261.

DeFries, R., Hansen, A., Newton, A. C. and Hansen, M. C. (2005) Increasing isolation of protected areas in tropical forests over the past twenty years. *Ecological Applications* **15**, 19–26.

Dekker, R. W. R. J. (1990) The distribution and status of nesting grounds of the maleo *Macrocephalon maleo* in Sulawesi, Indonesia. *Biological Conservation* **51**, 139–150.

DeMartini, E. E., Parrish, F. A. and Parrish, J. D. (1996) Interdecadal changes in reef fish populations at French Frigate Shoals and Midway Atoll, Northwestern Hawaiian Islands. *Bulletin of Marine Science* **58**, 804–825.

Dexter, N. (2003) Stochastic models of foot and mouth disease in feral pigs in the Australian semi-arid rangelands. *Journal of Applied Ecology* **40**, 293–306.

DeYoung, R. W., Demarais, S., Honeycutt, R. L., Rooney, A. P., Gonzales, R. A. and Gee, K. L. (2003) Genetic consequences of white-tailed deer (*Odocoileus virginianus*) restoration in Mississippi. *Molecular Ecology* **12**, 3237–3252.

Diamond, J. M. (1980) Patchy distributions of tropical birds. In: *Conservation Biology: An Evolutionary–Ecological Perspective* (eds. M. E. Soulé and B. A. Wilcox), pp. 57–74. Sinauer, New York.

Diamond, J. M. (1984) 'Normal' extinctions of isolated populations. In: *Extinctions* (ed. M. H. Nitecki), pp. 191–246. University of Chicago Press, Chicago.

Diamond, J. M. (1989) The present, past and future of human-caused extinctions. *Philosophical Transactions of the Royal Society B: Biological Sciences* **325**, 469–477.

Diamond, J. M., Bishop, K. D. and van Balen, S. (1987) Bird survival in an isolated Javan woodland: Island or mirror. *Conservation Biology* **1**, 132–142.

Dick, C. W. (2001) Genetic rescue of remnant tropical trees by an alien pollinator. *Proceedings of the Royal Society B: Biological Sciences* **268**, 2391–2396.

Didham, R. K., Hammond, P. M., Lawton, J. H., Eggleton, P. and Stork, N. E. (1998) Beetle species responses to tropical forest fragmentation. *Ecological Monographs* **68**, 295–323.

Didham, R. K., Tylianakis, J. M., Hutchison, M. A., Ewers, R. M. and Gemmell, N. J. (2005) Are invasive species the drivers of ecological change? *Trends in Ecology and Evolution* **20**, 470–474.

Dirzo, R. and Raven, P. H. (2003) Global state of biodiversity and loss. *Annual Review of Environmental Resources* **28**, 137–167.

Donald, P. F. (2004) Biodiversity impacts of some agricultural commodity production systems. *Conservation Biology* **18**, 17–37.

Douglas, M., Finlayson, C. M. and Storrs, M. J. (1998) Weed management in tropical wetlands of the

Northern Territory, Australia. In: *Wetlands in a Dry Land: Understanding for Management* (ed. W. D. Williams), pp. 239–251. Environment Australia, Biodiversity Group, Canberra, Australia.

Draulans, D. and Van Krunkelsven, E. (2002) The impact of war on forest areas in the Democratic Republic of Congo. *Oryx* **36**, 35–40.

Drew, J. A. (2005) Use of traditional ecological knowledge in marine conservation. *Conservation Biology* **19**, 1286–1293.

Driml, S. (1994) *Protection for Profit. Economic and Financial Values of the Great Barrier Reef World Heritage Area and Other Protected Areas. Research publication No. 35*. Great Barrier Reef Marine Park Authority, Townsville, Queensland.

du Toit, J. T., Walker, B. H. and Campbell, B. M. (2004) Conserving tropical nature: Current challenges for ecologists. *Trends in Ecology and Evolution* **19**, 12–17.

Dubinsky, Z. and Stambler, N. (1996) Marine pollution and coral reefs. *Global Change Biology* **2**, 511–526.

Dudgeon, D. (1999) *Tropical Asian Streams. Zoobenthos, Ecology and Conservation*. Hong Kong University Press, Hong Kong.

Dudgeon, D. (2000a) Large-scale hydrological changes in tropical Asia: Prospects for riverine biodiversity. *BioScience* **50**, 793–806.

Dudgeon, D. (2000b) The ecology of tropical Asian rivers and streams in relation to biodiversity conservation. *Annual Review of Ecology and Systematics* **31**, 239–263.

Dudgeon, D., Arthington, A. H., Gessner, M. O., Kawabata, Z. I., Knowler, D. J., Leveque, C., Naiman, R. J., Prieur-Richard, A. H., Soto, D., Stiassny, M. L. J. and Sullivan, C. A. (2006) Freshwater biodiversity: Importance, threats, status and conservation challenges. *Biological Reviews* **81**, 163–182.

Dudley, S. F. J. and Simpfendorfer, C. A. (2006) Population status of 14 shark species caught in the protective gillnets off KwaZulu-Natal beaches, South Africa, 1978–2003. *Marine and Freshwater Research* **57**, 225–240.

Dukes, J. S. and Mooney, H. A. (1999) Does global change increase the success of biological invaders? *Trends in Ecology and Evolution* **14**, 135–139.

Dulvy, N. K., Sadovy, Y. and Reynolds, J. D. (2003) Extinction vulnerability in marine populations. *Fish and Fisheries* **4**, 25–64.

Dunn, R. R. (2005) Modern insect extinctions, the neglected majority. *Conservation Biology* **19**, 1030–1036.

Durst, P. B., Killmann, W. and Brown, C. (2004) Asia's new woods. *Journal of Forestry* **102**, 46–53.

Dymond, C. C., Field, R. D., Roswintiarti, O. and Guswanto (2005) Using satellite fire detection to calibrate components of the Fire Weather Index system in Malaysia and Indonesia. *Environmental Management* **35**, 426–440.

Eakin, C. M. (1996) Where have all the carbonates gone? A model comparison of calcium carbonate budgets before and after the 1982–1983 El Niño. *Coral Reefs* **15**, 109–119.

East, T., Kumpel, N. F., Milner-Gulland, E. J. and Rowcliffe, J. M. (2005) Determinants of urban bushmeat consumption in Rio Muni, Equatorial Guinea. *Biological Conservation* **126**, 206–215.

Eckert, S. A. and Stewart, B. S. (2001) Telemetry and satellite tracking of whale sharks, *Rhincodon typus*, in the Sea of Cortez, Mexico, and the north Pacific Ocean. *Environmental Biology of Fishes* **60**, 299–308.

Edinger, E. N., Jompa, J., Limmon, G. V., Widjatmoko, W. and Risk, M. J. (1998) Reef degradation and coral biodiversity in Indonesia: Effects of land-based pollution, destructive fishing practices and changes over time. *Marine Pollution Bulletin* **36**, 617–630.

Edinger, E. N., Limmon, G. V., Jompa, J., Widjatmoko, W., Heikoop, J. M. and Risk, M. J. (2000) Normal coral growth rates on dying reefs: Are coral growth rates good indicators of reef health? *Marine Pollution Bulletin* **5**, 404–425.

Ehrlich, P. R. and Ehrlich, A. H. (1996) *Betrayal of Science and Reason: How Anti-environmental Rhetoric Threatens Our Future*. Island Press, Washington, DC.

Ellison, A. M. (2000) Mangrove restoration: Do we know enough? *Restoration Ecology* **8**, 219–229.

Emanuel, K. A. (1987) The dependence of hurricane intensity on climate. *Nature* **326**, 483–485.

Epstein, P. R. (1998) Global warming and vector-borne disease. *Lancet* **351**, 1737–1737.

Epstein, P. R. (2000) Is global warming harmful to health? *Scientific American* **283**, 50–57.

Epstein, P. R., Diaz, H. F., Elias, S., Grabherr, G., Graham, N. E., Martens, W. J. M., Mosley-Thompson,

E. and Susskind, J. (1998) Biological and physical signs of climate change: Focus on mosquito-borne diseases. *Bulletin of the American Meteorological Society* 79, 409–417.

Espino-Barr, E., Ruiz-Luna, A. and Garcia-Boa, A. (2002) Changes in tropical fish assemblages associated with small-scale fisheries: A case study in the Pacific off central Mexico. *Reviews in Fish Biology and Fisheries* 12, 393–401.

Estrada, A., Coates-Estrada, R., Dadda, A. A. and Cammarano, P. (1998) Dung and carrion beetles in tropical rain forest fragments and agricultural habitats at Los Tuxtlas, Mexico. *Journal of Tropical Ecology* 14, 577–593.

Eudey, A. A. (1994) Temple and pet primates in Thailand. *Revue d'Écologie de la Terre et de la Vie* 49, 273–280.

Fa, J. E., Yuste, J. E. G. and Castelo, R. (2000) Bushmeat markets on Bioko Island as a measure of hunting pressure. *Conservation Biology* 14, 1602–1613.

Fa, J. E., Peres, C. A. and Meeuwig, J. (2002) Bushmeat exploitation in tropical forests: An intercontinental comparison. *Conservation Biology* 16, 232–237.

Fa, J. E., Currie, D. and Meeuwig, J. (2003) Bushmeat and food security in the Congo Basin: Linkages between wildlife and people's future. *Environmental Conservation* 30, 71–78.

Fa, J. E., Ryan, S. F. and Bell, D. J. (2005) Hunting vulnerability, ecological characteristics and harvest rates of bushmeat species in Afrotropical forests. *Biological Conservation* 121, 167–176.

Fabricius, K. E. (2005) Effects of terrestrial runoff on the ecology of corals and coral reefs: Review and synthesis. *Marine Pollution Bulletin* 50, 125–146.

Fabricius, K. E. and De'Ath, G. (2001) Biodiversity on the Great Barrier Reef: Large-scale patterns and turbidity-related local loss of soft coral taxa. In: *Oceanographic Processes of Coral Reefs: Physical and Biological Links in the Great Barrier Reef* (ed. E. Wolanski), pp. 127–144. CRC Press, London.

Fabricius, K. E. and De'Ath, G. (2004) Identifying ecological change and its causes: A case study on coral reefs. *Ecological Applications* 14, 1448–1465.

Fahrig, L. (2003) Effects of habitat fragmentation on biodiversity. *Annual Review of Ecology, Evolution, and Systematics* 34, 487–515.

FAO (1993) *Forest Resources Assessment 1990: Tropical Countries*. FAO Forestry Paper 112, Food and Agriculture Organization of the United Nations, Rome.

FAO (1999) *The State of World Fisheries and Aquaculture 1998*. Food and Agriculture Organization of the United Nations, Rome, Italy.

FAO (2004) *Trends and Current Status of the Contribution of the Forestry Sector to National Economies*. Food and Agricultural Organization of the United Nations, Forest Products and Economics Division, Rome.

FAO (2005) *State of the World's Forests*. Food and Agricultural Organization of the United Nations, Rome.

Fearnside, P. M. (1989) Extractive reserves in Brazilian Amazonia. *BioScience* 39, 387–393.

Fearnside, P. M. and Laurance, W. F. (2003) Comment on 'Determination of deforestation rates of the world's humid tropical forests'. *Science* 299, 1015a.

Fedorov, A. V. and Philander, S. G. (2000) Is El Niño changing? *Science* 288, 1997–2002.

Feely, R. A., Sabine, C. L., Lee, K., Berelson, W., Kleypas, J., Fabry, V. J. and Millero, F. J. (2004) Impact of anthropogenic CO_2 on the $CaCO_3$ system in the oceans. *Science* 305, 362–366.

Feinsinger, P., Margutti, L. and Oviedo, R. D. (1997) School yards and nature trails: Ecology education outside the university. *Trends in Ecology and Evolution* 12, 115–120.

Ferraz, G., Russell, G. J., Stouffer, P. C., Bierregaard, R. O., Pimm, S. L. and Lovejoy, T. E. (2003) Rates of species loss from Amazonian forest fragments. *Proceedings of the National Academy of Sciences of the USA* 100, 14069–14073.

Figueredo, C. C. and Giani, A. (2005) Ecological interactions between Nile tilapia (*Oreochromis niloticus* L.) and the phytoplanktonic community of the Furnas Reservoir (Brazil). *Freshwater Biology* 50, 1391–1403.

Fitt, W. K., Brown, B. E., Warner, M. E. and Dunne, R. P. (2001) Coral bleaching: Interpretation of thermal tolerance limits and thermal thresholds in tropical corals. *Coral Reefs* 20, 51–65.

Fitzsimmons, K. (2001) Environmental and conservation issues in tilapia aquaculture. In: *Tilapia: Production, Marketing and Technological Developments*. (ed. S. Singh), pp. 128–131. FAO-Infofish, Kuala Lumpur.

Fjedeså, J. and Lovett, J. C. (1997) Biodiversity and environmental stability. *Biodiversity and Conservation* 6, 315–323.

Flannery, T. (2005) *The Weather Makers. The History and Future Impact of Climate Change.* Text Publishing, Melbourne.

Fleming, L., Dewailly, E. and Baden, D. G. (2000) The epidemiologic of marine harmful algal blooms. *Epidemiology* 11, 143.

Flint, E. P. (1994) Changes in land use in South and Southeast Asia from 1880 to 1980: A data base prepared as part of a coordinated research program on carbon fluxes in the tropics. *Chemosphere* 29, 1015–1062.

Foley, J. A., DeFries, R., Asner, G. P., Barford, C., Bonan, G., Carpenter, S. R., Chapin, F. S., Coe, M. T., Daily, G. C., Gibbs, H. K., Helkowski, J. H., Holloway, T., Howard, E. A., Kucharik, C. J., Monfreda, C., Patz, J. A., Prentice, I. C., Ramankutty, N. and Snyder, P. K. (2005) Global consequences of land use. *Science* 309, 570–574.

Fong, Q. S. W. and Anderson, J. L. (2000) Assessment of the Hong Kong shark fin trade. *INFOFISH International* 1, 28–32.

Fong, Q. S. W. and Anderson, J. L. (2002) International shark fin markets and shark management: An integrated market preference-cohort analysis of the blacktip shark (*Carcharhinus limbatus*). *Ecological Economics* 40, 117–130.

Foster, S. J. and Vincent, A. C. J. (2005) Enhancing sustainability of the international trade in seahorses with a single minimum size limit. *Conservation Biology* 19, 1044–1050.

Foxcroft, L. C., Lotter, W. D., Runyoro, V. A. and Mattay, P. M. C. (2006) A review of the importance of invasive alien plants in the Ngorongoro Conservation Area and Serengeti National Park. *African Journal of Ecology* 44, 404–406.

Frankham, R., Ballou, J. D. and Biscoe, D. A. (2002) *Introduction to Conservation Genetics.* Cambridge University Press, Cambridge.

Fredericksen, N. J. and Fredericksen, T. S. (2002) Terrestrial wildlife responses to logging and fire in a Bolivian tropical humid forest. *Biodiversity and Conservation* 11, 27–38.

Freed, L. A., Cann, R. L., Goff, M. L., Kuntz, W. A. and Bodner, G. R. (2005) Increase in avian malaria at upper elevation in Hawai'i. *Condor* 107, 753–764.

Freeland, W. J. (1986) Populations of cane toad, *Bufo marinus*, in relation to time since colonization. *Australian Wildlife Research* 13, 321–329.

Freeland, W. J. (1990) Large herbivorous mammals – exotic species in northern Australia. *Journal of Biogeography* 17, 445–449.

Freese, L., Auster, P. J., Heifetz, J. and Wing, B. L. (1999) Effects of trawling on seafloor habitat and associated invertebrate taxa in the Gulf of Alaska. *Marine Ecology Progress Series* 182, 119–126.

Friedlander, A. M. and Parrish, J. D. (1997) Fisheries harvest and standing stock in a Hawaiian bay. *Fisheries Research* 32, 33–50.

Frisk, M. G., Miller, T. J. and Fogarty, M. J. (2001) Estimation and analysis of biological parameters in elasmobranch fishes: A comparative life history study. *Canadian Journal of Fisheries and Aquatic Sciences* 58, 969–981.

Fritts, T. H. and Rodda, G. H. (1998) The role of introduced species in the degradation of island ecosystems: A case history of Guam. *Annual Review of Ecology and Systematics* 29, 113–140.

Fryxell, J., Falls, J. B., Falls, E. A., Brooks, R. J., Dix, L. and Strickland, M. (2001) Harvest dynamics of mustelid carnivores in Ontario, Canada. *Wildlife Biology* 7, 151–159.

Fryxell, J. M., Lynn, D. H. and Chris, P. J. (2006) Harvest reserves reduce extinction risk in an experimental microcosm. *Ecology Letters* 9, 1025–1031.

Fuchs, E. J., Lobo, J. A. and Quesada, M. (2003) Effects of forest fragmentation and flowering phenology on the reproductive success and mating patterns of the tropical dry forest tree *Pachira quinata*. *Conservation Biology* 17, 149–157.

Fujii, S., Somiya, I., Nagare, H. and Serizawa, S. (2001) Water quality characteristics of forest rivers around Lake Biwa. *Water Science and Technology* 43, 183–192.

Furness, R. W. (2003) Impacts of fisheries on seabird communities. *Scientia Marina* 67, 33–45.

Futuyma, D. J. and Moreno, G. (1988) The evolution of ecological specialization. *Annual Review of Ecology and Systematics* 19, 207–233.

Garcia, S. M. and Newton, C. (1997) Global trends in fisheries management. In: *American Fisheries*

Society Symposium (eds. E. K. Pikitch, D. D. Hubert and M. P. Sissenwine), pp. 3–27. American Fisheries Society, Bethesda, MD.

Gardner, T. A., Cote, I. M., Gill, J. A., Grant, A. and Watkinson, A. R. (2003) Long-term region-wide declines in Caribbean corals. *Science* 301, 958–960.

Garnett, S. T. and Brook, B. W. (2007) Modelling to forestall extinction of Australian tropical birds. *Journal of Ornithology*, in press.

Gascon, C. (1993) Breeding habitat use by five Amazonian frogs at forest edge. *Biodiversity and Conservation* 2, 438–444.

Gascon, C., Lovejoy, T. E., Bierregaard, R. O., Malcolm, J. R., Stouffer, P. C., Vasconcelos, H. L., Laurance, W. F., Zimmerman, B., Tocher, M. and Borges, S. (1999) Matrix habitat and species richness in tropical forest remnants. *Biological Conservation* 91, 223–229.

Gascon, C., Williamson, G. B. and da Fonseca, G. A. B. (2000) Receding forest edges and vanishing reserves. *Science* 288, 1356–1358.

Gaston, K. J. and Blackburn, T. M. (1995) Birds, body size and the threat of extinction. *Philosophical Transactions of the Royal Society B: Biological Sciences* 347, 205–212.

Gaston, K. J. and Spicer, J. I. (1998) *Biodiversity: An Introduction*. Blackwell Science, Oxford.

Geist, H. J. and Lambin, E. F. (2002) Proximate causes and underlying driving forces of tropical deforestation. *BioScience* 52, 143–150.

Gentry, A. H. and Lopez-Parodi, J. (1980) Deforestation and increased flooding of the upper Amazon. *Science* 210, 1354–1356.

Gerber, L. R. and DeMaster, D. P. (1999) A quantitative approach to endangered species act classification of long-lived vertebrates: Application to the North Pacific humpback whale. *Conservation Biology* 13, 1203–1214.

Gerber, L. R., Botsford, L. W., Hastings, A., Possingham, H. P., Gaines, S. D., Palumbi, S. R. and Andelman, S. (2003) Population models for marine reserve design: A retrospective and prospective synthesis. *Ecological Applications* 13, S47-S64.

Gerwing, J. J. (2002) Degradation of forests through logging and fire in the eastern Brazilian Amazon. *Forest Ecology and Management* 157, 131–141.

Gewin, V. (2004) Troubled waters: The future of global fisheries. *PLoS Biology* 2, e113.

Ghazoul, J. (2002) Impact of logging on the richness and diversity of forest butterflies in a tropical dry forest in Thailand. *Biodiversity and Conservation* 11, 521–541.

Ghazoul, J. (2004) Alien abduction: Disruption of native plant–pollinator interactions by invasive species. *Biotropica* 36, 156–164.

Ghazoul, J. and McLeish, M. (2001) Reproductive ecology of tropical forest trees in logged and fragmented habitats in Thailand and Costa Rica. *Plant Ecology* 153, 335–345.

Giambelluca, T. W. (2002) Hydrology of altered tropical forest. *Hydrological Processes* 16, 1665–1669.

Giday, M., Asfaw, Z., Elmqvist, T. and Woldu, Z. (2003) An ethnobotanical study of medicinal plants used by the Zay people in Ethiopia. *Journal of Ethnopharmacology* 85, 43–52.

Gillespie, T. W. (2001) Application of extinction and conservation theories for forest birds in Nicaragua. *Conservation Biology* 15, 699–709.

Gilpin, M. E. and Diamond, J. M. (1980) Subdivision of nature reserves and the maintenance of species diversity. *Nature* 285, 567–568.

Gilpin, M. E. and Soulé, M. E. (1986) Minimum viable populations: Processes of species extinction. In: *Conservation Biology* (ed. M. E. Soulé), pp. 19–34. Sinauer, Sunderland, MA.

Giri, C., Defourny, P. and Shrestha, S. (2003) Land cover characterization and mapping of continental Southeast Asia using multi-resolution satellite sensor data. *International Journal of Remote Sensing* 24, 4181–4196.

Gleick, P. H. (1996) Water resources. In: *Encyclopaedia of Climate and Weather* (ed. S. H. Schneider), pp. 817–823. Oxford University Press, New York.

Glynn, P. (1996) Coral reef bleaching: Facts, hypotheses, and implications. *Global Change Biology* 2, 495–509.

Glynn, P. W. (1991) Coral reef bleaching in the 1980s and possible connections with global warming. *Trends in Ecology and Evolution* 6, 175–179.

Godley, B. J., Broderick, A. C. and Hays, G. C. (2001) Nesting of green turtles (*Chelonia mydas*) at Ascension Island, South Atlantic. *Biological Conservation* 97, 151–158.

Goerck, J. M. (1997) Patterns of rarity in the birds of the Atlantic forest of Brazil. *Conservation Biology* 11, 112–118.

Goldschmidt, T. (1996) *Darwin's Dreampond: Drama on Lake Victoria*. MIT Press, Cambridge, MA.

Goldstein, J. H., Daily, G. C., Friday, J. B., Matson, P. A. and Naylor, R. A. (2006) Business strategies for conservation on private lands: Koa forestry as a case study. *Proceedings of the National Academy of Sciences of the USA* 103, 10140–10145.

Gophen, M., Ochumba, P. B. O. and Kaufman, L. S. (1995) Some aspects of perturbation in the structure and biodiversity of the ecosystem of Lake Victoria (East Africa). *Aquatic Living Resources* 8, 27–41.

Goreau, T. J. (1990) Coral bleaching in Jamaica. *Nature* 343, 417.

Goreau, T. J. and Hayes, R. L. (1994) Coral bleaching and ocean 'hot spots'. *Ambio* 23, 176–180.

Goreau, T., McClanahan, T., Hayes, R. and Strong, A. (2000) Conservation of coral reefs after the 1998 global bleaching event. *Conservation Biology* 14, 5–17.

Gorog, A. J., Pamungkas, B. and Lee, R. J. (2005) Nesting ground abandonment by the maleo (*Macrocephalon maleo*) in North Sulawesi: Identifying conservation priorities for Indonesia's endemic megapode. *Biological Conservation* 126, 548–555.

Gossling, S. (2001) The consequences of tourism for sustainable water use on a tropical island: Zanzibar, Tanzania. *Journal of Environmental Management* 61, 179–191.

Goswami, B. N., Venugopal, V., Sengupta, D., Madhusoodanan, M. S. and Xavier P. K. (2006) Increasing trend of extreme rain events over India in a warming environment. *Science* 314, 1442–1445.

Gotelli, N. J. and Graves, G. R. (1990) Body size and the occurrence of avian species on land-bridge islands. *Journal of Biogeography* 17, 315–325.

Gowda, J. and Raffaele, E. (2004) Spine production is induced by fire: A natural experiment with three *Berberis* species. *Acta Oecologica* 26, 239–245.

Gracia, A. (1996) White shrimp (*Penaeus setiferus*) recruitment overfishing. *Marine and Freshwater Research* 47, 59–65.

Graham, N. A. J., Wilson, S. K., Jennings, S., Polunin, N. V. C., Bijoux, J. P. and Robinson, J. (2006) Dynamic fragility of oceanic coral reef ecosystems. *Proceedings of the National Academy of Sciences of the USA* 103, 8425–8429.

Greathead, D. J. (1968) Biological control of *Lantana*: A review and discussion of recent developments in East Africa. *Pest Articles and News Summaries* 14, 167–175.

Greathead, D. J. (1991) Biological control in the tropics – present opportunities and future prospects. *Insect Science and its Application* 12, 3–8.

Green, A. L., Birkeland, C. E. and Randall, R. H. (1999) Twenty years of disturbance and change in Fagatele Bay National Marine Sanctuary, American Samoa. *Pacific Science* 53, 376–400.

Green, E. P. and Bruckner, A. W. (2000) The significance of coral disease epizootiology for coral reef conservation. *Biological Conservation* 96, 347–361.

Green, P. T., Lake, P. S. and O'Dowd, D. J. (2004) Resistance of island rain forest to invasion by alien plants: Influence of microhabitat and herbivory on seedling performance. *Biological Invasions* 6, 1–9.

Greenberg, R., Bichier, P., Angon, A. C., MacVean, C., Perez, R. and Cano, E. (2000) The impact of avian insectivory on arthropods and leaf damage in some Guatemalan coffee plantations. *Ecology* 81, 1750–1755.

Grelle, C. E. V. (2005) Predicting extinction of mammals in the Brazilian Amazon. *Oryx* 39, 347–350.

Griffiths, A. D., Holland, D. C. and Whitehead, P. J. (2004) Impact of the exotic cane toad (*Bufo marinus*) on the survival of Lowland *Varanus* species in Kakadu National Park. Final Report: October 2004. Kakadu National Park and Department of Environment and Heritage, Jabiru, Australia.

Groombridge, B. and Luxmoore, R. (1991) *Pythons in Southeast Asia. A Review of Distribution, Status and Trade in Three Selected Species*. Convention on International Trade in Endangered Spcies (CITES) Secretariat, Lausanne.

Groomsbridge, B. and Jenkins, M. (1998) *Freshwater Biodiversity: A Preliminary Global Assessment*. World Conservation Monitoring Centre, Cambridge.

Grove, R. H., Damodaran, V., Sangwan, S. and (Eds) (1998) *Nature and the Orient: The Environmental History of South and Southeast Asia*. Oxford University Press, Delhi.

Guan, Y., Zheng, B. J., He, Y. Q., Liu, X. L., Zhuang, Z. X., Cheung, C. L., Luo, S. W., Li, P. H., Zhang, L. J., Guan, Y. J., Butt, K. M., Wong, K. L., Chan, K. W., Lim, W., Shortridge, K. F., Yuen, K. Y., Peiris, J. S. M. and Poon, L. L. M. (2003) Isolation and characterization of viruses related to the SARS coronavirus from animals in Southern China. *Science* 302, 276–278.

Guénette, S. and Pitcher, T. J. (1999) An age-structured model showing the benefits of marine reserves in controlling overexploitation. *Fisheries Research* 39, 295–303.

Guevara, S., Purata, S. E. and Vandermaarel, E. (1986) The role of remnant forest trees in tropical secondary succession. *Vegetatio* 66, 77–84.

Guo, J. (2006) The Galápagos Islands kiss their goat problem goodbye. *Science* 313, 1567–1567.

Gur, A. and Zamir, D. (2004) Unused natural variation can lift yield barriers in plant breeding. *PLoS Biology* 2, e245.

Gust, N. (2004) Variation in the population biology of protogynous coral reef fishes over tens of kilometres. *Canadian Journal of Fisheries and Aquatic Sciences* 61, 205–218.

Haila, Y. (2002) A conceptual genealogy of fragmentation research: From island biogeography to landscape ecology. *Ecological Applications* 12, 321–334.

Hales, S., Weinstein, P. and Woodward, A. (1999) Ciguatera fish poisoning, El Niño, and sea surface temperature. *Ecosystem Health* 5, 20–25.

Hall, M. A. (1998) An ecological view of the tuna-dolphin problem. *Reviews in Fish Biology and Fisheries* 8, 1–34.

Hall, S. J. and Mainprize, B. M. (2005) Managing by-catch and discards: How much progress are we making and how can we do better? *Fish and Fisheries* 6, 134–155.

Halpern, B. S. (2003) The impact of marine reserves: Do reserves work and does reserve size matter? *Ecological Applications* 13, S117-S137.

Halwart, M. (1994) The golden apple snail *Pomacea canaliculata* in Asian rice farming systems – present impact and future threat. *International Journal of Pest Management* 40, 199–206.

Hamann, A. and Curio, E. (1999) Interactions among frugivores and fleshy fruit trees in a Philippine submontane rain forest. *Conservation Biology* 13, 766–773.

Hamer, K. C., Hill, J. K., Lace, L. A. and Langan, A. M. (1997) Ecological and biogeographical effects of forest disturbance on tropical butterflies of Sumba, Indonesia. *Journal of Biogeography* 24, 67–75.

Hamer, K. C., Hill, J. K., Benedick, S., Mustaffa, N., Sherratt, T. N., Maryati, M. and Chey, V. K. (2003) Ecology of butterflies in natural and selectively logged forests of northern Borneo: The importance of habitat heterogeneity. *Journal of Applied Ecology* 40, 150–162.

Hampton, J. O., Spencer, P. B. S., Alpers, D. L., Twigg, L. E., Woolnough, A. P., Doust, J., Higgs, T. and Pluske, J. (2004) Molecular techniques, wildlife management and the importance of genetic population structure and dispersal: A case study with feral pigs. *Journal of Applied Ecology* 41, 735–743.

Hansen, J., Sato, M., Ruedy, R., Nazarenko, L., Lacis, A., Schmidt, G. A., Russell, G., Aleinov, I., Bauer, M., Bauer, S., Bell, N., Cairns, B., Canuto, V., Chandler, M., Cheng, Y., Del Genio, A., Faluvegi, G., Fleming, E., Friend, A., Hall, T., Jackman, C., Kelley, M., Kiang, N., Koch, D., Lean, J., Lerner, J., Lo, K., Menon, S., Miller, R., Minnis, P., Novakov, T., Oinas, V., Perlwitz, J., Rind, D., Romanou, A., Shindell, D., Stone, P., Sun, S., Tausnev, N., Thresher, D., Wielicki, B., Wong, T., Yao, M. and Zhang, S. (2005) Efficacy of climate forcings. *Journal of Geophysical Research* 110, D18104.

Hansen, J., Sato, M., Ruedy, R., Lo, K., Lea, D. W. and Medina-Elizade, M. (2006) Global temperature change. *Proceedings of the National Academy of Sciences of the USA* 103, 14288–14293.

Hansen, M. C. and DeFries, R. S. (2004) Detecting long-term global forest change using continuous fields of tree-cover maps from 8-km Advanced Very High Resolution Radiometer (AVHRR) data for the years 1982–99. *Ecosystems* 7, 695–716.

Hanski, I. (1989) Metapopulation dynamics: Does it help to have more of the same? *Trends in Ecology and Evolution* 4, 113–114.

Hanski, I. (1991) Single-species metapopulation dynamics – concepts, models and observations. *Biological Journal of the Linnean Society* 42, 17–38.

Hardtër, R., Chow, W. Y. and Hock, O. S. (1997) Intensive plantation cropping, a source of sustainable food and energy production in the tropical rain forest areas in southeast Asia. *Forest Ecology and Management* 91, 93–102.

Harper, L. H. (1987) *The Conservation of Ant-Following Birds in Central Amazonian Forest Fragments.* PhD thesis, State University of New York, Albany, New York.

Harper, L. H. (1989) The persistence of ant-following birds in small Amazonian forest fragments. *Acta Amazonica* 19, 249–263.

Harrington, G. N., Freeman, A. N. D. and Crome, F. H. J. (2001) The effects of fragmentation of an

Australian tropical rain forest on populations and assemblages of small mammals. *Journal of Tropical Ecology* 17, 225–240.

Harrison, R. D. (2000) Repercussions of El Niño: Drought causes extinction and the breakdown of mutualism in Borneo. *Proceedings of the Royal Society B: Biological Sciences* 267, 911–915.

Harrison, R. D., Hamid, A. A., Kenta, T., Lafrankie, J., Lee, H. S., Nagamasu, H., Nakashizuka, T. and Palmiotto, P. (2003) The diversity of hemi-epiphytic figs (*Ficus*; Moraceae) in a Bornean lowland rain forest. *Biological Journal of the Linnean Society* 78, 439–455.

Hartemink, A. E. (2005) Plantation agriculture in the tropics – environmental issues. *Outlook on Agriculture* 34, 11–21.

Hartley, S. E. and Jones, T. H. (2003) Plant diversity and insect herbivores: Effects of environmental changes in contrasting models. *Oikos* 101, 6–17.

Hartnett, D. C., Potgieter, A. F. and Wilson, G. W. T. (2004) Fire effects on mycorrhizal symbiosis and root system architecture in southern African savanna grasses. *African Journal of Ecology* 42, 328–337.

Hartshorn, G. and Bynum, N. (1999) Tropical forest synergies. *Science* 286, 2093–2094.

Harvell, C. D., Kim, K., Burkholder, J. M., Colwell, R. R., Epstein, P. R., Grimes, D. J., Hofmann, E. E., Lipp, E. K., Osterhaus, A. D. M. E., Overstreet, R. M., Porter, J. W., Smith, G. W. and Vasta, G. R. (1999) Emerging marine diseases – climate links and anthropogenic factors. *Science* 285, 1505–1510.

Harvell, C. D., Mitchell, C. E., Ward, J. R., Altizer, S., Dobson, A. P., Ostfeld, R. S. and Samuel, M. D. (2002) Climate warming and disease risks for terrestrial and marine biota. *Science* 296, 2158–2162.

Harvey, P. H. and Pagel, M. D. (1991) *The Comparative Method in Evolutionary Biology*. Oxford University Press, Oxford.

Harwood, J. and Stokes, K. (2003) Coping with uncertainty in ecological advice: Lessons from fisheries. *Trends in Ecology and Evolution* 18, 617–622.

Hastings, A. and Botsford, L. W. (2003) Comparing designs of marine reserves for fisheries and for biodiversity. *Ecological Applications* 13, S65–S70.

Haugaasen, T., Barlow, J. and Peres, C. A. (2003) Surface wildfires in central Amazonia: Short-term impact on forest structure and carbon loss. *Forest Ecology and Management* 179, 321–331.

Hawkins, J. P., Roberts, C. M., Van't Hof, T., De Meyer, K. and Tratalos, J. A., C. (1999) Effects of recreational scuba diving on Caribbean coral and fish communities. *Conservation Biology* 13, 888–897.

Hayes, M. L., Bonaventura, J., Mitchell, T. P., Prospero, J. M., Shinn, E. A., Van Dolah, F. and Barber, R. T. (2001) How are climate and marine biological outbreaks functionally linked? *Hydrobiologia* 460, 213–220.

Hayes, T. M. (2006) Parks, people, and forest protection: An institutional assessment of the effectiveness of protected areas. *World Development* 34, 2064–2075.

Hays, G. C. (2004) Good news for sea turtles. *Trends in Ecology and Evolution* 19, 349–351.

He, F. L. and Legendre, P. (1996) On species–area relations. *American Naturalist* 148, 719–737.

Heaney, L. R. (1991) A synopsis of climate and vegetational change in Southeast Asia. *Climate Change* 19, 53–61.

Heard, S. B. and Mooers, A. O. (1999) Phylogenetically patterned speciation rates and extinction risks change the loss of evolutionary history during extinctions. *Proceedings of the Royal Society B: Biological Sciences* 267, 613–620.

Hecht, S. B., Kandel, S., Gomes, I., Cuellar, N. and Rosa, H. (2006) Globalization, forest resurgence, and environmental politics in El Salvador. *World Development* 34, 308–323.

Hedgpeth, J. W. (1993) Foreign invaders. *Science* 261, 34–35.

Heino, M. and Godo, O. R. (2002) Fisheries-induced selection pressures in the context of sustainable fisheries. *Bulletin of Marine Science* 70, 639–656.

Heinsohn, R., Lacy, R. C., Lindenmayer, D. B., Marsh, H., Kwan, D. and Lawler, I. R. (2004) Unsustainable harvest of dugongs in Torres Strait and Cape York (Australia) waters: Two case studies using population viability analysis. *Animal Conservation* 7, 417–425.

Heltberg, R. (2001) Impact of the ivory trade ban on poaching incentives: A numerical example. *Ecological Economics* 36, 189–195.

Heppell, S. S. and Crowder, L. B. (1996) Analysis of a fisheries model for harvest of hawksbill sea turtles (*Eretmochelys imbricata*). *Conservation Biology* 10, 874–880.

Herrera, C. M. and Pellmyr, O. (2002) *Plant–Animal Interactions*. Blackwell Science, Oxford.

Heywood, V. H. and Stuart, S. N. (1992) Species extinctions in tropical forests. In: *Tropical Deforestation and Species Extinction* (eds. T. C. Whitmore and J. A. Sayer), pp. 91–117. Chapman & Hall, London.

Heywood, V. H., Mace, G. M., May, R. M. and Stuart, S. N. (1994) Uncertainties in extinction rates. *Nature* 368, 105–105.

Hilborn, R. (2002) Marine reserves and fisheries management. *Science* 295, 1233–1234.

Hill, J. K., Hamer, K. C., Lace, L. A. and Banham, W. M. T. (1995) Effects of selective logging on tropical forest butterflies on Buru, Indonesia. *Journal of Applied Ecology* 32, 754–760.

Hill, J. K., Hamer, K. C., Tangah, J. and Dawood, M. (2001) Ecology of tropical butterflies in rain forest gaps. *Oecologia* 128, 294–302.

Hilton, M. J. and Manning, S. S. (1995) Conversion of coastal habitats in Singapore: Indications of unsustainable development. *Environmental Conservation* 22, 307–322.

Hites, R. A., Foran, J. A., Carpenter, D. O., Hamilton, M. C., Knuth, B. A. and Schwager, S. J. (2004) Global assessment of organic contaminants in farmed salmon. *Science* 303, 226–229.

Hoare, P. (2004) A process for community and government cooperation to reduce the forest fire and smoke problem in Thailand. *Agriculture Ecosystems and Environment* 104, 35–46.

Hodgson, G. (1990) Sediment and the settlement of larvae of the reef coral *Pocillopora damicornis*. *Coral Reefs* 9, 41–43.

Hoegh-Guldberg, O. (1999) Climate change, coral bleaching and the future of the world's coral reef. *Marine and Freshwater Research* 50, 839–866.

Hoekstra, J. M., Boucher, T. M., Ricketts, T. H. and Roberts, C. (2005) Confronting a biome crisis: Global disparities of habitat loss and protection. *Ecology Letters* 8, 23–29.

Hoffmann, B. D., Andersen, A. N. and Hill, G. J. E. (1999) Impact of an introduced ant on native rain forest invertebrates: *Pheidole megacephala* in monsoonal Australia. *Oecologia* 120, 595–604.

Hoffmann, M., Kelly, H. and Evans, T. (2003) Simulating land-cover change in south-central Indiana: An agent-based model of deforestation and afforestation. In: *Complexity and Ecosystem Management: The Theory and Practice of Multi-agent Systems* (ed. M. A. Janssen), pp. 218–247. Edward Elgar Publishers, Cheltenham.

Hoffmann, W. A., Schroeder, W. and Jackson, R. B. (2003) Regional feedbacks among fire, climate, and tropical deforestation. *Journal of Geophysical Research* 108, 4721.

Hogan, Z. S. and May., B. P. (2002) Twenty-seven new microsatellites for the migratory Asian catfish family Pangasiidae. *Molecular Ecology Notes* 2, 38–41.

Hogan, Z. S., Moyle, P. B., May, B., Vander Zanden, M. J. and Baird, I. G. (2004) The imperiled giants of the Mekong. *American Scientist* 92, 228–237.

Holdo, R. M. and McDowell, L. R. (2004) Termite mounds as nutrient-rich food patches for elephants. *Biotropica* 36, 231–239.

Holdsworth, A. R. and Uhl, C. (1997) Fire in Amazonian selectively logged rain forest and the potential for fire reduction. *Ecological Applications* 7, 713–725.

Homewood, K. M. (2004) Policy, environment and development in African rangelands. *Environmental Science and Policy* 7, 125–143.

Houghton, J. (2004) *Global Warming: The Complete Briefing*. Cambridge University Press, Cambridge.

Houlahan, J. E., Findlay, C. S., Schmidt, B. R., Meyer, A. H. and Kuzmin, S. L. (2000) Quantitative evidence for global amphibian population declines. *Nature* 404, 752–755.

Howarth, R. W., Billen, G., Swaney, D., Townsend, A., Jaworski, N., Lajtha, K., Downing, J. A., Elmgren, R., Caraco, N., Jordan, T., Berendse, F., Freney, J., Kudeyarov, V., Murdoch, P. and Zhu, Z. L. (1996) Regional nitrogen budgets and riverine N and P fluxes for the drainages to the North Atlantic Ocean: Natural and human influences. *Biogeochemistry* 35, 75–139.

Howe, G. E., Marking, L. L., Bills, T. D., Rach, J. J. and Mayer, F. L. (1994) Effects of water temperature and pH on toxicity of terufos, trichlorofon, 4-nitrophenol and 2,4-dinitrophenol to the amphipod *Gammarus pseudolimnaeus* and rainbow trout (*Oncorhyncus mykiss*). *Environmental Toxicology and Chemistry* 13, 51–66.

Howe, H. F. (1984) Implications of seed dispersal by animals for tropical reserve management. *Biological Conservation* 30, 261–281.

Hoyle, M. and James, M. (2005) Global warming, human population pressure, and viability of the world's smallest butterfly. *Conservation Biology* 19, 1113–1124.

Hucke-Gaete, R., Osman, L. P., Moreno, C. A., Findlay, K. P. and Ljungblad, D. K. (2004) Discovery of a blue whale feeding and nursing ground in southern Chile. *Proceedings of the Royal Society B: Biological Sciences* **271**, S170–S173.

Hughes, J. B., Daily, G. C. and Ehrlich, P. R. (1997) Population diversity: Its extent and extinction. *Science* **278**, 689–692.

Hughes, T. P., Baird, A. H., Bellwood, D. R., Card, M., Connolly, S. R., Folke, C., Grosberg, R., Hoegh-Guldberg, O., Jackson, J. B. C., Kleypas, J., Lough, J. M., Marshall, P., Nystrom, M., Palumbi, S. R., Pandolfi, J. M., Rosen, B. and Roughgarden, J. (2003) Climate change, human impacts, and the resilience of coral reefs. *Science* **301**, 929–933.

Hunte, W. and Wittenberg, M. (1992) Effects of eutrophication and sedimentation on juvenile corals. 2. Settlement. *Marine Biology* **114**, 625–631.

Hunter, C. L. and Evans, C. W. (1995) Coral reefs in Kaneohe Bay, Hawaii: Two centuries of Western influence and two decades of data. *Bulletin of Marine Science* **57**, 501–515.

Hutchings, J. A. (2000) Collapse and recovery of marine fishes. *Nature* **406**, 882–885.

Huth, A. and Ditzer, T. (2001) Long-term impacts of logging in a tropical rain forest: A simulation study. *Forest Ecology and Management* **142**, 33–51.

Hyrenbach, K. D., Forney, K. A. and Dayton, P. K. (2000) Marine protected areas and ocean basin management. *Aquatic Conservation-Marine and Freshwater Ecosystems* **10**, 437–458.

Ickes, K. (2001) Hyper-abundance of native wild pigs (*Sus scrofa*) in a lowland dipterocarp rain forest of Peninsular Malaysia. *Biotropica* **33**, 682–690.

Ickes, K. and Thomas, S. C. (2003) Native, wild pigs (*Sus scrofa*) at Pasoh and their impacts on the plant community. In: *Pasoh: Ecology of a Lowland Rain Forest of Southeast Asia* (eds. T. Okuda, N. Manokaran, Y. Matsumoto, K. Niyama, S. C. Thomas and P. S. Asthton), pp. 507–520. Springer-Verlag, Tokyo.

Ickes, K., Dewalt, S. J. and Appanah, S. (2001) Effects of native pigs (*Sus scrofa*) on woody understorey vegetation in a Malaysian lowland rain forest. *Journal of Tropical Ecology* **17**, 191–206.

Inigo-Elias, E. E. and Ramos, M. A. (1991) The psittacine trade in Mexico. In: *Neotropical Wildlife Use and Conservation* (eds. J. G. Robinson and K. H. Redford), pp. 251–266. University of Chicago Press, Chicago.

IPCC (1996) *Climate Change 1995: The Science of Climate Change. Contribution of Working Group I to the Second Assessment Report of the Intergovernmental Panel on Climate Change.* Cambridge University Press, Cambridge.

IPPC (2001a) *Third Assessment Report of the Intergovernmental Panel on Climate Change (Working Group I and II).* Cambridge University Press, Cambridge.

IPPC (2001b) *Climate Change 2001: The Scientific Basis.* Cambridge University Press, Cambridge.

IPCC (2002) Workshop for Carbon Dioxide Capture and Storage (Working Group III: Mitigation of Climate Change). ECN – Environment Canada, Regina, Canada.

Islam, K. R., Ahmed, M. R., Bhuiyan, M. K. and Badruddin, A. (2001) Deforestation effects on vegetative regeneration and soil quality in tropical semi-evergreen degraded and protected forests of Bangladesh. *Land Degradation and Development* **12**, 45–56.

ITTO (2003) *Annual Review and Assessment of the World Timber Situation.* International Tropical Timber Organization (ITTO), Yokohama, Japan.

ITTO (2006) *Status of Tropical Forest Management 2005.* International Tropical Timber Organization (ITTO), Yokohama, Japan.

IUCN (2000) Communities and forest management in South Asia. In: *Regional Profile of the Working Group on Community Involvement in Forest Management* (ed. M. Poffenberger), p. 162. Asian Forest Network, Santa Barbara, CA.

IUCN (2005) *IUCN Red List of Threatened Species.* World Conservation Union, Gland (www.iucnredlist.org).

Jablonski, D., Roy, K. and Valentine, J. W. (2006) Out of the tropics: Evolutionary dynamics of the latitudinal diversity gradient. *Science* **314**, 102–106.

Jackson, J. B. C. (2001) What was natural in the coastal oceans? *Proceedings of the National Academy of Sciences of the USA* **98**, 5411–5418.

Jackson, J. B. C., Kirby, M. X., Berger, W. H., Bjorndal, K. A., Botsford, L. W., Bourque, B. J., Bradbury, R. H., Cooke, R., Erlandson, J., Estes, J. A., Hughes, T. P., Kidwell, S., Lange, C. B., Lenihan, H. S., Pandolfi, J. M., Peterson, C. H., Steneck, R. S., Tegner, M. J. and Warner, R. R. (2001) Historical overfishing and the recent collapse of coastal ecosystems. *Science* **293**, 629–638.

James, A. N., Green, M. J. B. and Paine, J. R. (1999) *Global Review of Protected Area Budgets and Staff.* WCMC, Cambridge.

James, A., Gaston, K. J. and Balmford, A. (2000) Why private institutions alone will not do enough to protect biodiversity. *Nature* **404**, 120.

James, H. F. (1995) Prehistoric extinctions and ecological changes on oceanic islands. *Ecological Studies* **115**, 88–102.

James, M., Gilbert, F. and Zalat, S. (2003) Thyme and isolation for the Sinai baton blue butterfly (*Pseudophilotes sinaicus*). *Oecologia* **134**, 445–453.

Jang, C. J., Nishigami, Y. and Yanagisawa, Y. (1996) Assessment of global forest change between 1986 and 1993 using satellite-derived terrestrial net primary productivity. *Environmental Conservation* **23**, 315–321.

Janzen, D. H. (2004) Setting up tropical biodiversity for conservation through non-damaging use: Participation by parataxonomists. *Journal of Applied Ecology* **41**, 181–187.

Jenkins, M. (2003) Prospects for biodiversity. *Science* **302**, 1175–1177.

Jenkins, P. D., Kilpatrick, C. W., Robinson, M. F. and Timmins, R. J. (2005) Morphological and molecular investigations of a new family, genus and species of rodent (Mammalia: Rodentia: Hystricognatha) from Lao PDR. *Systematics and Biodiversity* **2**, 419–454.

Jennings, G. (1992) *Raptor*. Doubleday, New York.

Jennings, S. and Polunin, N. (1997) Impacts of predator depletion by fishing on the biomass and diversity of non-target reef fish communities. *Coral Reefs* **16**, 71–82.

Jennings, S., Grandcourt, E. M. and Polunin, N. V. C. (1995) The effects of fishing on the diversity, biomass and trophic structure of Seychelles' reef fish communities. *Coral Reefs* **14**, 225–235.

Jennings, S., Reynolds, J. D. and Polunin, N. V. C. (1999) Predicting the vulnerability of tropical reef fishes to exploitation with phylogenies and life histories. *Conservation Biology* **13**, 1466–1475.

Jepson, P. (2001) Global biodiversity plan needs to convince local policy-makers. *Nature* **409**, 12.

Jepson, P., Jarvie, J. K., MacKinnon, K. and Monk, K. A. (2001) The end for Indonesia's lowland forests? *Science* **292**, 859–861.

Jerozolimski, A. and Peres, C. A. (2003) Bringing home the biggest bacon: A cross-site analysis of the structure of hunter–kill profiles in Neotropical forests. *Biological Conservation* **111**, 415–425.

Johannes, R. E. (1978) Traditional marine conservation methods in Oceania and their demise. *Annual Reviews in Ecology and Systematics* **9**, 349–364.

Johannes, R. E. (1998a) Government-supported, village-based management of marine resources in Vanuatu – managing marine resources in Palau, Micronesia. *Ocean and Coastal Management* **40**, 165–186.

Johannes, R. E. (1998b) The case for data-less marine resource management: Examples from tropical nearshore finfish-fisheries. *Trends in Ecology and Evolution* **13**, 243–246.

Johannes, R. E. (2002) The renaissance of community-based marine resource management in Oceania. *Annual Review of Ecology and Systematics* **33**, 317–340.

Johns, A. D. (1992) Species conservation in managed tropical forests. In: *Tropical Deforestation and Species Extinction* (eds. T. C. Whitmore and J. Sayer), pp. 15–54. Chapman & Hall, London.

Johnson, C. N., Isaac, J. L. and Fisher, D. O. (2007) Rarity of a top predator triggers continent-wide collapse of mammal prey: Dingoes and marsupials in Australia. *Proceedings of the Royal Society B: Biological Sciences* **274**, 341–346.

Johnson, N., Revenga, C. and Echeverria, J. (2001) Managing water for people and nature. *Science* **292**, 1071–1072.

Johnson, S. (2003) Estimating the extent of illegal trade of tropical forest products. *International Forestry Review* **5**, 247–252.

Johnston, F. H., Kavanagh, A. M., Bowman, D. and Scott, R. K. (2002) Exposure to bushfire smoke and asthma: An ecological study. *Medical Journal of Australia* **176**, 535.

Jones, G. P., McCormick, M. I., Srinivasan, M. and Eagle, J. V. (2004) Coral decline threatens fish biodiversity in marine reserves. *Proceedings of the National Academy of Sciences of the USA* **101**, 8251–8253.

Jones, M. J., Sullivan, M. S., Marsden, S. J. and Linsley, M. D. (2001) Correlates of extinction risk of birds from two Indonesian islands. *Biological Journal of the Linnean Society* **73**, 65–79.

Jones, R. J. and Hoegh-Guldberg, O. (1999) Effects of cyanide on coral photosynthesis: Implications for identifying the cause of coral bleaching and for assessing the environmental effects of cyanide fishing. *Marine Ecology Progress Series* 177, 83–91.

Jones, R. J. and Steven, A. L. (1997) Effects of cyanide on corals in relation to cyanide fishing on reefs. *Marine and Freshwater Research* 48, 517–522.

Jones, R. J., Bowyer, J., Hoegh-Guldberg, O. and Blackall, L. L. (2004) Dynamics of a temperature-related coral disease outbreak. *Marine Ecology Progress Series* 281, 63–77.

Jorgenson, J. P. (2000) Wildlife conservation and game harvest by Maya hunters in Quintana Roo, Mexico. In: *Hunting for Sustainability in Tropical Forests* (eds. J. G. Robinson and E. L. Bennett). Columbia University Press, New York.

Jules, E. S. and Shahani, P. (2003) A broader ecological context to habitat fragmentation: Why matrix habitat is more important than we thought. *Journal of Vegetation Science* 14, 459–464.

Junk, W. J. (1993) Wetlands of tropical South America. In: *Wetlands of the World I: Inventory, Ecology and Management* (eds. D. F. Whigham, D. Dykyjová and S. Hejný), pp. 679–739. Kluwer Academic Publishers, Dordrecht.

Junk, W. J. (2002) Long-term environmental trends and the future of tropical wetlands. *Environmental Conservation* 29, 414–435.

Jha, S. and Bawa, K. S. (2006) Population growth, human development, and deforestation in biodiversity hotspots. *Conservation Biology* 20, 906-912.

Kaimowitz, D. (2004) The great flood myth. *New Scientist* 182, 18.

Kaplan, S. and Kaplan, R. (1989) *The Experience of Nature: A Psychological Perspective*. Cambridge University Press, Cambridge.

Kapos, V. (1989) Effects of isolation on the water status of forest patches in the Brazilian Amazon. *Journal of Tropical Ecology* 5, 173–185.

Kappel, C. V. (2005) Losing pieces of the puzzle: Threats to marine, estuarine, and diadromous species. *Frontiers in Ecology and the Environment* 3, 275–282.

Karesh, W. B., Cook, R. A., Bennett, E. L. and Newcomb, J. (2005) Wildlife trade and global disease emergence. *Emerging Infectious Diseases* 11, 1000–1002.

Karr, J. R. (1982a) Avian extinction on Barro-Colorado Island, Panama – a reassessment. *American Naturalist* 119, 220–239.

Karr, J. R. (1982b) Population variability and extinction in the avifauna of a tropical land-bridge island. *Ecology* 63, 1975–1978.

Katsigris, E., Bull, G. Q., White, A., Barr, C., Barney, K., Bun, Y., Kahrl, F., King, T., Lankin, A., Lebedev, A., Shearman, R., Sheingauz, A., Su, Y. F. and Weyerhaeuser, H. (2004) The China forest products trade: Overview of Asia-Pacific supplying countries, impacts and implications. *International Forestry Review* 6, 237–253.

Kattan, G. H. (1992) Rarity and vulnerability: The birds of the Cordillera Central of Colombia. *Conservation Biology* 6, 64–70.

Kattan, G. H., Alvarezlopez, H. and Giraldo, M. (1994) Forest fragmentation and bird extinctions: San Antonio 80 years later. *Conservation Biology* 8, 138–146.

Kauffman, J. B., Sanford, R. L. J., Cummings, D. L., Salcedo, I. H. and Sampaio, E. V. S. B. (1993) Biomass and nutrient dynamics associated with slash fires in Neotropical dry forests. *Ecology* 74, 140–151.

Kaufman, L. and Ochumba, P. (1993) Evolutionary and conservation biology of cichlid fishes as revealed by faunal remnants in northern Lake Victoria. *Conservation Biology* 7, 719–730.

Kaufman, L. S., Chapman, L. J. and Chapman, C. A. (1997) Evolution in fast forward: Haplochromine fishes of the Lake Victoria region. *Endeavour* 21, 23–30.

Kawanishi, K. and Sunquist, M. E. (2004) Conservation status of tigers in a primary rainforest of Peninsular Malaysia. *Biological Conservation* 120, 329–344.

Keller, M. and Reiners, W. A. (1994) Soil atmosphere exchange of nitrous oxide, nitric oxide, and methane under secondary succession of pasture to forest in the atlantic lowlands of Costa Rica. *Global Biogeochemical Cycles* 8, 399–409.

Keppler, F., Hamilton, J. T. G., Brass, M. and Rockmann, T. (2006) Methane emissions from terrestrial plants under aerobic conditions. *Nature* 439, 187–191.

Kermen, J. and Janota-Bassalik, L. (1987) A note on the forest soil as a biological filter in the sanitary purification of municipal waste water evaluated on the basis of *Escherichia coli* titre. *Acta Microbiologica Polonica* 36, 109–118.

Kerr, J. T. and Burkey, T. V. (2002) Endemism, diversity, and the threat of tropical moist forest extinctions. *Biodiversity and Conservation* 11, 695–704.

Kevan, P. G., Hussein, N. T., Hussey, N. and Wahid, M. B. (1986) The use of *Elaeidobius kamerunicus* for pollination of oil palm. *Planter* 62, 89–99.

Khan, I. (1991) Effect of urban and industrial wastes on species diversity of the diatom community in a tropical river, Malaysia. *Hydrobiologia* 224, 175–184.

Kidson, C., Indaratna, K. and Looareesuwan, S. (2000) The malaria cauldron of Southeast Asia: Conflicting strategies of contiguous nation states. *Parasitologia* 42, 101–110.

Kiew, R. (1991) The limestone flora. In: *The State of Nature Conservation in Malaysia* (ed. R. Kiew), pp. 42–50. Malayan Nature Society, Kuala Lumpur.

Kilpatrick, A. M. (2006) Facilitating the evolution of resistance to avian malaria in Hawaiian birds. *Biological Conservation* 128, 475–485.

King, M. and Faasili, U. (1999) Community-based management of subsistence fisheries in Samoa. *Fisheries Management and Ecology* 6, 133–144.

Kinnaird, M. F. and O'Brien, T. G. (1998) Ecological effects of wildfire on lowland rainforest in Sumatra. *Conservation Biology* 12, 954–956.

Kinnaird, M. F. and O'Brien, T. G. (1999) Breeding ecology of the Sulawesi red-knobbed hornbill *Aceros cassidix*. *Ibis* 141, 60–69.

Kinnaird, M. F. and O'Brien, T. G. (2001) Who's scratching whom? Reply. *Conservation Biology* 15, 1459–1460.

Kinnaird, M. F., Sanderson, E. W., O'Brien, T. G., Wibisono, H. T. and Woolmer, G. (2003) Deforestation trends in a tropical landscape and implications for endangered large mammals. *Conservation Biology* 17, 245–257.

Kirkman, H. (1978) Decline of seagrass in northern areas of Moreton Bay, Queensland. *Aquatic Botany* 5, 63–76.

Kiss, A. (2004) Is community-based ecotourism a good use of biodiversity conservation funds? *Trends in Ecology and Evolution* 19, 232–237.

Klein, A.-M., Steffan-Dewenter, I. and Tscharntke, T. (2003a) Pollination of *Coffea canephora* in relation to local and regional agroforestry management. *Journal of Applied Ecology* 40, 837–845.

Klein, A.-M., Steffan-Dewenter, I. and Tscharntke, T. (2003b) Fruit set of highland coffee increases with the diversity of pollinating bees. *Proceedings of the Royal Society B: Biological Sciences* 270, 955–961.

Klein, B. C. (1989) Effects of forest fragmentation on dung and carrion beetle communities in central Amazonia. *Ecology* 70, 1715–1725.

Kleypas, J. A., Buddemeier, R. W., Archer, D., Gattuso, J. P., Langdon, C. and Opdyke, B. N. (1999) Geochemical consequences of increased atmospheric carbon dioxide on coral reefs. *Science* 284, 118–120.

Kling, H. J., Mugidde, R. and Hecky, R. E. (2001) Recent changes in the phytoplankton community of Lake Victoria in response to eutrophication. In: *The Great Lakes of the World (GLOW): Food-web, Health and Integrity* (eds. M. Munawar and R. E. Hecky), pp. 47–65. Backhuys, Leiden.

Klink, C. A. and Machado, R. B. (2005) Conservation of the Brazilian Cerrado. *Conservation Biology* 19, 707–713.

Knowlton, N. (2001) The future of coral reefs. *Proceedings of the National Academy of Sciences of the USA* 98, 5419–5425.

Koh, L. P., Dunn, R. R., Sodhi, N. S., Colwell, R. K., Proctor, H. C. and Smith, V. S. (2004a) Species coextinctions and the biodiversity crisis. *Science* 305, 1632–1634.

Koh, L. P., Sodhi, N. S. and Brook, B. W. (2004b) Ecological correlates of extinction proneness in tropical butterflies. *Conservation Biology* 18, 1571–1578.

Kolar, C. S. and Lodge, D. M. (2001) Progress in invasion biology: Predicting invaders. *Trends in Ecology and Evolution* 16, 199–204.

Komdeur, J. (1994) Conserving the Seychelles warbler *Acrocephalus sechellensis* by translocation from Cousin island to the islands of Aride and Cousine. *Biological Conservation* 67, 143–152.

Kong, L., Yuen, B. and Sodhi, N. S. (1997) Nature and nurture, danger and delight: Urban women's experiences of the natural world. *Landscape Research* 22, 245–266.

Kong, L., Yuen, B., Sodhi, N. S. and Briffett, C. (1999) The construction and experience of nature: Perspectives of urban youths. *Tijdschrift Voor Economische en Sociale Geografie* 90, 3–16.

Korontzi, S. (2005) Seasonal patterns in biomass burning emissions from southern African vegetation fires for the year 2000. *Global Change Biology* 11, 1680–1700.

Kovats, R. S., Campbell-Lendrum, D. H., McMichael, A. J., Woodward, A. and Cox, J. S. (2001) Early effects of climate change: Do they include changes in vector-borne disease? *Philosophical Transactions of the Royal Society B: Biological Sciences* 356, 1057–1068.

Kowarik, I. (1995) Clonal growth in *Ailanthus altissima* on a natural site in West Virginia. *Journal of Vegetation Science* 6, 853–856.

Krammer, E. A. (1997) Measuring landscape changes in remnant tropical dry forest. In: *Tropical Forest Remnants: Ecology, Management and Conservation of Fragmented Communities* (eds. W. F. Laurance and R. O. Bierregaard), pp. 400–409. University of Chicago Press, Chicago.

Krantz, G. W. and Poinar, G. O. (2004) Mite, nematodes and the multimillion dollar weevil. *Journal of Natural History* 38, 135–141.

Kremen, C., Razafimahatratra, V., Guillery, R. P., Rakotomalala, J., Weiss, A. and Ratsisompatrarivo, J. S. (1999) Designing the Masoala National Park in Madagascar based on biological and socioeconomic data. *Conservation Biology* 13, 1055–1068.

Kremen, C., Niles, J. O., Dalton, M. G., Daily, G. C., Ehrlich, P. R., Fay, J. P., Grewal, D. and Guillery, R. P. (2000) Economic incentives for rain forest conservation across scales. *Science* 288, 1828–1832.

Kroeze, C. and Seitzinger, S. P. (1998) Nitrogen inputs to rivers, estuaries and continental shelves and related nitrous oxide emissions in 1990 and 2050: A global model. *Nutrient Cycling in Agroecosystems* 52, 195–212.

Krogh, M. and Reid, D. (1996) Bycatch in the protective shark meshing programme off south-eastern New South Wales, Australia. *Biological Conservation* 77, 219–226.

Kummer, D. M. and Turner, B. L. I. (1994) The human causes of deforestation in Southeast Asia: The recurrent patterns is that of large-scale logging for exports, followed by agricultural expansion. *BioScience* 44, 323–328.

Kupfer, J. A., Malanson, G. P. and Franklin, S. B. (2006) Not seeing the ocean for the islands: The mediating influence of matrix-based processes on forest fragmentation effects. *Global Ecology and Biogeography* 15, 8–20.

Kurtén, B. and Anderson, A. (1980) *Pleistocene Mammals of North America*. Columbia University Press, New York.

Kushmaro, A., Rosenberg, E., Fine, M., Ben Haim, Y. and Loya, Y. (1998) Effect of temperature on bleaching of the coral *Oculina patagonica* by *Vibrio* AK-1. *Marine Ecology Progress Series* 171, 131–137.

Kwan, D., Marsh, H. and Delean, S. (2006) Factors influencing the sustainability of customary dugong hunting by a remote indigenous community. *Environmental Conservation* 33, 164–171.

Lach, L., Britton, D. K., Rundell, R. J. and Cowie, I. D. (2000) Food preference and reproductive plasticity in an invasive freshwater snail. *Biological Invasions* 2, 279–288.

Lacki, M. J. and Lancia, R. A. (1983) Effects of wild pigs on beech growth in Great Smoky Mountains National Park, USA. *Journal of Wildlife Management* 50, 655–659.

Lafferty, K. D. and Gerber, L. R. (2002) Good medicine for conservation biology: The intersection of epidemiology and conservation theory. *Conservation Biology* 16, 593–604.

Laidlaw, R. K. (2000) Effects of habitat disturbance and protected areas on mammals of Peninsular Malaysia. *Conservation Biology* 14, 1639–1648.

La Marca, E., Lips, K. R., Lotters, S., Puschendorf, R., Ibanez, R., Rueda-Almonacid, J. V., Schulte, R., Marty, C., Castro, F., Manzanilla-Puppo, J., Garcia-Perez, J. E., Bolanos, F., Chaves, G., Pounds, J. A., Toral, E. and Young, B. E. (2005) Catastrophic population declines and extinctions in neotropical harlequin frogs (Bufonidae: Atelopus). *Biotropica* 37, 190–201.

Lamb, D., Erskine, P. D. and Parrotta, J. A. (2005) Restoration of degraded tropical forest landscapes. *Science* 310, 1628–1632.

Lambert, F. (1991) The conservation of fig-eating birds in Malaysia. *Biological Conservation* 58, 31–40.

Larsen, T. H., Williams, N. M. and Kremen, C. (2005) Extinction order and altered community structure rapidly disrupt ecosystem functioning. *Ecology Letters* 8, 538–547.

Laurance, S. G. and Laurance, W. F. (1999) Tropical wildlife corridors: Use of linear rainforest remnants by arboreal mammals. *Biological Conservation* 91, 231–239.

Laurance, W. F. (1990) Comparative responses of five arboreal marsupials to tropical forest fragmentation. *Journal of Mammalogy* 71, 641–653.

Laurance, W. F. (1999) Reflections on the tropical deforestation crisis. *Biological Conservation* **91**, 109–117.

Laurance, W. F. (2004) The perils of payoff: Corruption as a threat to global biodiversity. *Trends in Ecology and Evolution* **19**, 399–401.

Laurance, W. F. (2006a) A change in climate. *Tropinet* **17**, 1–3.

Laurance, W. F. (2006b) The value of trees. *New Scientist* **190**, 24–24.

Laurance, W. F. and Williamson, G. B. (2001) Positive feedbacks among forest fragmentation, drought, and climate change in the Amazon. *Conservation Biology* **15**, 1529–1535.

Laurance, W. F., Ferreira, L. V., Rankin-De Merona, J. M. and Laurance, S. G. (1998a) Rain forest fragmentation and the dynamics of Amazonian tree communities. *Ecology* **79**, 2032–2040.

Laurance, W. F., Laurance, S. G. and Delamonica, P. (1998b) Tropical forest fragmentation and greenhouse gas emissions. *Forest Ecology and Management* **110**, 173–180.

Laurance, W. F., Vasconcelos, H. L. and Lovejoy, T. E. (2000) Forest loss and fragmentation in the Amazon: Implications for wildlife conservation. *Oryx* **34**, 39–45.

Laurance, W. F., Albernaz, A. K. M. and Da Costa, C. (2001a) Is deforestation accelerating in the Brazilian Amazon? *Environmental Conservation* **28**, 305–311.

Laurance, W. F., Perez-Salicrup, D., Delamonica, P., Fearnside, P. M., D'Angelo, S., Jerozolinski, A., Pohl, L. and Lovejoy, T. E. (2001b) Rain forest fragmentation and the structure of Amazonian liana communities. *Ecology* **82**, 105–116.

Laurance, W. F., Lovejoy, T. E., Vasconcelos, H. L., Bruna, E. M., Didham, R. K., Stouffer, P. C., Gascon, C., Bierregaard, R. O., Laurance, S. G. and Sampaio, E. (2002) Ecosystem decay of Amazonian forest fragments: A 22-year investigation. *Conservation Biology* **16**, 605–618.

Lawton, J. H. and May, R. M. (1995) *Extinction Rates*. Oxford University Press, New York.

Lebedys, A. (2004) *Trends and Current Status of the Contribution of the Forest Sector to National Economies*. Forest Finance Working Paper, Food and Agriculture Organization of the United Nations, Rome.

Lee, C. E. (2002) Evolutionary genetics of invasive species. *Trends in Ecology and Evolution* **17**, 386–391.

Leck, C. F. (1979) Avian extinctions in an isolated tropical wet-forest preserve, Ecuador. *Auk* **96**, 343–352.

Lee, K. A. and Klasing, K. C. (2004) A role for immunology in invasion biology. *Trends in Ecology and Evolution* **19**, 523–529.

Lee, R. J., O'Brien, T. G., Kinnaird, M. F. and Dwiyahreni, A. A. (1999) *Impact of Wildlife hunting in Sulawesi, Indonesia, Technical Memorandum 4*. PKA/Wildlife Conservation Society Indonesia Program, Bogor.

Lee, T. M., Soh, M. C. K., Sodhi, N., Koh, L. P. and Lim, S. L. H. (2005) Effects of habitat disturbance on mixed species bird flocks in a tropical sub-montane rainforest. *Biological Conservation* **122**, 193–204.

Lee, T. M., Sodhi, N. S. and Prawiradilaga, D. M. (2007) Birds, local people, and protected areas in Sulawesi, Indonesia. In: *Biodiversity and Human Livelihoods in Protected Areas: Case Studies from the Malay Archipelago* (eds. N. S. Sodhi, G. Acciaioli, M. Erb and A. K. J. Tan), pp. 78–94. Cambridge University Press, Cambridge.

Legesse, D. and Ayenew, T. (2006) Effect of improper water and land resource utilization on the central Main Ethiopian Rift lakes. *Quaternary International* **148**, 8–18.

Lens, L., Van Dongen, S., Norris, K., Githiru, M. and Matthysen, E. (2002) Avian persistence in fragmented rainforest. *Science* **298**, 1236–1238.

Lever, C. (2001) *The Cane Toad: The History and Ecology of a Successful Colonist*. Westbury Academic and Scientific Publishing, Otley, UK.

Lever, C. (2004) The impact of traditional Chinese medicine on threatened species. *Oryx* **38**, 13–14.

Levine, J. S., Cofer, W. R., Cahoon, D. R. and L., W. E. (1995) Biomass burning: A driver for global change. *Environmental Science and Technology* **29**, 120A–125A.

Levins, R. (1970) Extinction. In: *Some Mathematical Questions in Biology* (ed. M. Gerstenhaber), pp. 77–107. American Mathematical Society, Providence, RI.

Lewis, J. B. (1977) Processes of organic production on coral reefs. *Biological Reviews of the Cambridge Philosophical Society* **52**, 305–347.

Lewison, R. L. and Crowder, L. B. (2003) Estimating fishery bycatch and effects on a vulnerable seabird population. *Ecological Applications* 13, 743–753.

Lewison, R. L., Crowder, L. B., Read, A. J. and Freeman, S. A. (2004a) Understanding impacts of fisheries bycatch on marine megafauna. *Trends in Ecology and Evolution* 19, 598–604.

Lewison, R. L., Freeman, S. A. and Crowder, L. B. (2004b) Quantifying the effects of fisheries on threatened species: The impact of pelagic longlines on loggerhead and leatherback sea turtles. *Ecology Letters* 7, 221–231.

Ley, J. A., Halliday, I. A., Tobin, A. J., Garrett, R. N. and Gribble, N. A. (2002) Ecosystem effects of fishing closures in mangrove estuaries of tropical Australia. *Marine Ecology Progress Series* 245, 223–238.

Lin, H. J., Dai, X. X., Shao, K. T., Su, H. M., Lo, W. T., Hsieh, H. L., Fang, L. S. and Hung, J. J. (2006) Trophic structure and functioning in a eutrophic and poorly flushed lagoon in southwestern Taiwan. *Marine Environmental Research* 62, 61–82.

Lindenmayer, D. B., Franklin, J. F. and Fischer, J. (2006) General management principles and a checklist of strategies to guide forest biodiversity conservation. *Biological Conservation* 131, 433–445.

Ling, S., Kumpel, N. and Albrechtsen, L. (2002) No new recipes for bushmeat. *Oryx* 36, 330–330.

Linthicum, K. J., Anyamba, A., Tucker, C. J., Kelley, P. W., Myers, M. F. and Peters, C. J. (1999) Climate and satellite indicators to forecast Rift Valley fever epidemics in Kenya. *Science* 285, 397–400.

Liow, L. H. (2000) Mangrove conservation in Singapore: A physical or a psychological impossibility? *Biodiversity and Conservation* 9, 309–332.

Lizaso, J. L. S., Goñi, R., Reñones, O., García Charton, J. A., Galzin, R., Bayle, J. T., Sánchez Jerez, P., Pérez Ruzafa, A. and Ramos, A. A. (2000) Density dependence in marine protected populations: A review. *Environmental Conservation* 27, 144–158.

Lockwood, J. L. (1999) Using taxonomy to predict success among introduced avifauna: Relative importance of transport and establishment. *Conservation Biology* 13, 560–567.

Lockwood, J. L., Brooks, T. M. and McKinney, M. L. (2000) Taxonomic homogenization of the global avifauna. *Animal Conservation* 3, 27–35.

Lodge, D. M. (2001) Lakes. In: *Future Scenarios of Global Biodiversity: Scenarios for the 21st Century* (eds. III, F. S. Chaplin, O. E. Sala and E. Huber-Sannwald), pp. 277–313. Springer-Verlag, New York.

Loibooki, M., Hofer, H., Campbell, K. L. I. and East, M. L. (2002) Bushmeat hunting by communities adjacent to the Serengeti National Park, Tanzania: The importance of livestock ownership and alternative sources of protein and income. *Environmental Conservation* 29, 391–398.

Loiselle, B. A. and Hoppes, W. G. (1983) Nest predation in insular and mainland lowland rainforest in Panama. *Condor* 85, 93–95.

Long, A. J. (1994) The importance of tropical montane cloud forests for endemic and threatened birds. In: *Tropical Montane Cloud Forests* (eds. L. S. Hamilton, O. J. James and F. N. Scatena), pp. 79–106. Springer-Verlag, New York.

Longhurst, A. R. and Pauly, D. (1987) *Ecology of Tropical Oceans.* Academic Press, San Diego.

Loreau, M., Oteng-Yeboah, A., Arroyo, M. T. K., Babin, D., Barbault, R., Donoghue, M., Gadgil, M., Hauser, C., Heip, C., Larigauderie, A., Ma, K., Mace, G., Mooney, H. A., Perrings, C., Raven, P., Sarukhan, J., Schei, P., Scholes, R. J. and Watson, R. T. (2006) Diversity without representation. *Nature* 442, 245–246.

Lotze, H. K., Lenihan, H. S., Bourque, B. J., Bradbury, R. H., Cooke, R. G., Kay, M. C., Kidwell, S. M., Kirby, M. X., Peterson, C. H. and Jackson, J. B. C. (2006) Depletion, degradation, and recovery potential of estuaries and coastal seas. *Science* 312, 1806–1809.

Lovejoy, T. E. (2006) Protected areas: A prism for a changing world. *Trends in Ecology and Evolution* 21, 329–333.

Lovejoy, T. E., Bierregaard, R. O. J., Rylands, A. B., Malcolm, J. R. and Quintela, C. E. (1997) Edge and other effects of isolation on Amazon forest fragments. In: *Tropical Forest Remnants: Ecology, Management and Conservation of Fragmented Communities* (eds. W. F. Laurance and R. O. J. Bierregaard), pp. 257–285. University of Chicago Press, Chicago.

Loya, Y. (2004) The coral reefs of Eilat – past, present and future: Three decades of coral community structure studies. In: *Coral Reef Health and Disease* (eds. E. Rosenberg and Y. Loya), pp. 1–34. Springer, Berlin.

Lubchenco, J., Palumbi, S. R., Gaines, S. D. and Andelman, S. (2003) Plugging a hole in the ocean: The emerging science of marine reserves. *Ecological Applications* 13, S3-S7.

Lundberg, G., Kottelat, M., Smith, G. R., Stiassny, M. L. J. and Gill, A. C. (2000) So many fishes, so little time: An overview of recent ichthyological discovery in continental waters. *Annals of the Missouri Botanical Gardens* **87**, 26–62.

MacArthur, R. H. and Wilson, E. O. (1967) *The Theory of Island Biogeography.* Princeton University Press, Prineton, NJ.

McClanahan, T. R. (1994) Kenyan coral reef lagoon fish: Effects of fishing, substrate complexity, and sea urchins. *Coral Reefs* **13**, 231–241.

McClenachan, L., Jackson, J. B. C. and Newman, M. J. H. (2006) Conservation implications of historic sea turtle nesting beach loss. *Frontiers in Ecology and the Environment* **4**, 290–296.

McCrary, J. K., van den Berghe, E. P., McKaye, K. R. and Lopez Perez, L. J. (2001) Tilapia cultivation: A threat to native fish species in Nicaragua. *Encuentro* **58**, 3–19.

McCurry, J. (1997) Physicians add their warnings to Kyoto summit. *Lancet* **350**, 1825–1825.

Mace, G. M., Balmford, A., Boitani, L., Cowlishaw, G., Dobson, A. P., Faith, D. P., Gaston, K. J., Humphries, C. J., Vane-Wright, R. I., Williams, P. H., Lawton, J. H., Margules, C. R., May, R. M., Nicholls, A. O., Possingham, H. P., Rahbek, C. and van Jaarsveld, A. S. (2000) It's time to work together and stop duplicating conservation efforts. *Nature* **405**, 393–393.

McGowan, P. J. K. and Garson, P. J. (2002) The Galliformes are highly threatened: Should we care? *Oryx* **36**, 311–312.

Mack, R. N., Simberloff, D., Lonsdale, W. M., Evans, H., Clout, M. and Bazzaz, F. A. (2000) Biotic invasions: Causes, epidemiology, global consequences, and control. *Ecological Applications* **10**, 689–710.

McKinney, M. L. (1997) Extinction vulnerability and selectivity: Combining ecological and paleontological views. *Annual Review of Ecology and Systematics* **28**, 495–516.

McKinney, M. L. (1999) High rates of extinction and threat in poorly studied taxa. *Conservation Biology* **13**, 1273–1281.

McKinney, M. L. and Lockwood, J. L. (1999) Biotic homogenization: A few winners replacing many losers in the next mass extinction. *Trends in Ecology and Evolution* **14**, 450–453.

McLain, D. K., Moulton, M. P. and Sanderson, J. G. (1999) Sexual selection and extinction: The fate of plumage-dimorphic and plumage-monomorphic birds introduced onto islands. *Evolutionary Ecology Research* **1**, 549–565.

McNeil, W. H. (1976) *Plagues and People.* Anchor Press/Doubleday, New York.

McPhee, R. D. E. (1999) *Extinctions in Near Time: Causes, Contexts, and Consequences.* Plenum, New York.

McPherson, J. M. and Vincent, A. C. J. (2004) Assessing East African trade in seahorse species as a basis for conservation under international controls. *Aquatic Conservation-Marine and Freshwater Ecosystems* **14**, 521–538.

Magrath, W. and Arens, P. (1989) The costs of soil erosion on Java: A natural resource accounting approach. In: *Environment Department Working Paper No. 18.* Policy Planning and Research Staff, World Bank, Washington, DC.

Magsalay, P., Brooks, T., Dutson, G. and Timmins, R. (1995) Extinction and conservation on Cebu. *Nature* **373**, 294–294.

Maguire, L. A. (1996) Making the role of values in conservation explicit: Values and conservation biology. *Conservation Biology* **10**, 914–916.

Mainka, S. A. and Mills, J. A. (1995) Wildlife and traditional Chinese medicine – supply and demand for wildlife species. *Journal of Zoo and Wildlife Medicine* **26**, 193–200.

Majluf, P., Babcock, E. A., Riveros, J. C., Schreiber, M. A. and Alderete, W. (2002) Catch and bycatch of sea birds and marine mammals in the small-scale fishery of Punta San Juan, Peru. *Conservation Biology* **16**, 1333–1343.

Malcolm, J. (1997) Biomass and diversity of small mammals in Amazonian forest fragments. In: *Tropical Forest Remnants: Ecology, Management and Conservation of Fragmented Communities* (eds. W. F. Laurance and R. O. Bierregaard), pp. 207–221. University of Chicago Press, Chicago.

Malcolm, J. R., Liu, C. R., Neilson, R. P., Hansen, L. and Hannah, L. (2006) Global warming and extinctions of endemic species from biodiversity hotspots. *Conservation Biology* **20**, 538–548.

Malhi, Y. and Grace, J. (2000) Tropical forests and atmospheric carbon dioxide. *Trends in Ecology and Evolution* **15**, 332–337.

Maliao, R. J., Webb, E. L. and Jensen, K. R. (2004) A survey of stock of the donkey's ear abalone,

Haliotis asinina L. in the Sagay Marine Reserve, Philippines: Evaluating the effectiveness of marine protected area enforcement. *Fisheries Research* **66**, 343–353.

Malmqvist, B. and Rundle, S. (2002) Threats to the running water ecosystems of the world. *Environmental Conservation* **29**, 134–153.

Mann, M. E., Bradley, R. S. and Hughes, M. K. (1998) Global-scale temperature patterns and climate forcing over the past six centuries. *Nature* **392**, 779–787.

Manne, L. L., Brooks, T. M. and Pimm, S. L. (1999) Relative risk of extinction of passerine birds on continents and islands. *Nature* **399**, 258–261.

Marini, M. A. (2001) Effects of forest fragmentation on birds of the Cerrado region, Brazil. *Bird Conservation International* **11**, 13–25.

Marod, D., Kutintara, U., Tanaka, H. and Nakashizuka, T. (2002) The effects of drought and fire on seed and seedling dynamics in a tropical seasonal forest in Thailand. *Plant Ecology* **161**, 41–57.

Marsh, D. M. and Pearman, P. B. (1997) Effects of habitat fragmentation on the abundance of two species of Leptodactylid frogs in an Andean montane forest. *Conservation Biology* **11**, 1323–1328.

Marsh, H., Harris, A. H. N. and Lawler, I. R. (1997) The sustainability of the indigenous dugong fishery in Torres Strait, Australia/Papua New Guinea. *Conservation Biology* **11**, 1375–1386.

Marsh, H., Eros, C., Corkeron, P. and Breen, B. (1999) A conservation strategy for dugongs: Implications of Australian research. *Marine and Freshwater Research* **50**, 979–990.

Marsh, H., Penrose, H. and Eros, C. (2003) A future for the dugong? In: *Marine Mammals. Fisheries, Tourism and Management Issues* (eds. N. J. Gales, M. A. Hindell and R. Kirkwood), pp. 383–399. CSIRO Publishing, Melbourne.

Marsh, H., De'Ath, G., Gribble, N. and Lane, B. (2005) Historical marine population estimates: Triggers or targets for conservation? The dugong case study. *Ecological Applications* **15**, 481–492.

Marte, C. L. (2003) Larviculture of marine species in Southeast Asia: Current research and industry prospects. *Aquaculture* **227**, 293–304.

Martin, P. (2005) *Twilight Of The Mammoths: Ice Age Extinctions and the Rewilding of America.* University of California Press, Berkeley.

Martin-Smith, K. M., Samoilys, M. A., Meeuwig, J. J. and Vincent, A. C. J. (2004) Collaborative development of management options for an artisanal fishery for seahorses in the central Philippines. *Ocean and Coastal Management* **47**, 165–193.

Mathooko, J. M. (2001) Disturbance of a Kenya Rift Valley stream by the daily activities of local people and their livestock. *Hydrobiologia* **458**, 131–139.

Matson, P. A., Vitousek, P. M., Ewel, J. J., Mazzarino, M. J. and Robertson, G. P. (1987) Nitrogen transformations following tropical forest felling and burning on a volcanic soil. *Ecology* **68**, 491–502.

Matson, P. A., Parton, W. J., Power, A. G. and Swift, M. J. (1997) Agricultural intensification and ecosystem properties. *Science* **277**, 504–509.

Matsuda, B. M. (1997) Conservation biology, values, and advocacy. *Conservation Biology* **11**, 1449–1450.

Matsuishi, T., Muhoozi, L., Mkumbo, O., Budeba, Y., Njiru, M., Asila, A., Othina, A. and Cowx, I. G. (2006) Are the exploitation pressures on the Nile perch fisheries resources of Lake Victoria a cause for concern? *Fisheries Management and Ecology* **13**, 53–71.

Matthews, E. (2001) *Understanding the FRA 2000.* World Resources Institute Forest Briefing No. 1, World Resources Institute, Washington.

Maxwell, A. L. (2004) Fire regimes in north-eastern Cambodian monsoonal forests, with a 9300-year sediment charcoal record. *Journal of Biogeography* **31**, 225–239.

Mayfield, M. M., Boni, M. E., Daily, G. C. and Ackerly, D. (2005) Species and functional diversity of native and human-dominated plant communities. *Ecology* **86**, 2365–2372.

Mayr, E. and Diamond, J. (2001) *The Birds of Northern Melansia: Speciation, Ecology and Biogeography.* Oxford University Press, Oxford.

Meehl, G. A., Washington, W. M., Collins, W. D., Arblaster, J. M., Hu, A. X., Buja, L. E., Strand, W. G. and Teng, H. Y. (2005) How much more global warming and sea level rise? *Science* **307**, 1769–1772.

Meekan, M. G., Bradshaw, C. J. A., Press, M., McLean, C., Richards, A., Quasnichka, S. and Taylor, J. G. (2006) Population size and structure of whale sharks (*Rhincodon typus*) at Ningaloo Reef, Western Australia. *Marine Ecology Progress Series* **319**, 275–285.

Meijaard, E. and Nijman, V. (2000) The local extinction of the proboscis monkey *Nasalis larvatus* in Pulau Kaget Nature Reserve, Indonesia. *Oryx* **34**, 66–70.

Meijaard, E., Sheil, D., Nasi, R., Augeri, D., Rosenbaum, B., Iskandar, D., Setyawati, T., Lammertink, M., Rachmatika, I., Wong, A., Soehartono, T., Stanley, S. and O'Brien, T. (2005) *Life After Logging: Reconciling Wildlife Conservation and Production Forestry in Indonesian Borneo.* CIFOR, Bogor, Indonesia.

Meltzer, D. G. A. (1993) Historical survey of disease problems in wildlife populations – southern African mammals. *Journal of Zoo and Wildlife Medicine* **24**, 237–244.

Mendelsohn, R. and Balick, M. (1995) Private property and rainforest conservation. *Conservation Biology* **9**, 1322–1323.

Mesquita, R. C. G., Delamonica, P. and Laurance, W. F. (1999) Effect of surrounding vegetation on edge-related tree mortality in Amazonian forest fragments. *Biological Conservation* **91**, 129–134.

Messel, H. and Vorlicek, G. C. (1986) Population dynamics and status of *Crocodylus porosus* in the tidal waterways of northern Australia. *Australian Wildlife Research* **13**, 71–111.

Michalski, F. and Peres, C. A. (2005) Anthropogenic determinants of primate and carnivore local extinctions in a fragmented forest landscape of southern Amazonia. *Biological Conservation* **124**, 383–396.

Milberg, P. and Tyrberg, T. (1993) Naive birds and noble savages: A review of man-caused prehistoric extinctions of island birds. *Ecography* **16**, 229–250.

Millennium Ecosystem Assessment (2005) *Ecosystems and Human Well-being: Synthesis.* Island Press, Washington, DC.

Miller, G. H., Fogel, M. L., Magee, J. W., Gagan, M. K., Clarke, S. J. and Johnson, B. J. (2005) Ecosystem collapse in Pleistocene Australia and a human role in megafaunal extinction. *Science* **309**, 287–290.

Miller, J. R. (2005) Biodiversity conservation and the extinction of experience. *Trends in Ecology and Evolution* **20**, 430–434.

Miller, L. and Douglas, B. C. (2006) On the rate and causes of twentieth century sea-level rise. *Philosophical Transactions of the Royal Society A: Mathematical, Physical and Engineering Sciences* **364**, 805–820.

Mills, A. J. and Fey, M. V. (2005) Interactive response of herbivores, soils and vegetation to annual burning in a South African savanna. *Austral Ecology* **30**, 435–444.

Mills, J. A. (1993) Tiger bone trade in South Korea. *Cat News* **19**, 13–16.

Mills, M. S. L. (2004) Bird community responses to savanna fires: Should managers be concerned? *South African Journal of Wildlife Research* **34**, 1–11.

Milner-Gulland, E. J. and Bennett, E. L. (2003) Wild meat: The bigger picture. *Trends in Ecology and Evolution* **18**, 351–357.

Milon, J. W. (2000) Pastures, fences, tragedies and marine reserves. *Bulletin of Marine Science* **66**, 901–916.

Milton, D. A. (2001) Assessing the susceptibility to fishing of populations of rare trawl bycatch: Sea snakes caught by Australia's Northern Prawn Fishery. *Biological Conservation* **101**, 281–290.

Mimura, N. (1999) Vulnerability of island countries in the South Pacific to sea level rise and climate change. *Climate Research* **12**, 137–143.

Mistry, J. and Berardi, A. (2005) Assessing fire potential in a Brazilian savanna nature reserve. *Biotropica* **37**, 439–451.

Mistry, J., Berardi, A., Andrade, V., Kraho, T., Kraho, P. and Leonardos, O. (2005) Indigenous fire management in the Cerrado of Brazil: The case of the Kraho of Tocantins. *Human Ecology* **33**, 365–386.

Mitchell, J. F. B., Lowe, J., Wood, R. A. and Vellinga, M. (2006) Extreme events due to human-induced climate change. *Philosophical Transactions of the Royal Society A: Mathematical, Physical and Engineering Sciences* **364**, 2117–2133.

Moegenburg, S. M. and Levey, D. J. (2002) Prospects for conserving biodiversity in Amazonian extractive reserves. *Ecology Letters* **5**, 320–324.

Mokaya, S. K., Mathooko, J. M. and Leichtfried, M. (2004) Influence of anthropogenic activities on water quality of a tropical stream ecosystem. *African Journal of Ecology* **42**, 281–288.

Møller, A. P. (1995) Patterns of fluctuating asymmetry in sexual ornaments of birds from marginal and central populations. *American Naturalist* **145**, 316–327.

Mölsä, H., Reynolds, J. E., Coenen, E. J. and Lindqvist, O. V. (1999) Fisheries research towardss resource management on Lake Tanganyika. *Hydrobiologia* 407, 1–24.

Mooney, H. A. and Cleland, E. E. (2001) The evolutionary impact of invasive species. *Proceedings of the National Academy of Sciences of the USA* 98, 5446–5451.

Moran, E. F. (1988) Following the Amazonian Highways. In: *People of the Rain Forest* (eds. J. S. Denslow and C. Padoch), pp. 155–162. University of California Press, Berkeley, CA.

Morris, A. V., Roberts, C. M. and Hawkins, J. P. (2000) The threatened status of groupers (Epinephelinae). *Biodiversity and Conservation* 9, 919–942.

Morrow, E. H. and Fricke, C. (2004) Sexual selection and the risk of extinction in mammals. *Proceedings of the Royal Society B: Biological Sciences* 271, 2395–2401.

Morton, B. (1989) Life-history characteristics and sexual strategy of *Mytilopsis sallei* (Bivalvia, Dreissenacea), introduced into Hong Kong. *Journal of Zoology* 219, 469–485.

Morton, B. (1994) Hong Kong's coral communities: Status, threats and management plans. *Marine Pollution Bulletin* 29, 74–83.

Morton, D.C., DeFries, R.S., Shimabukuro,Y.E., Anderson, L.O., Arai, E., Espirito-Santo, F.D., Freitas, R. and Moristte, J. (2006) Cropland expansion changes deforestation dynamics in the southern Brazilian Amazon. *Proceedings of National Academy of Sciences of the USA* 103, 14637–14641.

Morwood, M. J., Soejono, R. P., Roberts, R. G., Sutikna, T., Turney, C. S. M., Westaway, K. E., Rink, W. J., Zhao, J. X., van den Bergh, G. D., Due, R. A., Hobbs, D. R., Moore, M. W., Bird, M. I. and Fifield, L. K. (2004) Archaeology and age of a new hominid from Flores in eastern Indonesia. *Nature* 431, 1087–1091.

Mosquera, I., Côté, I. M., Jennings, S. and Reynolds, J. D. (2000) Conservation benefits of marine reserves for fish populations. *Animal Conservation* 3, 321–332.

Moss, B. (2000) Biodiversity in freshwaters: An issue of species preservation or system functioning? *Environmental Conservation* 27, 1–4.

Moyle, P. B. and Cech, J. J. (2004) *Fishes: An Introduction to Ichthyology*, 5th edn. Prentice Hall, Upper Saddle River, NJ.

Mrosovsky, N. (2000) Sustainable Use of Hawksbill Turtles: Contemporary Issues in Conservation. In: *Key Centre for Tropical Wildlife Management*, p. 107. Northern Territory University, Darwin.

Mrosovsky, N. (2003) Predicting extinction: Fundamental flaws in IUCN's Red List system, exemplified by the case of sea turtles. (http://members.seaturtle.org/mrosovsky/)

Mulliken, T. A., Broad, S. R. and Thomsen, J. B. (1992) The wild bird trade – an overview. In: *Perceptions, Conservation and Management of Wild Birds in Trade* (eds. J. B. Thomsen, S. R. Edwards and T. A. Mulliken), pp. 1–41. TRAFFIC International, Cambridge.

Mulrennan, M. E. and Woodroffe, C. D. (1998) Saltwater intrusion into the coastal plains of the lower Mary River, Northern Territory, Australia. *Journal of Environmental Management* 54, 169–188.

Mumby, P. J., Edwards, A. J., Arias-Gonzalez, J. E., Lindeman, K. C., Blackwell, P. G., Gall, A., Gorczynska, M. I., Harbourne, A. R., Pescod, C. L., Renken, H., Wabnitz, C. C. C. and Llewellyn, G. (2004) Mangroves enhance the biomass of coral reef fish communities in the Caribbean. *Nature* 427, 533–536.

Munday, P. L. (2004) Habitat loss, resource specialization, and extinction on coral reefs. *Global Change Biology* 10, 1642–1647.

Munn, C. A. and Terborgh, J. W. (1979) Multi-species territoriality in neotropical foraging flocks. *Condor* 81, 338–347.

Munro, J. L. (1983) Coral reef fish and fisheries of the Caribbean Sea. In: *Caribbean Coral Reef Fishery Resources* (ed. J. L. Munro), ICLARM Studies and Reviews 7, pp. 1–9. International Centre for Living Aquatic Resources Management, Manila, Philippines.

Murawski, S. A., Brown, R., Lai, H. L., Rago, P. R. and Hendrickson, L. (2000) Large-scale closed areas as a fishery management tool in temperate marine systems: The Georges Bank experience. *Bulletin of Marine Science* 66, 775–798.

Murcia, C. (1995) Edge effects in fragmented forests: Implications for conservation. *Trends in Ecology and Evolution* 10, 58–62.

Musick, J. A., Burgess, G., Cailliet, G., Camhi, M. and Fordham, S. (2000) Management of sharks and their relatives (Elasmobranchii). *Fisheries* 25, 9–13.

Myers, N. (1986) Tropical deforestation and a mega-extinction spasm. In: *Conservation Biology: The Science of Scarcity and Diversity* (ed. M. E. Soulé) pp. 394–409. Sinauer, Sunderland, MA.

Myers, N. (1988) Threatened biotas: 'Hotspots' in tropical forests. *Environmentalist* 8, 187–208.

Myers, N. (1990) The biodiversity challenge: Expanded hotspots analysis. *Environmentalist* 10, 243–256.

Myers, N. (1991) Tropical forests – present status and future outlook. *Climate Change* 19, 3–32.

Myers, N. (1996) Environmental services of biodiversity. *Proceedings of the National Academy of Sciences of USA* 93, 2764–2769.

Myers, N. (1997) Ecosystem services: What we know and what we expensively don't know. *AAAS Annual Meeting and Science Innovation Exposition* 163, A6.

Myers, N. (1998) Lifting the veil on perverse subsidies. *Nature* 392, 327–328.

Myers, N. (1999) Environmental scientists: Advocates as well? *Environmental Conservation* 26, 163–165.

Myers, N. and Knoll, A. H. (2001) The biotic crisis and the future of evolution. *Proceedings of the National Academy of Sciences of the USA* 98, 5389–5392.

Myers, N., Mittermeier, R. A., Mittermeier, C. G., da Fonseca, G. A. B. and Kent, J. (2000) Biodiversity hotspots for conservation priorities. *Nature* 403, 853–858.

Nagl, S., Tichy, H., Mayer, W. E., Takezaki, N., Takahata, N. and Klein, J. (2000) The origin and age of haplochromine fishes in Lake Victoria, East Africa. *Proceedings of the Royal Society B: Biological Sciences* 267, 1049–1061.

Naidoo, R. and Adamowicz, W. L. (2005) Economic benefits of biodiversity exceed costs of conservation at an African rainforest reserve. *Proceedings of the National Academy of Sciences of the USA* 102, 16712–16716.

National Research Council (1990) *Decline of the Sea Turtles: Causes and Prevention*. National Academy Press, Washington, DC.

Nauta, T. A., Bongco, A. E. and Santos-Borja, A. C. (2003) Set-up of a decision support system to support sustainable development of the Laguna de Bay, Philippines. *Marine Pollution Bulletin* 47, 211–219.

Naylor, R. and Ehrlich, P. (1997) The value of natural pest control services in agriculture. In: *Nature's Services: Societal Dependence on Natural Ecosystems* (ed. G. Daily), pp. 151–174. Island Press, Washington, DC.

Naylor, R. (1996) Invasions in agriculture: Assessing the cost of the golden apple snail in Asia. *Ambio* 25, 443–448.

Naylor, R. L. and Ehrlich, P. R. (1997) Natural pest control services and agriculture. In: *Nature's Services* (ed. G. C. Daily), pp. 151–174. Island Press, Washington, DC.

Naylor, R. T. (2005) The underworld of ivory. *Crime Law and Social Change* 42, 261–295.

Nee, S. and May, R. M. (1997) Extinction and the loss of evolutionary history. *Science* 278, 692–694.

Neilan, W., Catterall, C. P., Kanowski, J. and McKenna, S. (2006) Do frugivorous birds assist rainforest succession in weed dominated oldfield regrowth of subtropical Australia? *Biological Conservation* 129, 393–407.

Neill, C., Piccolo, M. C., Cerri, C. C., Steudler, P. A., Melillo, J. M. and Brito, M. (1997) Net nitrogen mineralization and net nitrification rates in soils following deforestation for pasture across the southwestern Brazilian Amazon Basin landscape. *Oecologia* 110, 243–252.

Neubert, M. G. (2003) Marine reserves and optimal harvesting. *Ecology Letters* 6, 843–849.

Neumann, R. P. (2001) Africa's 'last wilderness' reordering space for political and economic control in colonial Tanzania. *Africa* 71, 641–665.

Newmark, W. D. (1991) Tropical forest fragmentation and the local extinction of understory birds in the eastern Usambara mountains, Tanzania. *Conservation Biology* 5, 67–78.

Newmark, W. D. (2006) A 16-year study of forest disturbance and understory bird community structure and composition in Tanzania. *Conservation Biology* 20, 122–134.

Newsome, A. E. and Noble, I. R. (1986) Ecological and physiological characters of invading species. In: *Ecology of Biological Invasions* (eds. R. H. Groves and J. J. Burdon), pp. 1–20. Cambridge University Press, Cambridge.

Ng, P. K. L., Chou, L. M. and Lam, T. J. (1993) The status and impact of introduced freshwater animals in Singapore. *Biological Conservation* 64, 19–24.

Niesten, E. T., Rice, R. E., Ratay, S. M. and Paratore, K. (2004) *Commodities and Conservation: The Need for Greater Habitat Protection in the Tropics*. Centre for Applied Biodiversity Science, Conservation International, Washington DC.

Nijman, V. (2005) Decline of the endemic Hose's langur *Presbytis hosei* in Kayan Mentarang National Park, east Borneo. *Oryx* 39, 223–226.

Niklaus, P. A., Leadley, P. W., Schmid, B. and Korner, C. (2001) A long-term field study on biodiversity × elevated CO_2 interactions in grassland. *Ecological Monographs* 71, 341–356.

Nilsson, C. and Berggren, K. (2000) Alterations of riparian ecosystems caused by river regulation. *BioScience* 50, 783–792.

Njiforti, H. L. (1996) Preferences and present demand for bushmeat in north Cameroon: Some implications for wildlife conservation. *Environmental Conservation* 23, 149–155.

Norgrove, L. and Hulme, D. (2006) Confronting conservation at Mount Elgon, Uganda. *Development and Change* 37, 1093–1116.

Norris, D. R. (2004) Mosquito-borne diseases as a consequence of land use change. *EcoHealth* 1, 19–24.

Norris, K. and Harper, N. (2004) Extinction processes in hot spots of avian biodiversity and the targeting of pre-emptive conservation action. *Proceedings of the Royal Society B: Biological Sciences* 271, 123–130.

Noss, R. F. (1996) Conservation biology, values, and advocacy. *Conservation Biology* 10, 904–904.

Novotny, V. (1999) Diffuse pollution from agriculture – a worldwide outlook. *Water Science and Technology* 39, 1–13.

Nowell, K., Chyi, W.-L. and Pei, C.-J. (1992) *The Horns of a Dilemma: The Market for Rhino Horn in Taiwan*. TRAFFIC International, Cambridge.

Oakwood, M. (2004) The effect of cane toads on a marsupial carnivore, the northern quoll, *Dasyurus hallucatus*. Department of Environment and Heritage, Darwin.

O'Brien, T. G. and Kinnaird, M. F. (1996) Changing populations of birds and mammals in North Sulawesi. *Oryx* 30, 150–156.

O'Brien, T. G. and Kinnaird, M. F. (2003) Caffeine and conservation. *Science* 300, 587–587.

O'Brien, T. G., Kinnaird, M. F., Nurcahyo, A., Prasetyaningrum, M. and Iqbal, M. (2003) Fire, demography and the persistence of siamang (*Symphalangus syndactylus*: Hylobatidae) in a Sumatran rainforest. *Animal Conservation* 6, 115–121.

Odour, G. (1996) Biological pest control and invasives. In: *Proceedings of the Norway/UN Conference on Alien Species* (eds. O. T. Sandlund, P. J. Schei and A. Viken), pp. 116–122. Directorate for Nature Management and Norwegian Institute for Nature Research, Trondheim.

O'Dowd, D. J., Green, P. T. and Lake, P. S. (2003) Invasional 'meltdown' on an oceanic island. *Ecology Letters* 6, 812–817.

Ogden, J. C. (1980) Faunal relationships in Caribbean seagrass beds. In: *Handbook of Seagrass Biology: An Ecosystem Perspective* (eds. R. C. Phillips and C. P. McRoy), pp. 173–198. Garland STPM, New York.

OgutuOhwayo, R., Hecky, R. E., Cohen, A. S. and Kaufman, L. (1997) Human impacts on the African Great Lakes. *Environmental Biology of Fishes* 50, 117–131.

Ohsawa, M. (1995) Latitudinal comparison of altitudinal changes in forest structure, leaf-type, and species richness in humid monsoon Asia. *Vegetatio* 121, 3–10.

Olschewski, R., Tscharntke, T., Benitez, P. C., Schwarze, S. and Klein, A. M. (2006) Economic evaluation of pollination services comparing coffee landscapes in Ecuador and Indonesia. *Ecology and Society* 11, Article No. 7.

Olson, D. M. and Dinerstein, E. (1998) The global 200: A representation approach to conserving the Earth's most biologically valuable ecoregions. *Conservation Biology* 12, 502–515.

Ooi, J. B. (1976) *Peninsular Malaysia*. Longman, London.

Orme, C. D. L., Davies, R. G., Burgess, M., Eigenbrod, F., Pickup, N., Olson, V. A., Webster, A. J., Ding, T. S., Rasmussen, P. C., Ridgely, R. S., Stattersfield, A. J., Bennett, P. M., Blackburn, T. M., Gaston, K. J. and Owens, I. P. F. (2005) Global hotspots of species richness are not congruent with endemism or threat. *Nature* 436, 1016–1019.

Orr, J. C., Fabry, V. J., Aumont, O., Bopp, L., Doney, S. C., Feely, R. A., Gnanadesikan, A., Gruber, N., Ishida, A., Joos, F., Key, R. M., Lindsay, K., Maier-Reimer, E., Matear, R., Monfray, P., Mouchet, A., Najjar, R. G., Plattner, G. K., Rodgers, K. B., Sabine, C. L., Sarmiento, J. L., Schlitzer, R., Slater, R. D., Totterdell, I. J., Weirig, M. F., Yamanaka, Y. and Yool, A. (2005) Anthropogenic ocean acidification over the twenty-first century and its impact on calcifying organisms. *Nature* 437, 681–686.

Otway, N. M., Bradshaw, C. J. A. and Harcourt, R. G. (2004) Estimating the rate of quasi-extinction of the Australian grey nurse shark (*Carcharias taurus*) population using deterministic age- and stage-classified models. *Biological Conservation* 119, 341–350.

Owens, I. P. F. and Bennett, P. M. (2000) Ecological basis of extinction risk in birds: Habitat loss versus human persecution and introduced predators. *Proceedings of the National Academy of Sciences of the USA* 97, 12144–12148.

Pace, M. L., Cole, J. J., Carpenter, S. R. and Kitchell, J. F. (1999) Trophic cascades revealed in diverse ecosystems. *Trends in Ecology and Evolution* 14, 483–488.

Page, S. E. and Rieley, J. O. (1998) Tropical peatlands: a review of their natural resource functions with particular reference to Southeast Asia. *International Peat Journal* 8, 95–106.

Page, S. E., Siegert, F., Rieley, J. O., Boehm, H. D. V., Jaya, A. and Limin, S. (2002) The amount of carbon released from peat and forest fires in Indonesia during 1997. *Nature* 420, 61–65.

Pagiola, S. (2002) Paying for water services in Central America: Learning from Costa Rica. In: *Selling Forest Environmental Services* (eds. S. Pagiola, J. Bishop and N. Landell-Mills), pp. 37–62. Earthscan, London.

Palmer, M., Bernhardt, E., Chornesky, E., Collins, S., Dobson, A., Duke, C., Gold, B., Jacobson, R., Kingsland, S., Kranz, R., Mappin, M., Martinez, M. L., Micheli, F., Morse, J., Pace, M., Pascual, M., Palumbi, S., Reichman, O. J., Simons, A., Townsend, A. and Turner, M. (2004) Ecology for a crowded planet. *Science* 304, 1251–1252.

Pandolfi, J. M., Bradbury, R. H., Sala, E., Hughes, T. P., Bjorndal, K. A., Cooke, R. G., McArdle, D., McClenachan, L., Newman, M. J. H., Paredes, G., Warner, R. R. and Jackson, J. B. C. (2003) Global trajectories of the long-term decline of coral reef ecosystems. *Science* 301, 955–958.

Pardon, L. G., Brook, B. W., Griffiths, A. D. and Braithwaite, R. W. (2003) Determinants of survival for the northern brown bandicoot under a landscape-scale fire experiment. *Journal of Animal Ecology* 72, 106–115.

Parra-Olea, G., Martínez-Meyer, E. and de León, G. F. P. (2005) Forecasting climate change effects on salamander distribution in the highlands of central Mexico. *Biotropica* 37, 202–208.

Parshad, V. R. (1999) Rodent control in India. *Integrated Pest Management Reviews* 4, 97–126.

Pascual, M., Ahumada, J. A., Chaves, L. F., Rodo, X. and Bouma, M. (2006) Malaria resurgence in the East African highlands: Temperature trends revisited. *Proceedings of the National Academy of Sciences of USA* 103, 5829–5834.

Pataki, D. E. (2002) Atmospheric CO_2, climate and evolution – lessons from the past. *New Phytologist* 154, 10–12.

Paterson, R. (1979) Shark meshing takes a heavy toll of harmless marine animals. *Australian Fisheries* 38, 17–23.

Paterson, R. A. (1990) Effects of long-term anti-shark measures on target and non-target species in Queensland, Australia. *Biological Conservation* 52, 147–159.

Pattanavibool, A. and Dearden, P. (2002) Fragmentation and wildlife in montane evergreen forests, northern Thailand. *Biological Conservation* 107, 155–164.

Patterson, B. D. (2004) *The Lions of Tsavo: Exploring the Legacy of Africa's Notorious Man-eaters.* McGraw-Hill, New York.

Patterson, B. D., Kasiki, S. M., Selempo, E. and Kays, R. W. (2004) Livestock predation by lions (*Panthera leo*) and other carnivores on ranches neighbouring Tsavo National Parks, Kenya. *Biological Conservation* 119, 507–516.

Patz, J. A. and Olson, S. H. (2006) Climate change and health: Global to local influences on disease risk. *Annals of Tropical Medicine and Parasitology* 100, 535–549.

Pauly, D. (1998) Tropical fishes: Patterns and propensities. *Journal of Fish Biology* 53, 1–17.

Pauly, D. and Christensen, V. (1995) Primary production required to sustain global fisheries. *Nature* 374, 255–257.

Pauly, D., Silvestre, G. and Smith, I. R. (1989) On development, fisheries and dynamite: A brief review of tropical fisheries management. *Natural Resource Modeling* 3, 307–329.

Pauly, D., Christensen, V., Dalsgaard, J., Froese, R. and Torres, F. (1998) Fishing down marine food webs. *Science* 279, 860–863.

Pauly, D., Christensen, V., Guenette, S., Pitcher, T. J., Sumaila, U. R., Walters, C. J., Watson, R. and Zeller, D. (2002) Towards sustainability in world fisheries. *Nature* 418, 689–695.

Paxton, J. (2003) Shark nets in the spotlight. *Nature Australia* 27, 84.

Payne, J. (1995) Links between vertebrates and the conservation of Southeast Asian rainforests. In:

Ecology, Conservation and Management of Southeast Asian rainforests (eds. R. B. Primack and T. E. Lovejoy), pp. 54–65. Yale University Press, New Haven, CT.

Pearson, H. (2003) Lost forest fuels malaria. *Nature Science Update* (www.nature.com/ nsu/031124/031124-12.html).

Pearson, R. G. and Endean, R. (1969) A preliminary study of the coral predator *Acanthaster planci* (L.) (Asteroidea) on the Great Barrier Reef. *Department of Harbours and Marine, Queensland, Fisheries Notes* **3**, 27–55.

Pearson, R. G. (1981) Recovery and recolonization of coral reefs. *Marine Ecology Progress Series* **4**, 105–122.

Peh, K. S. H. (2007) Potential effects of climate change on altitudinal distributions in tropical birds. *Condor* **109**, 437–440.

Peh, K. S. H., de Jong, J., Sodhi, N. S., Lim, S. L. H. and Yap, C. A. M. (2005) Lowland rainforest avifauna and human disturbance: Persistence of primary forest birds in selectively logged forests and mixed-rural habitats of southern Peninsular Malaysia. *Biological Conservation* **123**, 489–505.

Pemberton, C. E. (1957) Progress in the biological control of undesirable plants in Hawaii. In: *Proceedings of the Ninth Pacific Science Congress*, pp. 124–126. Secretariat of Congress, Department of Science, Bangkok, Thailand.

Pereira, H. M. and Cooper, H. D. (2006) Towards the global monitoring of biodiversity change. *Trends in Ecology and Evolution* **21**, 123–129.

Peres, C. A. (2000) Effects of subsistence hunting on vertebrate community structure in Amazonian forests. *Conservation Biology* **14**, 240–253.

Peres, C. A. (2001) Synergistic effects of subsistence hunting and habitat fragmentation on Amazonian forest vertebrates. *Conservation Biology* **15**, 1490–1505.

Peres, C. A., Baider, C., Zuidema, P. A., Wadt, L. H. O., Kainer, K. A., Gomes-Silva, D. A. P., Salomao, R. P., Simoes, L. L., Franciosi, E. R. N., Valverde, F. C., Gribel, R., Shepard, G. H., Kanashiro, M., Coventry, P., Yu, D. W., Watkinson, A. R. and Freckleton, R. P. (2003) Demographic threats to the sustainability of Brazil nut exploitation. *Science* **302**, 2112–2114.

Perfecto, I., Vandermeer, J. H., Bautista, G. L., Nunez, G. I., Greenberg, R., Bichier, P. and Langridge, S. (2004) Greater predation in shaded coffee farms: The role of resident neotropical birds. *Ecology* **85**, 2677–2681.

Peters, H. A. (2001) *Clidemia hirta* invasion at the Pasoh Forest Reserve: An unexpected plant invasion in an undisturbed tropical forest. *Biotropica* **33**, 60–68.

Phat, N. K., Knorr, W. and Kim, S. (2004) Appropriate measures for conservation of terrestrial carbon stocks – analysis of trends of forest management in Southeast Asia. *Forest Ecology and Management* **191**, 283–299.

Phillips, B. L., Brown, G. P. and Shine, R. (2003) Assessing the potential impact of cane toads on Australian snakes. *Conservation Biology* **17**, 1738–1747.

Phillips, B. L., Brown, G. P., Webb, J. K. and Shine, R. (2006) Invasion and the evolution of speed in toads. *Nature* **439**, 803–803.

PHPA (1998) *Species Recovery Plan Yellow-crested Cockatoo*. PHPA/LIPI/Birdlife International-IP, Bogor, Indonesia.

Pimentel, D., Harvey, C., Resosudarmo, P., Sinclair, K., Kurz, D., McNair, M., Crist, S., Shpritz, L., Fitton, L., Saffouri, R. and Blair, R. (1995) Environmental and economic costs of soil erosion and conservation benefits. *Science* **267**, 117–1123.

Pimentel, D., Lach, L., Zuniga, R. and Morrison, D. (2000) Environmental and economic costs of nonindigenous species in the United States. *BioScience* **50**, 53–65.

Pimm, S. L. (1996) Lessons from a kill. *Biodiversity and Conservation* **5**, 1059–1067.

Pimm, S. L. and Raven, P. (2000) Extinction by numbers. *Nature* **403**, 843–845.

Pimm, S. L., Jones, H. L. and Diamond, J. (1988) On the risk of extinction. *American Naturalist* **132**, 757–785.

Pimm, S. L., Lawton, J. H. and Cohen, J. E. (1991) Food web patterns and their consequences. *Nature* **350**, 669–674.

Pimm, S. L., Russell, G. J., Gittleman, J. L. and Brooks, T. M. (1995) The future of biodiversity. *Science* **269**, 347–350.

Piotto, D., Montagnini, F., Ugalde, L. and Kanninen, M. (2003) Performance of forest plantations in small and medium-sized farms in the Atlantic lowlands of Costa Rica. *Forest Ecology and Management* **175**, 195–204.

Piotto, D., Viquez, E., Montagnini, F. and Kanninen, M. (2004) Pure and mixed forest plantations with native species of the dry tropics of Costa Rica: A comparison of growth and productivity. *Forest Ecology and Management* **190**, 359–372.

Ploetz, R. C. (2000) Panama disease: A classic and destructive disease of banana. *Plant Health Progress* DOI:10.1094/PHP-2000-1204-01-HM.

Plumptre, A. J., Bizumuremyi, J.-B., Uwimana, F. and Ndaruhebeye, J-D. (1997) The effects of the Rwandan civil war on poaching of ungulates in the Parc National des Volcans. *Oryx* **31**, 265–273.

Poiner, I. R. and Peterken, C. (1996) Seagrasses. In: *The State of the Marine Environment Report for Australia: The Marine Environment – Technical Annex: 1.* (eds. L. P. Zann and P. Kailola), pp. 107–117. Great Barrier Marine Park Authority, Townsville, Australia.

Polis, G. A., Sears, A. L. W., Huxel, G. R., Strong, D. R. and Maron, J. (2000) When is a trophic cascade a trophic cascade? *Trends in Ecology and Evolution* **15**, 473–475.

Polunin, N. V. C. and Roberts, C. M. (eds.) (1996) *Reef Fisheries.* Chapman & Hall, London.

Pomeroy, R. S. (1995) Community-based and co-management institutions for sustainable coastal fisheries management in southeast Asia. *Ocean and Coastal Management* **27**, 143–162.

Poonswad, P., Sukkasem, C., Phataramata, S., Hayeemuida, S., Plongmai, K., Chuailua, P., Thiensongrusame, P. and Jirawatkavi, N. (2005) Comparison of cavity modification and community involvement as strategies for hornbill conservation in Thailand. *Biological Conservation* **122**, 385–393.

Posey, D. A. (1999) *Cultural and Spiritual Values of Biodiversity.* Intermediate Technology Publications, Nairobi, and United Nations Environment Programme, London.

Postel, S. L. (1998) Water for food production: Will there be enough in 2025? *BioScience* **48**, 629–637.

Postel, S. and Richter, B. (2003) *Rivers for Life: Managing Water for People and Nature.* Island Press, Washington, DC.

Pounds, J. A., Bustamante, M. R., Coloma, L. A., Consuegra, J. A., Fogden, M. P. L., Foster, P. N., La Marca, E., Masters, K. L., Merino-Viteri, A., Puschendorf, R., Ron, S. R., Sanchez-Azofeifa, G. A., Still, C. J. and Young, B. E. (2006) Widespread amphibian extinctions from epidemic disease driven by global warming. *Nature* **439**, 161–167.

Pounds, J. A., Fogden, M. P. L. and Campbell, J. H. (1999) Biological response to climate change on a tropical mountain. *Nature* **398**, 611–615.

Powell, A. H. and Powell, G. V. N. (1987) Population dynamics of male euglossine bees in Amazonian forest fragments. *Biotropica* **19**, 176–179.

Powell, G. V. N. (1985) Sociobiology and adaptive significance of interspecific foraging flocks in the Neotropics. *Ornithology Monographs* **36**, 1013–1021.

Pratchett, M. S. (2001) Influence of coral symbionts on feeding preferences of crown-of-thorns starfish *Acanthaster planci* in the west Pacific. *Marine Ecology Progress Series* **214**, 111–119.

Pratchett, M. S. (2005) Dynamics of an outbreak population of *Acanthaster planci* at Lizard Island, northern Great Barrier Reef (1995–1999). *Coral Reefs* **24**, 453–462.

Pratt, D. G., Macmillan, D. C. and Gordon, I. J. (2004) Local community attitudes to wildlife utilization in the changing economic and social context of Mongolia. *Biodiversity and Conservation* **13**, 591–613.

Prawiradilaga, D. M. (1997) The maleo *Macrocephalon maleo* on Buton. *Bulletin of the British Ornithologists' Club* **117**, 237.

Preen, A. (1995) Impacts of dugong foraging on seagrass habitats: Observational and experimental evidence for cultivation grazing. *Marine Ecology Progress Series* **124**, 201–213.

Prior, L. D., Brook, B. W., Williams, R. J., Werner, P. A., Bradshaw, C. J. A. and Bowman, D. M. J. S. (2006) Environmental and allometric drivers of tree growth rates in a north Australian savanna. *Forest Ecology and Management.* **234**, 164–180.

Pulliam, H. R. (1988) Sources, sinks, and population regulation. *American Naturalist* **132**, 652–661.

Purvis, A., Agapow, P. M., Gittleman, J. L. and Mace, G. M. (2000) Nonrandom extinction and the loss of evolutionary history. *Science* **288**, 328–330.

Putz, F. E., Dykstra, D. P. and Heinrich, R. (2000) Why poor logging practices persist in the tropics. *Conservation Biology* **14**, 951–956.

Rabinowitz, A. (1995) Helping a species go extinct: The Sumatran rhino in Borneo. *Conservation Biology* 9, 482–488.

Rahel, F. J. (2002) Homogenization of freshwater faunas. *Annual Review of Ecology and Systematics* 33, 291–315.

Rainboth, W. J. (1999) *Fishes of the Cambodian Mekong.* Food and Agriculture Organization of the United Nations (FAO), Rome.

Rajamani, L., Cabanban, A. S. and Rahman, R. A. (2006) Indigenous use and trade of dugong (*Dugong dugon*) in Sabah, Malaysia. *Ambio* 35, 266–268.

Ramstein, G., Fluteau, F., Besse, J. and Joussaume, S. (1997) Effect of orogeny, plate motion and land sea distribution on Eurasian climate change over the past 30 million years. *Nature* 386, 788–795.

Randrianasolo, A., Miller, J. S. and Consiglio, T. K. (2002) Application of IUCN criteria and Red List categories to species of five Anacardiaceae genera in Madagascar. *Biodiversity and Conservation* 11, 1289–1300.

Rao, M., Rabinowitz, A. and Khaing, S. T. (2002) Status review of the protected-area system in Myanmar, with recommendations for conservation planning. *Conservation Biology* 16, 360–368.

Rao, M. H., Myint, T., Zaw, T. and Htun, S. (2005) Hunting patterns in tropical forests adjoining the Hkakaborazi National Park, north Myanmar. *Oryx* 39, 292–300.

Raup, D. M. (1991a) *Extinction: Bad Genes or Bad Luck?* W.W. Norton, New York.

Raup, D. M. (1991b) A kill curve for Phanerozoic marine apecies. *Paleobiology* 17, 37–48.

Ravelo, A. C., Andreasen, D. H., Lyle, M., Lyle, A. O. and Wara, M. W. (2004) Regional climate shifts caused by gradual global cooling in the Pliocene epoch. *Nature* 429, 263–267.

Read, A. J., Drinker, P. and Northridge, S. (2006) Bycatch of marine mammals in US and global fisheries. *Conservation Biology* 20, 163–169.

Reaser, J. K., Pomerance, R. and Thomas, P. O. (2000) Coral bleaching and global climate change: Scientific findings and policy recommendations. *Conservation Biology* 14, 1500–1511.

Reed, J. M. (1999) The role of behaviour in recent avian extinctions and endangerments. *Conservation Biology* 13, 232–241.

Reeves, R. R. (2002) The origins and character of 'aboriginal subsistence' whaling: A global review. *Mammal Review* 32, 71–106.

Refisch, J. and Koné, I. (2005) Impact of commercial hunting on monkey populations in the Tai region, Côte d'Ivoire. *Biotropica* 37, 136–144.

Reid, D. D. and Krogh, M. (1992) Assessment of catches from protective shark meshing off New South Wales beaches between 1950 and 1990. *Australian Journal of Marine and Freshwater Research* 43, 283–296.

Reid, W. V. (1992) How many species will there be? In: *Deforestation and Species Extinction* (eds. T. Whitmore and J. Sayer), pp. 55–73. Chapman & Hall, London.

Reid, W. V. (1998) Biodiversity hotspots. *Trends in Ecology and Evolution* 13, 275–280.

Renjifo, L. M. (1999) Composition changes in a subAndean avifauna after long-term forest fragmentation. *Conservation Biology* 13, 1124–1139.

Rhymer, J. M. and Simberloff, D. (1996) Extinction by hybridization and introgression. *Annual Review of Ecology and Systematics* 27, 83–109.

Ribon, R., Simon, J. E. and De Mattos, G. T. (2003) Bird extinctions in Atlantic forest fragments of the Vicosa region, southeastern Brazil. *Conservation Biology* 17, 1827–1839.

Richards, J. F. and Flint, E. P. (1994) A century of land use change in South and Southeast Asia. In: *Effects of Land Use Change on Atmospheric CO_2 Concentrations: South and Southeast Asia as a Case Study* (ed. D. H. Dale), pp. 15–66. Springer-Verlag, New York.

Richardson, D. M., Macdonald, I. A. W., Hoffmann, J. H. and Henderson, L. (1997) Alien plant invasions. In: *Vegetation of Southern Africa* (eds. R. M. Cowling, D. M. Richardson and S. M. Pierce), pp. 534–570. Cambridge University Press, Cambridge.

Ricketts, T. H., Daily, G. C., Ehrlich, P. R. and Michener, C. D. (2004) Economic value of tropical forest to coffee production. *Proceedings of the National Academy of Sciences of the USA* 34, 12579–12582.

Riley, J. (2002) Mammals on the Sangihe and Talaud Islands, Indonesia, and the impact of hunting and habitat loss. *Oryx* 36, 288–296.

Roberts, C. M. (1997a) Ecological advice for the global fisheries crisis. *Trends in Ecology and Evolution* 12, 35–38.

Roberts, C. M. (1997b) Connectivity and management of Caribbean coral reefs. *Science* 278, 1454–1457.

Roberts, C. M. (2003) Our shifting perspectives on the oceans. *Oryx* 37, 166–177.

Roberts, C. M., Bohnsack, J. A., Gell, F., Hawkins, J. P. and Goodridge, R. (2001) Effects of marine reserves on adjacent fisheries. *Science* 294, 1920.

Roberts, C. M., McClean, C. J., Veron, J. E. N., Hawkins, J. P., Allen, G. R., McAllister, D. E., Mittermeier, C. G., Schueler, F. W., Spalding, M., Wells, F., Vynne, C. and Werner, T. B. (2002) Marine biodiversity hotspots and conservation priorities for tropical reefs. *Science* 295, 1280–1284.

Robertson, J. M. Y. and van Schaik, C. P. (2001) Causal factors underlying the dramatic decline of the Sumatran orang-utan. *Oryx* 35, 26–38.

Robinson, J. G. and Bennett, E. L. (2002) Will alleviating poverty solve the bushmeat crisis? *Oryx* 36, 332–332.

Robinson, J. G. and Bennett, E. L. (2004) Having your wildlife and eating it too: An analysis of hunting sustainability across tropical ecosystems. *Animal Conservation* 7, 397–408.

Robinson, J. G. and Thorbjarnarson, J. (2000) Sustainable use of hawksbill turtles: Contemporary issues in conservation. *Nature* 404, 704–704.

Robinson, J. G., Redford, K. H. and Bennett, E. L. (1999) Wildlife harvest in logged tropical forests. *Science* 284, 595–596.

Robinson, J. M. (2001) The dynamics of avicultural markets. *Environmental Conservation* 28, 76–85.

Rodrigues, A. S. L. and Gaston, K. J. (2001) How large do reserve networks need to be? *Ecology Letters* 4, 602–609.

Rodrigues, A. S. L., Brooks, T. M. and Gaston, K. J. (2004a) Integrating phylogenetic diversity in the selection of priority areas for conservation: Does it really make a difference? In: *Phylogeny and Conservation* (eds. A. Purvis, J. L. Gittleman and T. M. Brooks), pp. 101–119. Cambridge University Press, Cambridge.

Rodrigues, A. S. L., Andelman, S. J., Bakarr, M. I., Boitani, L., Brooks, T. M., Cowling, R. M., Fishpool, L. D. C., da Fonseca, G. A. B., Gaston, K. J., Hoffmann, M., Long, J. S., Marquet, P. A., Pilgrim, J. D., Pressey, R. L., Schipper, J., Sechrest, W., Stuart, S. N., Underhill, L. G., Waller, R. W., Watts, M. E. J. and Yan, X. (2004b) Effectiveness of the global protected area network in representing species diversity. *Nature* 428, 640–643.

Rodriguez, I. (2004) Indigenous versus scientific knowledge: The conflict over the use of fire in Canaima National Park, Venezuela. *Interciencia* 29, 121–129.

Rodriguez, J. P. (2000) Impact of the Venezuelan economic crisis on wild populations of animals and plants. *Biological Conservation* 96, 151–159.

Roessig, J. M., Woodley, C. M., Cech, J. J. and Hansen, L. J. (2004) Effects of global climate change on marine and estuarine fishes and fisheries. *Reviews in Fish Biology and Fisheries* 14, 251–275.

Roman, J. and Palumbi, S. R. (2003) Whales before whaling in the North Atlantic. *Science* 301, 451.

Root, T. L., Price, J. T., Hall, K. R., Schneider, S. H., Rosenzweig, C. and Pounds, J. A. (2003) Fingerprints of global warming on wild animals and plants. *Nature* 421, 57–60.

Rose, C. S. and Risk, M. J. (1985) Increase in *Cliona delitrix* infestation of *Montastrea cavernosa* heads on an organically polluted portion of the Grand Cayman fringing reef. *Pubblicazioni della Stazione Zoologica di Napoli. I: Marine Ecology* 6, 345–362.

Rosenbaum, B., O'Brien, T. G., Kinnaird, M. and Supriatna, J. (1998) Population densities of Sulawesi crested black macaques (*Macaca nigra*) on Bacan and Sulawesi, Indonesia: Effects of habitat disturbance and hunting. *American Journal of Primatology* 44, 89–106.

Rosenmeier, M. F., Brenner, M., Kenney, W. F., Whitmore, T. J. and Taylor, C. M. (2004) Recent eutrophication in the Southern Basin of Lake Peten Itza, Guatemala: Human impact on a large tropical lake. *Hydrobiologia* 511, 161–172.

Rossiter, N. A., Setterfield, S. A., Douglas, M. M. and Hutley, L. B. (2003) Testing the grass-fire cycle: Alien grass invasion in the tropical savannas of northern Australia. *Diversity and Distributions* 9, 169–176.

Roughgarden, J., Gaines, S. and Possingham, H. (1988) Recruitment dynamics in complex life cycles. *Science*, 241, 1460–1466.

Rouphael, A. B. and Inglis, G. J. (1997) Impacts of recreational scuba diving at sites with different reef topographies. *Biological Conservation* 82, 329–336.

Rowcliffe, M. (2002) Bushmeat and the biology of conservation. *Oryx* 36, 331.

Rowe, S. and Hutchings, J. A. (2003) Mating systems and the conservation of commercially exploited marine fish. *Trends in Ecology and Evolution* 18, 567–572.

Ruangpanit, N. (1995) Tropical seasonal forests in monsoon Asia: Emphasis on continental Southeast Asia. *Vegetatio* 121, 31–40.

Rudd, M. A., Tupper, M. H., Folmer, H. and van Kooten, G. C. (2003) Policy analysis for tropical marine reserves: Challenges and directions. *Fish and Fisheries* 4, 65–85.

Ruitenbeek, H. J. (1992) *Mangrove Management: An Economic Analysis of Management Options with a Focus on Bintuni Bay, Irian Jaya.* Environmental Mangement Development in Indonesia Project, Environmental Reports No. 8.

Runion, G. B. (2003) Climate change and plant pathosystems – future disease prevention starts here. *New Phytologist* 159, 531–533.

Ruttenberg, B. I. (2001) Effects of artisanal fishing on marine communities in the Galápagos Islands. *Conservation Biology* 15, 1691–1699.

Saarnak, C. F. (2001) A shift from natural to human-driven fire regime: Implications for trace-gas emissions. *Holocene* 11, 373–375.

Saberwal, V. K., Gibbs, J. P., Chellam, R. and Johnsingh, A. J. T. (1994) Lion-human conflict in the Gir Forest, India. *Conservation Biology* 8, 501–507.

Sabine, C. L., Feely, R. A., Gruber, N., Key, R. M., Lee, K., Bullister, J. L., Wanninkhof, R., Wong, C. S., Wallace, D. W. R., Tilbrook, B., Millero, F. J., Peng, T. H., Kozyr, A., Ono, T. and Rios, A. F. (2004) The oceanic sink for anthropogenic CO_2. *Science* 305, 367–371.

Sadovy, Y. (1994) Grouper stocks of the western central Atlantic: The need for management and management needs. *Proceedings of the Gulf and Caribbean Fisheries Institute* 43, 43–64.

Sadovy, Y. J. and Vincent, A. C. J. (2002) Ecological issues and the trades in live reef fishes. In: *Coral Reef Fishes: Dynamics and Diversity in a Complex Ecosystem* (ed. P. F. Sale), pp. 391–420. Academic Press, San Diego, CA.

Sadovy, Y. J., Donaldson, T. J., Graham, T. R., McGilvray, F., Muldoon, G. J., Phillips, M. J., Rimmer, M. A., Smith, A. and Yeeting, B. (2003) *While Stocks Last: The Live Reef Food Fish Trade.* Asian Development Bank, Manila, Philippines.

Safford, R. J. and Jones, C. G. (1997) Did organochlorine pesticide use cause declines in Mauritian forest birds? *Biodiversity and Conservation* 6, 1445–1451.

Saila, S. A. (1983) Importance and assessment of discards in commercial fisheries. *FAO Fisheries Circular* 765, 1–62.

Sakai, A. K., Allendorf, F. W., Holt, J. S., Lodge, D. M., Molofsky, J., With, K. A., Baughman, S., Cabin, R. J., Cohen, J. E., Ellstrand, N. C., McCauley, D. E., O'Neil, P., Parker, I. M., Thompson, J. N. and Weller, S. G. (2001) The population biology of invasive species. *Annual Review of Ecology and Systematics* 32, 305–332.

Sala, E., Aburto-Oropeza, O., Paredes, G., Barrera, J. C. and Dayton, P. K. (2002) A general model for designing networks of marine reserves. *Science* 298, 1991–1993.

Sala, O. E., Chaplin, F. S., III, Armesto, J. J., Berlow, E., Bloomfield, J., Dirzo, R., Huber-Sanwald, E., Huenneke, L. F., Jackson, R. B., Kinzig, A., Leemans, R., Lodge, D. M., Mooney, H. A., Oesterheld, M., Poff, N. L., Sykes, M. T., Walker, B. H., Walker, M. and Wall, D. (2000) Global biodiversity scenarios for the year 2100. *Science* 287, 1770–1774.

Salafsky, N., Dugelby, F. L. and Terborgh, J. W. (1993) Can extractive reserves save the rainforest? An ecological and socioeconomic comparison of non-timber forest product extraction systems in Petén, Guatemala, and West Kalimantan, Indonesia. *Conservation Biology* 7, 39–52.

Sánchez-Azofeifa, G. A., Quesada-Mateo, C., Gonzalez-Quesada, P., Dayanandan, S. and Bawa, K. S. (1999) Protected areas and conservation of biodiversity in the tropics. *Conservation Biology* 13, 407–411.

Sánchez-Pérez, J. M. and Trémolières, M. (1997) Variation in nutrient levels of the groundwater in the Upper Rhine alluvial forests as a consequence of hydrological regime and soil texture. *Global Ecology and Biogeography Letters* 6, 211–217.

Sánchez-Pérez, J. M., Trémolières, M. and Carbiener, R. (1991a) A site of natural purification for phosphates and nitrates carried by the Rhine flood waters – the alluvial ash-elm forest. *Comptes Rendus de L'Académie des Sciences Serie III – Sciences de la Vie* 312, 395–402.

Sánchez-Pérez, J. M., Trémolières, M., Schnitzler, A. and Carbiener, R. (1991b) Evolution de la qualité physico-chimique des eaux de la frange superficielle de la nappe phréatique en fonction du cycle saisonnier et des stades de succession des forêts alluviales rhénanes (*Querco ulmetum minoris* Issl. 24). *Acta Oecologica* **12**, 581–601.

Santilli, M., Moutinho, P., Schwartzman, S., Nepstad, D., Curran, L. and Nobre, C. (2005) Tropical deforestation and the Kyoto Protocol. *Climatic Change* **71**, 267–276.

Sastry, N. (2002) Forest fires, air pollution, and mortality in Southeast Asia. *Demography* **39**, 1–23.

Saunders, C. D., Brook, A. T. and Myers, O. E. (2006) Using psychology to save biodiversity and human well-being. *Conservation Biology* **20**, 702–705.

Saunders, D. A., Hobbs, R. J. and Margules, C. R. (1991) Biological consequences of ecosystem fragmentation: A review. *Conservation Biology* **5**, 18–32.

Saunders, K. (1997) Guarding against illegal foreign fishing activity in Australia's northern waters. *Western Fisheries*, 27–30.

Scanlan, J. C., Berman, D. M. and Grant, W. E. (2006) Population dynamics of the European rabbit (*Oryctolagus cuniculus*) in north eastern Australia: Simulated responses to control. *Ecological Modelling* **196**, 221–236.

Schafer, E. H. (1963) *The Golden Peaches of Samarkand: A Study of T'ang Exotics*. University of California Press, Berkeley, CA.

Scheren, P. A. G. M., Zanting, H. A. and Lemmens, A. M. C. (2000) Estimation of water pollution sources in Lake Victoria, East Africa: Application and elaboration of the rapid assessment methodology. *Journal of Environmental Management* **58**, 235–248.

Schlaepfer, M. A. and Gavin, T. A. (2001) Edge effects on lizards and frogs in tropical forest fragments. *Conservation Biology* **15**, 1079–1090.

Schlesinger, W. H. (1991) *Biogeochemistry: An Analysis of Global Change*. Academic Press, San Diego.

Schultz, P. W. (2000) Empathizing with nature: The effects of perspective taking on concern for environmental issues. *Journal of Social Issues* **56**, 391–406.

Schweithelm, J. (1998) *The Fire This Time. An Overview of Indonesia's Forest Fires in 1997/1998*. WWF Indonesia, Jakarta.

Scott, J. M., Kepler, C. B., van Riper III, C. and Fefer, S. I. (1988) Conservation of Hawaii's vanishing avifauna. *BioScience* **38**, 238–253.

Sebens, K. P. (1994) Biodiversity of coral reefs: What are we losing and why? *American Zoologist* **34**, 115–133.

Sechrest, W., Brooks, T. M., da Fonseca, G. A. B., Konstant, W. R., Mittermeier, R. A., Purvis, A., Rylands, A. B. and Gittleman, J. L. (2002) Hotspots and the conservation of evolutionary history. *Proceedings of the National Academy of Sciences of the USA* **99**, 2067–2071.

Seidenschwarz, F. (1986) Comparison of riverside herb communities with weed vegetation in the tropical lowlands of Peru. *Amazoniana-Limnologia et Oecologia Regionalis Systemae Fluminis Amazonas* **10**, 79–111.

Seidensticker, J., Christie, S. and Jackson, P. (1999) *Riding the Tiger: Tiger Conservation in Human-dominated Landscapes*. Cambridge University Press, Cambridge, United Kingdom.

Seitre, R. and Seitre, J. (1992) Causes of land bird extinction in French Polynesia. *Oryx* **26**, 215–222.

Sekercioglu, C. H. (2002) Impacts of birdwatching on human and avian communities. *Environmental Conservation* **29**, 282–289.

Sekercioglu, C. H. (2006) Increasing awareness of avian ecological function. *Trends in Ecology and Evolution* **21**, 464–471.

Sekercioglu, C. H., Daily, G. C. and Ehrlich, P. R. (2004) Ecosystem consequences of bird declines. *Proceedings of the National Academy of Sciences of the USA* **101**, 18042–18047.

Sekercioglu, C. H., Ehrlich, P. R., Daily, G. C., Aygen, D., Goehring, D. and Sandi, R. F. (2002) Disappearance of insectivorous birds from tropical forest fragments. *Proceedings of the National Academy of Sciences of the USA* **99**, 263–267.

Seneca Creek Associates, LLC and Wood Resources International (2004) 'Illegal' Logging and Global Wood Markets: The Competitive Impacts on the U.S. Wood Products Industry. Poolesville, MD, Seneca Creek Associates.

Shaanker, R. U., Ganeshaiah, K. N., Krishnan, S., Ramya, R., Meera, C., Aravind, N. A., Kumar, A., Rao, D., Vanaraj, G., Ramachandra, J., Gauthier, R., Ghazoul, J., Poole, N. and Reddy, B. V. C. (2004) Livelihood gains and ecological costs of non-timber forest product dependence: Assessing the roles of dependence, ecological knowledge and market structure in three contrasting human and ecological settings in south India. *Environmental Conservation* **31**, 242–253.

Shaffer, M. L. (1981) Minimum population sizes for species conservation. *BioScience* **31**, 131–134.

Shahabuddin, G., Herzner, G. A., Aponte, C. R. and Gomez, M. D. C. (2000) Persistence of a frugivorous butterfly species in Venezuelen forest fragments: The role of movement and habitat quality. *Biodiversity and Conservation* **9**, 1623–1641.

Shanker, K., Hiremath, A. and Bawa, K. (2005) Linking biodiversity conservation and livelihoods in India. *PLoS Biology* **3**, e1878.

Shanley, P. and Luz, L. (2003) The impacts of forest degradation on medicinal plant use and implications for health care in eastern Amazonia. *BioScience* **53**, 573–584.

Shapiro, A. M. (2002) The Californian urban butterfly fauna is dependent on alien plants. *Diversity and Distributions* **8**, 31–40.

Sheil, D. and Lawrence, A. (2004) Tropical biologists, local people and conservation: New opportunities for collaboration. *Trends in Ecology and Evolution* **19**, 634–638.

Shepherd, C. R. and Magnus, N. (2004) *Nowhere to Hide: The Trade in Sumatran Tiger. A TRAFFIC Southeast Asia Report*. WWF, Fauna and Flora International, Wildlife Conservation Society.

Shepherd, S., McComb, A. J., Bulthuis, D. A., Neverauskas, V., Steffensen, D. A. and West, R. (1989) Decline of seagrass. In: *Biology of Seagrasses: A Treatise on the Biology of Seagrasses with Special Reference to the Australian Region* (eds. A. W. D. Larkum, A. J. McComb and S.Shepherd), pp. 346–393. Elsevier, Amsterdam.

Sheppard, C. R. C. (1999) Coral decline and weather patterns over 20 years in the Chagos Archipelago, Central Indian Ocean. *Ambio* **28**, 472–478.

Sheppard, C. R. C. (2003) Predicted recurrences of mass coral mortality in the Indian Ocean. *Nature* **425**, 294–297.

Sheuyange, A., Oba, G. and Weladji, R. B. (2005) Effects of anthropogenic fire history on savanna vegetation in northeastern Namibia. *Journal of Environmental Management* **75**, 189–198.

Shi, H., Singh, A., Kant, S., Zhu, Z. L. and Waller, E. (2005) Integrating habitat status, human population pressure, and protection status into biodiversity conservation priority setting. *Conservation Biology* **19**, 1273–1285.

Shine, R., Ambariyanto, Harlow, P. S. and Mumpuni (1999) Reticulated pythons in Sumatra: Biology, harvesting and sustainability. *Biological Conservation* **87**, 349–357.

Shivji, M. S., Chapman, D. D., Pikitch, E. K. and Raymond, P. W. (2005) Genetic profiling reveals illegal international trade in fins of the great white shark, *Carcharodon carcharias*. *Conservation Genetics* **6**, 1035–1039.

Short, F. T. and Neckles, H. A. (1999) The effects of global climate change on seagrasses. *Aquatic Botany* **63**, 169–196.

Shukla, J. B. and Dubey, B. (1996) Effect of changing habitat on species: Application to Keoladeo National Park, India. *Ecological Modelling* **86**, 91–99.

Siegert, F., Ruecker, G., Hinrichs, A. and Hoffmann, A. A. (2001) Increased damage from fires in logged forests during droughts caused by El Niño. *Nature* **414**, 437–440.

Sieving, K. E. and Karr, J. R. (1997) Avian extinction and persistence mechanisms in lowland Panama. In: *Tropical Forest Remnants: Ecology, Management and Conservation of Fragmented Communities* (eds. W. F. Laurance and R. O. Bierregaard), pp. 156–170. University of Chicago Press, Chicago.

Sieving, K. E. (1992) Nest predation and differential insular extinction among selected forest birds of central Panama. *Ecology* **73**, 2310–2328.

Simberloff, D. (1986) Are we on the verge of a mass extinction in tropical rainforests? In: *Dynamics of Extinction* (ed. D. K. Elliott), pp. 165–180. Wiley-Interscience, New York.

Simberloff, D. (1992a) Do species–area curves predict extinction in fragmented forests? In: *Tropical Deforestation and Species Extinction* (eds. T. C. Whitmore and J. A. Sayer), pp. 75–89. Chapman & Hall, London.

Simberloff, D. (1992b) Species–area relationships, fragmentation, and extinction in tropical forests. *Malayan Nature Journal* **45**, 398–413.

Simberloff, D. (1995) Habitat fragmentation and population extinction of birds. *Ibis* **137**, S105-S111.

Simberloff, D. (1997) The biology of invasions. In: *Strangers in Paradise: Impact and Management of Nonindigenous Species in Florida* (eds. D. Simberloff, D. C. Schmitz and T. C. Brown), pp. 3–17. Island Press, Washington, DC.

Simberloff, D. (2001) Biological invasions – how are they affecting us, and what can we do about them? *Western North American Naturalist* **61**, 308–315.

Simberloff, D. (2003a) Confronting introduced species: A form of xenophobia? *Biological Invasions* **5**, 179–192.

Simberloff, D. (2003b) Eradication-preventing invasions at the outset. *Weed Science* **51**, 247–253.

Simberloff, D. (2003c) How much information on population biology is needed to manage introduced species? *Conservation Biology* **17**, 83–92.

Simberloff, D. (2006) Invasional meltdown six years later: Important phenomenon, unfortunate metaphor, or both? *Ecology Letters* **9**, 912–919.

Simberloff, D. S. and Abele, L. G. (1976a) Island biogeography and conservation: Strategy and limitations. *Science* **193**, 1032.

Simberloff, D. S. and Abele, L. G. (1976b) Island biogeography theory and conservation practice. *Science* **191**, 285–286.

Simberloff, D. and Cox, J. (1987) Consequences and costs of conservation corridors. *Conservation Biology* **1**, 63–71.

Simberloff, D. and Gibbons, L. (2004) Now you see them, now you don't – population crashes of established introduced species. *Biological Invasions* **6**, 161–172.

Simberloff, D. and Von Holle, B. (1999) Positive interactions of non-indigenous species: Invasional meltdown? *Biological Invasions* **1**, 21–32.

Simberloff, D., Parker, I. M. and Windle, P. N. (2005) Introduced species policy, management, and future research needs. *Frontiers in Ecology and the Environment* **3**, 12–20.

Simpfendorfer, C. (1992) Biology of tiger sharks (*Galeocerdo cuvier*) caught by the Queensland Shark Meshing Program off Townsville, Australia. *Australian Journal of Marine and Freshwater Research* **43**, 33–43.

Sinclair, A. R. E., Mduma, S. A. R. and Arcese, P. (2002) Protected areas as biodiversity benchmarks for human impact: Agriculture and the Serengeti avifauna. *Proceedings of the Royal Society B: Biological Sciences* **269**, 2401–2405.

Sist, P., Sheil, D., Kartawinata, K. and Priyadi, H. (2003) Reduced-impact logging in Indonesian Borneo: Some results confirming the need for new silvicultural prescriptions. *Forest Ecology and Management* **179**, 415–427.

Sizer, N. and Plouvier, D. (2000) *Increased investment and trade by transnational logging companies in Africa, the Carribean and the Pacific: Implications for the sustainable management and conservation of tropical forests.* WWF, WRI, Safran, Brussels.

Sizer, N. and Tanner, E. V. J. (1999) Responses of woody plant seedlings to edge formation in a lowland tropical rainforest, Amazonia. *Biological Conservation* **91**, 135–142.

Sladek Nowlis, J. and Roberts, C. M. (1999) Fisheries benefits and the optimum design of marine reserves. *Fishery Bulletin* **97**, 604–616.

Slik, J. W. F., Verburg, R. W. and Kessler, P. J. A. (2002) Effects of fire and selective logging on the tree species composition of lowland dipterocarp forest in East Kalimantan, Indonesia. *Biodiversity and Conservation* **11**, 85–98.

Smethurst, D. and Nietschmann, B. (1999) The distribution of manatees (*Trichechus manatus*) in the coastal waterways of Tortuguero, Costa Rica. *Biological Conservation* **89**, 267–274.

Smith, A. P. (1997) Deforestation, fragmentation, and reserve design in western Madagascar. In: *Tropical Forest Remnants: Ecology, Management and Conservation of Fragmented Communities* (eds. W. F. Laurance and R. O. Bierregaard), pp. 415–441. University of Chicago Press, Chicago.

Smith, B. D., Haque, A., Hossain, M. S. and Khan, A. (1998) River dolphins in Bangladesh: Conservation and the effects of water development. *Environmental Management* **22**, 323–335.

Smith, C. W. I. (2000) Impact of alien plants on Hawai'i's native biota. PhD thesis, University of Hawaii, Manoa.

Smith, J., Obidzinski, K., Subarudi, and Suramenggala, I. (2003) Illegal logging, collusive corruption and fragmented governments in Kalimantan, Indonesia. *International Forestry Review* **5**, 293–312.

Smith, R. J., Muir, R. D. J., Walpole, M. J., Balmford, A. and Leader-Williams, N. (2003) Governance and the loss of biodiversity. *Nature* **426**, 67–70.

Smith, S. D., Huxman, T. E., Zitzer, S. F., Charlet, T. N., Housman, D. C., Coleman, J. S., Fenstermaker, L. K., Seemann, J. R. and Nowak, R. S. (2000) Elevated CO_2 increases productivity and invasive species success in an arid ecosystem. *Nature* **408**, 79–82.

Smith, S. E., Au, D. W. and Show, C. (1998) Intrinsic rebound potential of 26 species of Pacific sharks. *Marine and Freshwater Research* **49**, 663–678.

So, N., Van Houdt, J. K. J. and Volckaert, F. A. M. (2006) Genetic diversity and population history of the migratory catfishes *Pangasianodon hypophthalmus* and *Pangasius bocourti* in the Cambodian Mekong River. *Fisheries Science* 72, 469–476.

Soares-Filho, B. S., Nepstad, D. C., Curran, L. M., Cerqueira, G. C., Garcia, R. A., Ramos, C. A., Voll, E., McDonald, A., Lefebvre, P. and Schlesinger, P. (2006) Modelling conservation in the Amazon basin. *Nature* 440, 520–523.

Sodhi, N. S. (2002) A comparison of bird communities of two fragmented and two continuous Southeast Asian rainforests. *Biodiversity and Conservation* 11, 1105–1119.

Sodhi, N. S. and Brook, B. W. (2006) *Southeast Asia Biodiversity in Crisis.* Cambridge University Press, Cambridge.

Sodhi, N. S. and Liow, L. H. (2000) Improving conservation biology research in Southeast Asia. *Conservation Biology* 14, 1211–1212.

Sodhi, N. S. and Sharp, I. (2006) *Winged Invaders: Pest Birds of the Asia Pacific.* SNP Reference, Singapore.

Sodhi, N. S., Peh, K. S. H., Lee, T. M., Turner, I. M., Tan, H. T. W., Prawiradilaga, D. M. and Darjono (2003) Artificial nest and seed predation experiments on tropical southeast Asian islands. *Biodiversity and Conservation* 12, 2415–2433.

Sodhi, N. S., Liow, L. H. and Bazzaz, F. A. (2004a) Avian extinctions from tropical and subtropical forests. *Annual Review of Ecology, Evolution, and Systematics* 35, 323–345.

Sodhi, N. S., Koh, L. P., Brook, B. W. and Ng, P. K. L. (2004b) Southeast Asian biodiversity: An impending disaster. *Trends in Ecology and Evolution* 19, 654–660.

Sodhi, N. S., Soh, M. C. K., Prawiradilaga, D. M., Darjono and Brook, B. W. (2005a) Persistence of lowland rainforest birds in a recently logged area in central Java. *Bird Conservation International* 15, 173–191.

Sodhi, N. S., Lee, T. M., Koh, L. P. and Dunn, R. R. (2005b) A century of avifaunal turnover in a small tropical rainforest fragment. *Animal Conservation* 8, 217–222

Sodhi, N. S., Koh, L. P., Brook, B. W. and Ng, P. K. L. (2005c) Response to Hau *et al*: Beyond Singapore: Hong Kong and Asian biodiversity. *Trends in Ecology and Evolution* 20, 282–283.

Sodhi, N. S., Koh, L. P., Prawiradilaga, D. M., Tinulele, I., Putra, D. D. and Tan, T. H. T. (2005d) Land use and conservation value for forest birds in Central Sulawesi (Indonesia). *Biological Conservation* 122, 547–558.

Sodhi, N. S., Lee, T. M., Koh, L. P. and Prawiradilaga, D. M. (2006a) Long-term avifaunal impoverishment in an isolated tropical woodlot. *Conservation Biology* 20, 772–779.

Sodhi, N. S., Brooks, T. M., Koh, L. P., Acciaioli, G., Erb, M., Tan, A. K. J., Curran, L. M., Brosius, P., Lee, T. M., Patlis, J. M., Gumal, M. and Lee, R. J. (2006b) Biodiversity and human livelihood crises in the Malay Archipelago. *Conservation Biology* 20, 1811–1813.

Soh, M. C. K., Sodhi, N. S. and Lim, S. L. H. (2006) High sensitivity of montane bird communities to habitat disturbance in Peninsular Malaysia. *Biological Conservation* 129, 149–166.

Soulé, M. E., Bolger, D. T., Alberts, A. C., Wright, J., Sorice, M. and Hill, S. (1988) Reconstructed dynamics of rapid extinctions of Chaparral-requiring birds in urban habitat islands. *Conservation Biology* 2, 75–92.

Spalding, M. D., Ravilious, C. and Green, E. P. (2001) *World Atlas of Coral Reefs.* University of California Press, Berkeley, CA.

Spielman, D., Brook, B. W. and Frankham, R. (2004) Most species are not driven to extinction before genetic factors impact them. *Proceedings of the National Academy of Sciences of the USA* 101, 15261–15264.

Spitzer, K., Novotny, V., Tonner, M. and Leps, J. (1993) Habitat preferences, distribution and seasonality of the butterflies (Lepidoptera, Papilionoidea) in a montane tropical rain-forest, Vietnam. *Journal of Biogeography* 20, 109–121.

Spotila, J. R., Reina, R. D., Steyermark, A. C., Plotkin, P. T. and Paladino, F. V. (2000) Pacific leatherback turtles face extinction. *Nature* 405, 529–530.

Spratt, D. (1998) Symposium: Global change, climate change and human health/zoonoses. *International Journal for Parasitology* 28, 925–926.

Stanford, C. B. (1999) *The Hunting Apes: Meat Eating and the Origin of Human Behaviour.* Princeton University Press, Princeton, NJ.

Starling, F., Lazzaro, X., Cavalcanti, C. and Moreira, R. (2002) Contribution of omnivorous tilapia to eutrophication of a shallow tropical reservoir: Evidence from a fish kill. *Freshwater Biology* **47**, 2443–2452.

Stattersfield, A. J., Crosby, M. J., Long, A. J. and Wege, D. C. (1998) *Endemic Bird Areas of the World: Priorities for Biodiversity Conservation.* BirdLife International, Cambridge.

Steadman, D. W. (1995) Prehistoric extinctions of Pacific island birds – biodiversity meets zooarchaeology. *Science* **267**, 1123–1131.

Stensland, E., Carlen, I., Sarnblad, A., Bignert, A. and Berggren, P. (2006) Population size, distribution, and behaviour of indo-pacific bottlenose (*Tursiops aduncus*) and humpback (*Sousa chinensis*) dolphins off the south coast of Zanzibar. *Marine Mammal Science* **22**, 667–682.

Stevens, J. D., Bonfil, R., Dulvy, N. K. and Walker, P. A. (2000) The effects of fishing on sharks, rays, and chimaeras (chondrichthyans), and the implications for marine ecosystems. *ICES Journal of Marine Science* **57**, 476–494.

Stevick, P. T., Allen, J., Clapham, P. J., Friday, N., Katona, S. K., Larsen, F., Lien, J., Mattila, D. K., Palsboll, P. J., Sigurjonsson, J., Smith, T. D., Oien, N. and Hammond, P. S. (2003) North Atlantic humpback whale abundance and rate of increase four decades after protection from whaling. *Marine Ecology Progress Series* **258**, 263–273.

Stewart, B. S. and Wilson, S. G. (2005) Threatened fishes of the world: *Rhincodon typus* (Smith 1828) (Rhincodontidae). *Environmental Biology of Fishes* **74**, 184–185.

Stibig, H. J. and Malingreau, J. P. (2003) Forest cover of insular southeast Asia mapped from recent satellite images of coarse spatial resolution. *Ambio* **32**, 469–475.

Stiles, D. (2004) The ivory trade and elephant conservation. *Environmental Conservation* **31**, 309–321.

Still, C. J., Foster, P. N. and Schneider, S. H. (1999) Simulating the effects of climate change on tropical montane cloud forests. *Nature* **398**, 608–610.

Stobutzki, I., Miller, M. and Brewer, D. (2001a) Sustainability of fishery bycatch: A process for assessing highly diverse and numerous bycatch. *Environmental Conservation* **28**, 167–181.

Stobutzki, I. C., Miller, M. J., Jones, J. P. and Salini, J. P. (2001b) Bycatch diversity and variation in penaeid fisheries: The implications for monitoring. *Fisheries Research* **53**, 283–301.

Stockwell, C. A., Mulvey, M. and Vinyard, G. L. (1996) Translocations and the preservation of allelic diversity. *Conservation Biology* **10**, 1133–1141.

Stolle, F., Chomitz, K. M., Lambin, E. F. and Tomich, T. P. (2003) Land use and vegetation fires in Jambi Province, Sumatra, Indonesia. *Forest Ecology and Management* **179**, 277–292.

Stone, R. (1995) Fishermen threaten Galápagos. *Science* **267**, 611–612.

Stork, N. E. and Lyal, C. H. C. (1993) Extinction or 'co-extinction' rates? *Nature* **366**, 307.

Stouffer, P. C. and Bierregaard, R. O. (1995a) Effects of forest fragmentation on understory hummingbirds in Amazonian Brazil. *Conservation Biology* **9**, 1085–1094.

Stouffer, P. C. and Bierregaard, R. O. (1995b) Use of Amazonian forest fragments by understory insectivorous birds. *Ecology* **76**, 2429–2445.

Strayer, D. L., Eviner, V. T., Jeschke, J. M. and Pace, M. L. (2006) Understanding the long-term effects of species invasions. *Trends in Ecology and Evolution* **21**, 645–651.

Struhsaker, T. T., Struhsaker, P. J. and Siex, K. S. (2005) Conserving Africa's rain forests: Problems in protected areas and possible solutions. *Biological Conservation* **123**, 45–54.

Stuart, S. N., Chanson, J. S., Cox, N. A., Young, B. E., Rodrigues, A. S. L., Fischman, D. L. and Waller, R. W. (2004) Status and trends of amphibian declines and extinctions worldwide. *Science* **306**, 1783–1786.

Stutchbury, B. J. M. and Morton, E. S. (2001) *Behavioural Ecology of Tropical Birds.* Academic Press, London.

Subba Rao, D. V. (2002) Marine sciences of the seas around India between 1874 and 2000 and prospects for the new millennium. In: *Ocean Yearbook* (eds. E. M. Borgese, A. Chircop and M. McConnell), pp. 195–227. University of Chicago Press, Chicago.

Subba Rao, D. V. (2005) Comprehensive review of the records of the biota of the Indian Seas and introduction of non-indigenous species. *Aquatic Conservation-Marine and Freshwater Ecosystems* **15**, 117–146.

Sukumar, R. (2003) *The Living Elephants: Evolutionary Ecology, Behaviour, and Conservation.* Oxford University Press, New York.

Sun, X. F., Katsigris, E. and White, A. (2004) Meeting China's demand for forest products: An overview of import trends, ports of entry, and supplying countries, with emphasis on the Asia-Pacific region. *International Forestry Review* 6, 227–236.

Sverdrup-Jensen, S. (2002) *Fisheries in the Lower Mekong Basin: Status and Perspectives. MRC Technical Paper No. 6*. Mekong River Commission, Phnom Penh, Cambodia.

Swaine, M. D. (1992) Characteristics of dry forests in West Africa and the influence of fire. *Journal of Vegetation Science* 3, 365–374.

Syed, R. A., Law, I. H. and Corley, R. H. V. (1982) Insect pollination of oil palm: Introduction, establishment and pollinating efficiency of *Elaeidobius kamerunicus* in Malaysia. *Planter* 58, 547–561.

Tabarelli, M., Mantovani, W. and Peres, C. A. (1999) Effects of habitat fragmentation on plant guild structure in the montane Atlantic forest of southeastern Brazil. *Biological Conservation* 91, 119–127.

Talbott, K. and Brown, M. (1998) Forest plunder in Southeast Asia: An environmental security nexus in Burma and Cambodia. *Environmental Change and Security Project Report* 4 (Spring), 53–60.

Taylor, D., Saksena, P., Sanderson, P. G. and Kucera, K. (1999) Environmental change and rain forests on the Sunda shelf of Southeast Asia: Drought, fire and the biological cooling of biodiversity hotspots. *Biodiversity and Conservation* 8, 1159–1177.

Teo, D. H. L., Tan, H. T. W., Corlett, R. T., Wong, C. M. and Lum, S. K. Y. (2003) Continental rain forest fragments in Singapore resist invasion by exotic plants. *Journal of Biogeography* 30, 305–310.

Terborgh, J. and Winter, B. (1980) Some causes of extinction. In: *Conservation Biology: An Evolutionary-Ecological Perspective* (eds. M. E. Soulé and B. A. Wilcox), pp. 119–133. Sinauer, New York.

Terborgh, J. (1974) Preservation of natural diversity: The problem of extinction-prone species. *BioScience* 24, 715–722.

Terborgh, J. (1992) Maintenance of diversity in tropical forests. *Biotropica* 24, 283–292.

Terborgh, J., Lopez, L. and Jose-Tello, S. (1997) Bird communities in transition: The Lago Guri Islands. *Ecology* 78, 1494–1501.

Thayer, G. W., Engel, D. W. and Bjorndal, K. A. (1982) Evidence for shortcircuiting of the detritus cycle of seagrass beds by the green turtle, *Chelonia mydas* L. *Journal of Experimental Marine Biology and Ecology* 62, 173–183.

Thibault, M. and Blaney, S. (2003) The oil industry as an underlying factor in the bushmeat crisis in Central Africa. *Conservation Biology* 17, 1807–1813.

Thibault, J. C., Martin, J. L., Penloup, A. and Meyer, J. Y. (2002) Understanding the decline and extinction of monarchs (Aves) in Polynesian Islands. *Biological Conservation* 108, 161–174.

Thiollay, J.-M. (1995) The role of traditional agroforests in the conservation of rain and forest bird diversity in Sumatra. *Conservation Biology* 9, 335–353.

Thiollay, J.-M. (1997) Distribution and abundance patterns of bird community and raptor populations in the Andaman archipelago. *Ecography* 20, 67–82.

Thiollay, J.-M. (1989) Area requirements for the conservation of rain forest raptors and game birds in French Guiana. *Conservation Biology* 3, 128–137.

Thomas, J. A. (1991) Rare species conservation: Case of European butterflies. In: *The Scientific Management of Temperate Communities for Conservation* (eds. I. Spellerberg and B. Goldsmith), pp. 149–197. Blackwell Scientific Publications, Oxford.

Thomas, J. A. and Morris, M. G. (1994) Patterns, mechanisms and rates of decline among UK invertebrates. *Philosophical Transactions of the Royal Society B: Biological Sciences* 344, 47–54.

Thomas, J. A., Bourn, N. A. D., Clarke, R. T., Stewart, K. E., Simcox, D. J., Pearman, G. S., Curtis, R. and Goodger, B. (2001) The quality and isolation of habitat patches both determine where butterflies persist in fragmented landscapes. *Proceedings of the Royal Society of London Series B: Biological Sciences* 268, 1791–1796.

Thomas, J. A., Telfer, M. G., Roy, D. B., Preston, C. D., Greenwood, J. J. D., Asher, J., Fox, R., Clarke, R. T. and Lawton, J. H. (2004) Comparative losses of British butterflies, birds, and plants and the global extinction crisis. *Science* 303, 1879–1881.

Thorrold, S. R. and Hare, J. A. (2002) Otolith applications in reef fish ecology. In: *Coral Reef Fishes. Dynamics and Diversity in a Complex Ecosystem* (ed. P. F. Sale), pp. 243–264. Academic Press, San Diego.

Tilman, D., May, R. M., Lehman, C. L. and Nowak, M. A. (1994) Habitat destruction and the extinction debt. *Nature* 371, 65–66.

Tilman, D., Fargione, J., Wolff, B., D'Antonio, C., Dobson, A., Howarth, R., Schindler, D., Schlesinger, W. H., Simberloff, D. and Swackhamer, D. (2001) Forecasting agriculturally driven global environmental change. *Science* 292, 281–284.

Tisdell, C. A. (1982) *Wild Pigs: Environmental Pest or Economic Resource?* Pergamon Press, Sydney.

Tomascik, T. and Sander, F. (1987a) Effects of eutrophication on reefbuilding corals. 2. Structure of scleractinian coral communities on fringing reefs, Barbados, West Indies. *Marine Biology* 94, 53–75.

Tomascik, T. and Sander, F. (1987b) Effects of eutrophication on reefbuilding corals. 3. Reproduction of the reef-building coral *Porites porites. Marine Biology* 94, 77–94.

Tomich, T. P., Thomas, D. E. and van Noordwijk, M. (2004) Environmental services and land-use change in Southeast Asia: From recognition to regulation or reward? *Agriculture, Ecosystems and Environment* 104, 229–244.

Trauernicht, C. and Ticktin, T. (2005) The effects of non-timber forest product cultivation on the plant community structure and composition of a humid tropical forest in southern Mexico. *Forest Ecology and Management* 219, 269–278.

Traveset, A. and Richardson, D. M. (2006) Biological invasions as disruptors of plant reproductive mutualisms. *Trends in Ecology and Evolution* 21, 208–216.

Tröeng, S. and Rankin, E. (2005) Long-term conservation efforts contribute to positive green turtle *Chelonia mydas* nesting trend at Tortuguero, Costa Rica. *Biological Conservation* 121, 111–116.

Tsutsui, N. D., Suarez, A. V., Holway, D. A. and Case, T. J. (2000) Reduced genetic variation and the success of an invasive species. *Proceedings of the National Academy of Sciences of the USA* 97, 5948–5953.

Tulloch, D. G. (1974) The feral swamp buffaloes of Australia's Northern Territory. In: *The Husbandry and Health of the Domestic Buffalo* (ed. W. R. Cockrill), pp. 493–505. Food and Agricultural Organization, Rome.

Tupper, M. H. and Juanes, F. (1999) Effects of a marine reserve on recruitment of grunts (Pisces: Haemulidae) at Barbados, West Indies. *Environmental Biology of Fishes* 55, 53–63.

Tupper, M. H. (2002) Marine reserves and fisheries management. *Science* 295, 1233.

Turner, A. and Antón, M. (1997) *The Big Cats and Their Fossil Relatives: An Illustrated Guide to Their Evolution and Natural History.* Columbia University Press, New York.

Turner, I. M. (1996) Species loss in fragments of tropical rain forest: A review of the evidence. *Journal of Applied Ecology* 33, 200–209.

Turner, I. M. and Corlett, R. T. (1996) The conservation value of small, isolated fragments of lowland tropical rain forest. *Trends in Ecology and Evolution* 11, 330–333.

Turner, I. M., Chua, K. S., Ong, J. S. Y., Soong, B. C. and Tan, H. T. W. (1996) A century of plant species loss from an isolated fragment of lowland tropical rain forest. *Conservation Biology* 10, 1229–1244.

Turner, I. M., Tan, H. T. W., Wee, Y. C., Ibrahim, A. B., Chew, P. T. and Corlett, R. T. (1994) A study of plant species extinction in Singapore: Lessons of the conservation of tropical biodiversity. *Conservation Biology* 8, 705–712.

Turner, I. M., Wong, Y. K., Chew, P. T. and Ibrahim, A. B. (1997) Tree species richness in primary and old secondary tropical forest in Singapore. *Biodiversity and Conservation* 6, 537–543.

Tylianakis, J. M., Tscharntke, T. and Lewis, O. T. (2007) Habitat modification alters structure of tropical host-parasitoid food webs. *Nature,* 445, 202–205.

Uhl, C. (1998) Perspectives on wildfire in the humid tropics. *Conservation Biology* 12, 942–943.

United Nations. (2004) *World Population to 2300.* United Nations, New York.

Valiela, I., Bowen, J. L. and York, J. K. (2001) Mangrove forests: One of the world's threatened major tropical environments. *BioScience* 51, 807–815.

Valladares, G., Salvo, A. and Cagnolo, L. (2006) Habitat fragmentation effects on trophic processes of insect–plant food webs. *Conservation Biology* 20, 212–217.

van Balen, S., Nijman, V. and Sozer, R. (1999) Distribution and conservation of the Javan Hawk-eagle *Spizaetus bartelsi. Bird Conservation International* 9, 333–349.

van Balen, S. B., Dirgayusa, I. W. A., Putra, I. M. W. A. and Prins, H. H. T. (2000) Status and distribution of the endemic Bali starling *Leucopsar rothschildi. Oryx* 34, 188–197.

van Beers, C. P. and de Moor, A. P. G. (1999) *Addicted to Subsidies.* Institute for Research on Public Expenditure, The Hague.

van Beers, C. and van den Bergh, J. (2001) Perseverance of perverse subsidies and their impact on trade and environment. *Ecological Economics* 36, 475–486.

Van Borm, S., Thomas, I., Hanquet, G., Lambrecht, N., Boschmans, M., Dupont, G., Decaestecker, M., Snacken, R. and van den Berg, T. (2005) Highly pathogenic H5N1 influenza virus in smuggled Thai eagles, Belgium. *Emerging Infectious Diseases* 11, 702–705.

Van der Kaay, H. J. (1998) Human diseases in relation to the degradation of tropical rainforests. *Rainforest Medical Bulletin* 5, no.2.

van Nieuwstadt, M. G. L., Sheil, D. and Kartawinata, K. (2001) The ecological consequences of logging in the burned forests of East Kalimantan, Indonesia. *Conservation Biology* 15, 1183–1186.

Van Riel, P., Jordaens, K., Martins, A. M. F. and Backeljau, T. (2000) Eradication of exotic species. *Trends in Ecology and Evolution* 15, 515–515.

van Riper III, C., van Riper, S. G., Goff, M. L. and Laird, M. (1986) The epizootiology and ecological significance of malaria in Hawaiian landbirds. *Ecological Monographs* 56, 327–344.

Van Waerebeek, K., Van Bressem, M. F., Felix, F., Alfaro Shigueto, J., Garcia Godos, A., Chavez Lisambart, L., Onton, K., Montes, D. and Bello, R. (1997) Mortality of dolphins and porpoises in coastal fisheries off Peru and southern Ecuador in 1994. *Biological Conservation* 81, 43–49.

Van Wilgen, B. W., Govender, N., Biggs, H. C., Ntsala, D. and Funda, X. N. (2004) Response of Savanna fire regimes to changing fire-management policies in a large African National Park. *Conservation Biology* 18, 1533–1540.

Vazquez, D. P. and Gittleman, J. L. (1998) Biodiversity conservation: Does phylogeny matter? *Current Biology* 8, R379–381.

Verschuren, D., Johnson, T. C., Kling, H. J., Edgington, D. N., Leavitt, P. R., Brown, E. T., Talbot, M. R. and Hecky, R. E. (2002) History and timing of human impact on Lake Victoria, East Africa. *Proceedings of the Royal Society B: Biological Sciences* 269, 289–294.

Vidal, O. (1993) Aquatic mammal conservation in Latin America – problems and perspectives. *Conservation Biology* 7, 788–795.

Vigilante, T. and Bowman, D. M. J. S. (2004) Effects of individual fire events on the flower production of fruit-bearing tree species, with reference to Aboriginal people's management and use, at Kalumburu, North Kimberley, Australia. *Australian Journal of Botany* 52, 405–415.

Vigliola, L., Doherty, P. J., Meekan, M. G., Drown, D., Jones, M. E. and Barber, P. H. (2007) Genetic identity determines risk of post-settlement mortality of a marine fish. *Ecology* 88, 1263–1277.

Vincent, A. C. J. (1996) *The International Trade in Seahorses*. TRAFFIC International, Cambridge.

Vine, P. J. (1973) Crown of thorns (*Acanthaster planci*) plagues: The natural causes theory. *Atoll Research Bulletin* 166, 1–10.

Viswanathan, P. N. and Krishna Murti, C. R. (1989) Effects of environmental chemicals on biota and ecosystems in tropical, arid and cold regions. 1. Effects of temperature and humidity on ecotoxicology of chemicals. In: *Ecotoxicology and Climate. SCOPE 38* (eds. P. Bordeau, J. A. Haines, W. Klein and C. R. K. Murti), pp. 139–154. John Wiley, Chichester.

Vitousek, P. M. (1994) Beyond global warming: Ecology and global change. *Ecology* 75, 1861–1876.

Vitousek, P. M. and Walker, L. R. (1989) Biological invasion by *Myrica faya* in Hawaii – plant demography, nitrogen fixation, and ecosystem effects. *Ecological Monographs* 59, 247–265.

Vitousek, P. M., Aber, J. D., Howarth, R. W., Likens, G. E., Matson, P. A., Schindler, D. W., Schlesinger, W. H. and Tilman, D. G. (1997) Human alteration of the global nitrogen cycle: Sources and consequences. *Ecological Applications* 7, 737–750.

Vitousek, P. M., Dantonio, C. M., Loope, L. L. and Westbrooks, R. (1996) Biological invasions as global environmental change. *American Scientist* 84, 468–478.

von Hippel, F. A. and von Hippel, W. (2002) Sex, drugs and animal parts: Will Viagra save threatened species? *Environmental Conservation* 29, 277–281.

Vörösmarty, C. J., Green, P., Salisbury, J. and Lammers, R. B. (2000) Global water resources: Vulnerability from climate change and population growth. *Science* 289, 284–288.

Walker, D. I. and Ormond, R. F. G. (1982) Coral death from sewage and phosphate pollution at Aqaba, Red Sea. *Marine Pollution Bulletin* 13, 21–25.

Walker, T. I. (1998) Can shark resources be harvested sustainably? A question revisited with a review of shark fisheries. *Marine and Freshwater Research* 49, 553–572.

Walters, B. B. (2000) Local mangrove planting in the Philippines: Are fisherfolk and fishpond owners effective restorationists? *Restoration Ecology* 8, 237–246.

Warburton, N. H. (1997) Structure and conservation of forest avifauna in isolated rainforest remnants in tropical Australia. In: *Tropical Forest Remnants: Ecology Management and Conservation of*

Fragmented Communities (eds. W. F. Laurance and R. O. Bierregaard), pp. 190–206. University of Chicago Press, Chicago.

Ward, J. R. and Lafferty, K. D. (2004) The elusive baseline of marine disease: Are diseases in ocean ecosystems increasing? *PLoS Biology* **2**, e120.

Ward, J. V., Tockner, K., Arscott, D. B. and Claret, C. (2002) Riverine landscape diversity. *Freshwater Biology* **47**, 517–539.

Wasser, S. K., Shedlock, A. M., Comstock, K., Ostrander, E. A., Mutayoba, B. and Stephens, M. (2004) Assigning African elephant DNA to geographic region of origin: applications to the ivory trade. *Proceedings of the National Academy of Sciences of the USA* **101**, 14847–14852.

Watkinson, A. R., Freckleton, R. P., Robinson, R. A. and Sutherland, W. J. (2000) Predictions of biodiversity response to genetically modified herbicide-tolerant crops. *Science* **289**, 1554–1557.

Watson, J. E. M., Whittaker, R. J. and Dawson, T. P. (2004) Avifaunal responses to habitat fragmentation in the threatened littoral forests of south-eastern Madagascar. *Journal of Biogeography* **31**, 1791–1807.

Watson, M. and Woinsarski, J. (2003) *A preliminary assessment of impacts of cane toads on terrestrial vertebrate fauna in Kakadu National Park.* Kakadu Research Advisory Committee, Darwin, Australia.

Webb, G. J.W. and Manolis, S.C. (1993). Conserving Australia's crocodiles through commercial incentives. In: *Herpetology in Australia – a Diverse Discipline* (eds. D. Lunney and D. Ayers), pp. 250–256. Royal Zoological Society of New South Wales, Sydney.

Webb, G., Manolis, S., Whitehead, P. and Letts, G. (1984) A Proposal for the Transfer of the Australian Population of *Crocodylus porosus* Schneider (1801), from Appendix I to Appendix II of CITES. Conservation Commission of the Northern Territory, Darwin, Australia.

Welcomme, R. L. (ed.) (1979) *Fisheries Ecology of Floodplain Rivers.* Longman, London.

Wells, S. and Edwards, A. (1989) Gone with the waves. *New Scientist* **124**, 47–51.

White, G. M., Boshier, D. H. and Powell, W. (2002) Increased pollen flow counteracts fragmentation in a tropical dry forest: An example from *Swietenia humilis* Zuccarini. *Proceedings of the National Academy of Sciences of the USA* **99**, 2038–2042.

Whitehead, H. (2002) Estimates of the current global population size and historical trajectory for sperm whales. *Marine Ecology Progress Series* **242**, 295–304.

Whitehead, H., Christal, J. and Dufault, S. (1997) Past and distant whaling and the rapid decline of sperm whales off the Galápagos Islands. *Conservation Biology* **11**, 1387–1396.

Whitehead, P. J. and Tschirner, K. (1991) Lead shot ingestion and lead poisoning of magpie geese *Anseranas semipalmata* foraging in a northern Australian hunting reserve. *Biological Conservation* **58**, 99–118.

Whitmore, T. C. (1980) The conservation of tropical rainforest. In: *Conservation Biology: An Evolutionary–Ecological Perspective* (eds. M. E. Soulé and B. A. Wilcox), pp. 303–318. Sinauer, New York.

Whitmore, T. C. (1997) Tropical forest disturbance, disappearance, and species loss. In: *Tropical Forest Remnants: Ecology, Management, and Conservation of Fragmented Communities* (eds. W. F. Laurance and R. O. J. Bierregaard), pp. 3–12. University of Chicago Press, Chicago.

Whitmore, T. C. and Sayer, J. A. (1992) *Tropical Deforestation and Species Extinction.* Chapman & Hall, London.

Whitten, A. J., Mustafa, M. and Henderson, G. S. (1987) *The Ecology of Sulawesi.* Gadjah Mada University Press, Yogyakarta, Indonesia.

Whitten, T., Castro, G., MacKinnon, K. and Platais, G. (2001) The World Bank and biodiversity conservation. *Oryx* **35**, 357–358.

Wickstrom, K. (2002) Marine reserves and fisheries management. *Science* **295**, 1233.

Wilcove, D. S., McLellan, C. H. and Dobson, A. P. (1986) Habitat fragmentaton in the temperate zone. In: *Conservation Biology* (ed. M. E. Soulé), pp. 237–256. Sinauer, Sunderland, MA.

Wiles, G. J., Bart, J., Beck, R. E. and Aguon, C. F. (2003) Impacts of the brown tree snake: Patterns of decline and species persistence in Guam's avifauna. *Conservation Biology* **17**, 1350–1360.

Wilkie, D. S. and Carpenter, J. F. (1999) Bushmeat hunting in the Congo Basin: An assessment of impacts and options for mitigation. *Biodiversity and Conservation* **8**, 927–955.

Wilkie, D. S., Starkey, M., Abernethy, K., Effa, E. N., Telfer, P. and Godoy, R. (2005) Role of prices and wealth in consumer demand for bushmeat in Gabon, Central Africa. *Conservation Biology* **19**, 268–274.

Williams, I. D. and Polunin, N. V. C. (2000) Differences between protected and unprotected reefs of the western Caribbean in attributes preferred by dive tourists. *Environmental Conservation* 27, 382–391.

Willis, E. O. and Oniki, Y. (1978) Birds and army ants. *Annual Review of Ecology and Systematics* 9, 243–263.

Willis, E. O. (1979) The composition of avian communities in remanescent woodlots in southern Brazil. *Papeis Avulsos de Zoologia* 33, 1–25.

Wilson, E. O. (1975) *Sociobiology: The New Synthesis.* Belknap Press, Cambridge, MA.

Wilson, E. O. (1988) *Biodiversity.* National Academy of Sciences/Smithsonian Institution, Washington, DC.

Wilson, E. O. (1989) Threats to biodiversity. *Scientific American* 261, 108, 112, 114, 116.

Wilson, E. O. (2000) A global biodiversity map. *Science* 289, 2279–2279.

Wilson, E. O. and Willis, E. O. (1975) Applied biogeography. In: *Ecology and Evolution of Communities* (eds. M. L. Cody and J. M. Diamond), pp. 522–534. Belknap Press, Cambridge, MA

Wilson, G. A., Nishi, J. S., Elkin, B. T. and Strobeck, C. (2005) Effects of a recent founding event and intrinsic population dynamics on genetic diversity in an ungulate population. *Conservation Genetics* 6, 905–916.

Wilson, S. G., Polovina, J. J., Stewart, B. S. and Meekan, M. G. (2006) Movements of whale sharks (*Rhincodon typus*) tagged at Ningaloo Reef, Western Australia. *Marine Biology* 148, 1157–1166.

Wilson, W. H., Francis, I., Ryan, K. and Davy, S. K. (2001) Temperature induction of viruses in symbiotic dinoflagellates. *Aquatic Microbial Ecology* 25, 99–102.

Windle, P. N. and Chavarría, G. (2005) The tragedy of the commons revisited: Invasive species. *Frontiers in Ecology and the Environment* 3, 107–108.

Wintner, S. P. (2000) Preliminary study of vertebral growth rings in the whale shark, *Rhincodon typus*, from the east coast of South Africa. *Environmental Biology of Fishes* 59, 441–451.

Wishart, M. J., Davies, B. R., Boon, P. J. and Pringle, C. M. (2000) Global disparities in river conservation: 'First World' values and 'Third World' realities. In: *Global Perspectives in River Conservation: Science, Policy and Practice* (ed. P. J. Boon), pp. 353–369. John Wiley and Sons, Chichester.

Witte, F., Goldschmidt, T., Goudswaard, P., Ligtvoet, W., van Oijen, M. and Wanink, J. H. (1992) Species extinction and concomitant ecological changes in Lake Victoria. *Netherlands Journal of Zoology* 42, 214–232.

Witte, F., Goudswaard, P. C., Katunzi, E. F. B., Mkumbo, O. C., Seehausen, O. and Wanink, J. H. (1999) Lake Victoria's ecological changes and their relationships with the riparian societies. In: *Ancient Lakes: Their Cultural and Biological Diversities* (eds. H. Kawanabe, G. Coulter and A. Roosevelt), pp. 189–202. Kenobi Productions, Ghent.

Woinarski, J. C. Z., Risler, J. and Kean, L. (2004) Response of vegetation and vertebrate fauna to 23 years of fire exclusion in a tropical *Eucalyptus* open forest, Northern Territory, Australia. *Austral Ecology* 29, 156–176.

Wolfe, N. D., Switzer, W. M., Carr, J. K., Bhullar, V. B., Shanmugam, V., Tamoufe, U., Prosser, A. T., Torimiro, J. N., Wright, A., Mpoudi-Ngole, E., McCutchan, F. E., Birx, D. L. and Folks, T. M. (2004) Naturally acquired simian retrovirus infections in central African hunters. *Lancet* 363, 932–937.

Wolfe, N. D., Daszak, P., Kilpatrick, A. M. and Burke, D. S. (2005) Bushmeat hunting deforestation, and prediction of zoonoses emergence. *Emerging Infectious Diseases* 11, 1822–1827.

Wong, T. C. M., Sodhi, N. S. and Turner, I. M. (1998) Artificial nest and seed predation experiments in tropical lowland rainforest remnants of Singapore. *Biological Conservation* 85, 97–104.

Woodroffe, R. and Ginsberg, J. R. (1998) Edge effects and the extinction of populations inside protected areas. *Science* 280, 2126–2128.

Woods, P. (1989) Effects of logging, drought, and fire on structure and composition of tropical forests in Sabah, Malaysia. *Biotropica* 21, 290–298.

Woodworth, B. L., Atkinson, C. T., LaPointe, D. A., Hart, P. J., Spiegel, C. S., Tweed, E. J., Henneman, C., LeBrun, J., Denette, T., DeMots, R., Kozar, K. L., Triglia, D., Lease, D., Gregor, A., Smith, T. and Duffy, D. (2005) Host population persistence in the face of introduced vector-borne diseases: Hawaii amakihi and avian malaria. *Proceedings of the National Academy of Sciences of the USA* 102, 1531–1536.

World Bank (2003) *World Development Indicators 2003.* World Bank Group (www.worldbank.org).

Worm, B., Lotze, H. K. and Myers, R. A. (2003) Predator diversity hotspots in the blue ocean. *Proceedings of the National Academy of Sciences of the USA* 100, 9884–9888.

WRI (2003) *World Resources 2002–2004: Decisions for the Earth: Balance, Voice, and Power.* United Nations Development Programme, World Bank, World Resources Institute.

Wright, S. J. (1981) Extinction-mediated competition: The *Anolis* lizards and insectivorous birds of the West Indies. *American Naturalist* 117, 181–192.

Wright, S. J. (2005) Tropical forests in a changing environment. *Trends in Ecology and Evolution* 20, 553–560.

Wright, S. J. and Muller-Landau, H. C. (2006) The uncertain future of tropical forest species. *Biotropica* 38, 443–445.

Xenopoulos, M. A., Lodge, D. M., Alcamo, J., Marker, M., Schulze, K. and Van Vuuren, D. P. (2005) Scenarios of freshwater fish extinctions from climate change and water withdrawal. *Global Change Biology* 11, 1557–1564.

Yamaguchi, M. (1986) *Acanthaster planci* infestations of reefs and coral assemblages in Japan: A retrospective analysis of control efforts. *Coral Reefs* 5, 23–30.

Yamamura, K. and Yokozawa, M. (2002) Prediction of a geographical shift in the prevalence of rice stripe virus disease transmitted by the small brown planthopper, *Laodelphax striatellus* (Fallen) (Hemiptera: Delphacidae), under global warming. *Applied Entomology and Zoology* 37, 181–190.

Yob, J. M., Field, H., Rashdi, A. M., Morrissy, C., van der Heide, B., Rota, P., Adzhar, A. B., White, J., Daniels, P., Jamaluddin, A. and Ksiazek, T. (2001) Nipah virus infection in bats (Order Chiroptera) in Peninsular Malaysia. *Emerging Infectious Diseases* 7, 439–441.

Zachos, J., Pagani, M., Sloan, L., Thomas, E. and Billups, K. (2001) Trends, rhythms, and aberrations in global climate 65 Ma to present. *Science* 292, 686–693.

Zakaria, M. and Nordin, M. (1998) Comparison of frugivory by birds in primary and logged lowland dipterocarp forests in Sabah, Malaysia. *Tropical Biodiversity* 5, 1–9.

Zavaleta, E. S., Shaw, M. R., Chiariello, N. R., Mooney, H. A. and Field, C. B. (2003) Additive effects of simulated climate changes, elevated CO_2, and nitrogen deposition on grassland diversity. *Proceedings of the National Academy of Sciences of the USA* 100, 7650–7654.

Zbinden, S. and Lee, D. R. (2005) Paying for environmental services: An analysis of participation in Costa Rica's PSA program. *World Development* 33, 255–272.

Zhang, H., Henderson-Sellers, A. and McGuffie, K. (2001) The compounding effects of tropical deforestation and greenhouse warming on climate. *Climatic Change* 49, 309–338.

Zieman, J. C., Fourqurean, J. W. and Frankovich, T. A. (1999) Seagrass die-off in Florida Bay: Long-term trends in abundance and growth of turtle grass *Thalassia testudinum*. *Estuaries* 22, 460–470.

Zuidema, P. A. and Boot, R. G. A. (2002) Demography of the Brazil nut tree (*Bertholletia excelsa*) in the Bolivian Amazon: Impact of seed extraction on recruitment and population dynamics. *Journal of Tropical Ecology* 18, 1–31.

Index

Echinochloa 104
eco-regions 31
ecolabelling 255
ecological 17, 31, 36, 40, 43, 49, 66,
   82–83, 92, 95, 106, 121, 144, 154–155,
   183, 204, 219, 222, 226, 228–229, 232,
   238, 247–248, 253, 260–261, 264, 266
ecological processes 43
ecology 61, 71, 88, 95, 107, 120, 128, 183
economic incentives 24, 266
economic losses 89, 144, 254
ecosystem 9, 14, 15, 31, 33, 40, 43, 47, 53,
   60–61, 67, 71, 89, 100, 101, 104, 105,
   109, 110, 153, 155, 168, 176, 181–182,
   187, 208, 228, 230, 238–239, 249, 258,
   260, 265–266
ecosystem functioning 187
ecosystem services 31, 33, 43, 47, 89, 182,
   249, 258, 260, 266
ecotourism 48, 101, 256, 261
Ecuador 61–62, 177, 216, 219, 220, 228,
   258, 261
edge effects 53, 54, 60, 64, 65, 66, 98, 213,
   223
edges 53–54, 56, 59, 60, 63, 64, 65–67, 69,
   97, 219
education 24, 87, 135–136, 139, 205,
   257–258, 260, 262
Egypt 107, 204
Eichhornia crassipes 104
Elaeidobius 43
Elaeis guineensis 20, 63, 231
elephant, Asiatic 61
elephantbirds 209
elephants 13, 69, 138, 145
Elephas maximus 6, 61, 69
Elettaria cardamomum 264
Eleutherodactylus chloronatus 61
Eleutherodactylus trepidotus 61
elevational changes 195
El Niño 47, 56, 72, 73, 82, 87–88, 105,
   148, 158, 189, 201
emissions 34–36, 40, 65, 82, 85, 87–88,
   205–206, 207, 249, 260
employment 48, 135, 254, 255
encephalitic 196
endangerment 103, 121, 130, 214, 216,
   232
Endemic Bird Areas 30
endemism 6, 27, 31, 148, 175, 182, 225,
   238, 247, 248
energy cycles 3
enforcement 25, 135, 136, 138, 155, 168,
   241, 247
Engraulis ringens 148
Enicurus scouleri 195

Enteroamoeba 107
enterprise 43, 48, 142
Epinephelinae 147, 150
epiphytes 225, 226
Equus 12
Equus caballus 100
eradication 105, 106, 108, 109, 110, 145
Eretmochelys imbricata 141
erosion 49, 223
Escherichia 107
establishment 12, 34, 43, 92, 94, 95, 102,
   109, 131, 140, 168, 178, 187, 214,
   246–247, 249, 255, 265
estuaries 183, 184, 185, 205
Etiella zinckenella 106
Eucalyptus 264
Euglandina rosea 106
Europe 1, 48, 49, 144
European rabbits 201
European Union 21, 87, 130
eutrophication 40–41, 96, 104, 163–164,
   169, 183
evaporation 37
evolution 47, 50, 52, 82, 94, 110, 189,
   200, 238
evolutionary 31, 33, 50, 190–191, 209,
   219, 226
Exclusive Economic Zone (EEZ) 153
exotic 13, 24, 39, 64, 66–67, 77, 85, 94,
   97, 100, 104–106, 109, 110, 176, 177
exports 1, 20, 22, 24, 48, 135, 251
extant 141, 144, 169, 221, 222, 225, 229,
   237, 239, 265
extinction 5, 27, 31, 54–56, 63, 82, 94–97,
   103, 106, 121, 125, 131, 133, 135, 139,
   147, 152, 156, 159, 162, 165, 169, 185,
   189, 199, 205, 208–214, 218–230, 232,
   238, 242, 253, 262, 265
extinction lags 214
extinction proneness 208, 219, 221, 222
extirpation 17, 64, 82, 137, 141, 173, 208,
   218, 230, 265

fairy bluebird 230
Falconidae 227
Falco punctatus 218, 225
farmers 18, 124
fauna 94, 98–99, 104–106, 127, 155, 209,
   218, 220, 223–225, 228, 265
fecundity 94, 137, 141, 147, 162, 168, 224
feedback loop 37, 67, 104, 135, 230
fertilizers 40, 56, 177, 179, 205, 260
Ficus 80, 82, 230
fig 230
fig wasps 82, 230
Fiji 102, 150, 160